# Viral Vectors in Veterinary Vaccine Development

Thiru Vanniasinkam • Suresh K. Tikoo •
Siba K. Samal

Editors

# Viral Vectors
# in Veterinary Vaccine
# Development

## A Textbook

*Editors*
Thiru Vanniasinkam
School of Biomedical Sciences
Charles Sturt University
Wagga Wagga, NSW, Australia

Suresh K. Tikoo
VIDO-InterVac & School of Public
Health
University of Saskatchewan
Saskatoon, SK, Canada

Siba K. Samal
Department of Veterinary Medicine
University of Maryland, College Park
College Park, MD, USA

ISBN 978-3-030-51929-2      ISBN 978-3-030-51927-8   (eBook)
https://doi.org/10.1007/978-3-030-51927-8

This Springer imprint is published by the registered company Springer Nature Switzerland AG.
The registered company address is: Gewerbestrasse 11, 6330 Cham, Switzerland

# Foreword

A viral vector vaccine consists of a nonpathogenic virus (the vector) that expresses protective antigen(s) of one or more heterologous pathogens. The pathogen gene sequence (typically encoding surface protein) is engineered by recombinant DNA methods to be inserted into the vector genome and expressed as part of the vector's transcriptional program. Following administration, a viral vector vaccine infects cells in the vaccine recipient. Some vectors are capable of limited vector replication in the recipient. Others are designed to be replication-defective and are restricted to a single cycle of infection. As described in this concise and very useful book, the viral vector vaccine approach is innovative and multifaceted, has resulted in a number of successful commercial veterinary vaccines to date, and has tremendous promise for further development.

The first demonstration of the efficacy of a viral vector vaccine was in 1983 when a recombinant vaccinia virus expresses the hemagglutinin protein of a human influenza A virus was shown to be immunogenic and protective in hamsters [5]. This quickly led to the use of vaccinia virus to express the surface glycoprotein of rabies virus [4], resulting in a licensed veterinary rabies vaccine that has had substantial success in rabies control [3]. Presently, a number of viruses are being developed as vaccine vectors. This book has excellent chapters describing veterinary vectors and vaccines – both commercial and experimental – based on adenovirus, poxvirus, herpesvirus, paramyxovirus, classical swine fever virus, rhabdovirus, coronavirus, and alphavirus. Additional important chapters discuss practical considerations in developing live vector vaccines, including manufacturing and regulatory issues.

Viral vector vaccines can be complicated and challenging to develop. Considerations include the numerous different available viral vectors and their diverse properties and strategies of construction and use; the biology and interplay of the vector, protective antigen, target host, and target pathogen; and the requirements for protective immunity. However, with this complexity comes the possibility to develop vaccines that are particularly well suited to their use. For example, the physical stability of vaccinia virus plus its ability to infect by the oral route led to the rabies vaccine mentioned above that can be delivered in bait left outdoors [3]. As a second example, herpesvirus of turkey provides a vector that is naturally nonpathogenic in poultry, is itself a vaccine against Marek's disease virus, and can be engineered to express one or more additional antigens of poultry pathogens to yield a single virus that is a bivalent or multivalent vaccine (chapter "Paramyxoviruses as Vaccine Vectors").

Both RNA and DNA viruses are being developed as vectors, with DNA vectors presently represented by a greater number of commercial vaccines. RNA viruses typically have small genomes, small sets of vector proteins and antigens, and less complex biology compared to DNA viruses. They are more limited in capacity for added foreign sequence and in some cases can have problems of point mutations and instability of the foreign sequence that need to be monitored. DNA viruses typically have substantially larger genomes with more complex organization and gene expression and a larger constellation of vector proteins and antigens. They typically have greater genetic stability and have a greater capacity such that several foreign antigens can be expressed by a single vector virus. In both RNA and DNA virus vectors, the foreign sequence usually is inserted as a supernumerary gene that does not function in the vector replication cycle. Alternatively, in a chimeric strategy, the encoded foreign protein is a functional substitution for a necessary vector protein, such as replacing the surface glycoprotein of a rhabdovirus vector with that of the target pathogen (chapter "Rhabdoviruses as Vaccine Vectors for Veterinary Pathogens"), or the creation of a West Nile virus vaccine for horses by the functional substitution of coding sequences for surface proteins of the attenuated yellow fever virus vector with their counterparts from the pathogen [2].

Because virus vector vaccines are infectious and often are replication-competent in vivo, they must be designed to be nonpathogenic but also must have sufficient infectivity and antigen expression to be immunogenic and protective. This balance can be difficult to achieve and is a major issue in vector development. Nonpathogenicity may be achieved in various ways. Some vector viruses are chosen because they have low virulence in their native hosts. Some vectors are attenuated or are replication-defective in a non-native host due to vector-host incompatibility. Some vectors have been attenuated by passage in cell culture under various conditions (e.g., extensive passage, suboptimal temperatures, cells from a non-native host, etc.), which can result in deletion of viral genes or genome regions that may not be needed in vitro but whose deletion is attenuating in vivo. These deletions may affect vector metabolism, tropism, virulence, and ability to suppress host responses. Attenuating point mutations also may appear in vector proteins or *cis*-acting nucleotide signals during passage or mutagenesis in vitro. Attenuation also may be achieved by direct manipulation of the vector genome by recombinant DNA methods, such as to delete or rearrange genes or introduce point mutations. Deletion of vector genes that are essential for replication results in replication-defective vectors that can be propagated in cell culture by complementation with vector proteins supplied by helper virus or engineered cells but which do not produce infectious virus in vivo (e.g., replication-defective adenovirus and alphavirus vectors, chapters "Adenovirus Vectors" and "Alphavirus-Based Vaccines", respectively).

Viral vector vaccines offer a number of important advantages, some of which remain to be fully investigated and developed. The vectors infect cells in the target host and express antigens intracellularly, inducing innate, cellular, antibody, mucosal, and systemic immunity. Broad immune stimulation enhances efficacy and provides immune regulatory crosstalk. Immunization typically is without need of an adjuvant (but see chapter "The Role of Adjuvants in the Application of Viral Vector Vaccines") and often at relatively low dose (in the case of replication-competent vectors) and often without need for multiple administrations. Expression of antigen in vivo presents antigenic sites in native form. In these important aspects, viral vaccine vectors mimic natural infections. Some viral vector vaccines can be designed to be bivalent or multivalent by expressing multiple antigens, and in some cases, the vector itself is a needed vaccine (e.g., herpesvirus of turkey noted above). Some vectors can be administered by multiple routes (e.g., topical versus oral versus parenteral) such that an optimal route can be chosen against mucosal versus systemic pathogens. Some vectors may have advantageous cell tropisms (e.g., efficient infection of antigen-presenting cells, with the potential for increased immunogenicity). Some vectors can be administered by convenient methods such as by spray, drinking water, or bait. The use of a virus vector that is not native to the target host avoids neutralization by maternal antibodies or immunity from prior infection with native viruses. Some replication-competent vectors can be produced relatively efficiently and cheaply. Using a vector to express isolated genes of a virulent pathogen avoids the need to handle the intact infectious pathogen.

Global human health is threatened by a number of emerging pathogens (primarily viruses) originating from animals, including HIV/AIDS, Ebola, avian influenza, Nipah and Hendra, SARS, the recent COVID-19, and others. A substantial proportion of older human pathogens likely also arose from animals. Thus, animal pathogens can have substantial impact on human health. As is abundantly illustrated in this book, veterinary application is a robust testing ground for new vaccines and new vaccine strategies. Some virus vector platforms, such as certain poxviruses and replication-defective adenovirus, are gaining substantial evaluation in both veterinary and human use. The increased availability of highly characterized vector platforms with clinical experience will facilitate and expedite their use for additional pathogens including newly emerging pathogens such as COVID-19. Some viral vector vaccines likely will be able to be used in both animals and humans. When feasible, vaccination of animals to control pathogens with zoonotic potential should reduce transmission (e.g., rabies). Thus, vaccines against pathogens of animal origin, and the use of the virus vector vaccine strategy, will be of increasing importance given the increasing threat to humans from animal pathogens.

Veterinary vaccine development is a complex and fascinating field of study because of the wide range of vaccine target species, target pathogens, and potential vaccines. As described in this book, there is a substantial number of commercial veterinary vaccines based on the viral vector vaccine strategy (also, see Ref. [1]). These include at least 13 commercial poxvirus-based vaccines targeting nine different pathogens (including rabies, canine distemper, and a poultry mycoplasma; chapter "Poxvirus Vectors"), several herpesvirus-based vaccines targeting at least five different pathogens (including Marek's disease, avian influenza, and infectious bursal disease; chapter "The Construction and Evaluation of Herpesvirus Vectors"), adenovirus-based rabies and foot-and-mouth disease vaccines (chapter "Adenovirus Vectors"), and others. We are still exploring the potential of the available vectors. A number of vectors have been improved by years of optimization. Continuing advances in the tools and understanding of molecular biology, pathogen-host interactions, and immunobiology will facilitate future increases in the number and effectiveness of viral vector vaccines. This timely book gives a valuable overview of the state of the art.

National Institute of Allergy and Infectious Diseases          Peter L. Collins
National Institute of Health
Bethesda, Maryland, USA
e-mail: PCOLLINS@niaid.nih.gov

## References

1. Current Veterinary Biologics Product Catalog – USDA APHIS. https://www.aphis.usda.gov/aphis/ourfocus/animalhealth/veterinary-biologics/CT_Vb_licensed_products
2. De Filette M, Ulbert S, Diamond M, Sanders NN. Recent progress in West Nile virus diagnosis and vaccination. Vet Res. 2012;43(16). http://www.veterinaryresearch.org/content/43/1/16
3. Desmettre P, Languet B, Chappuis G, Brochier B, Thomas I, Lecocq JP, Kieny M-P, Blancou J, Aubert M, Artois M, Pastoret P-P. Use of vaccinia rabies recombinant for oral vaccination of wildlife. Vet Microbiol. 1990;23:227–36.
4. Kieny MP, Lathe R, Drillen R, Spehner D, Skory S, Schmitt D, Wiktor T, Kaprowski H, Lecocq JP. Expression of rabies virus glycoprotein from a recombinant vaccinia virus. Nature. 1984;322:163–6.
5. Smith GL, Murphy BR, Moss B. Construction and characterization of an infectious vaccinia virus recombinant that expresses the influenza hemagglutinin gene and induces resistance to influenza virus infection in hamsters. Proc Natl Acad Sci U S A. 1983;80:7155–9.

# Contents

**Part IV   Application of Viral Vector Vaccines, Challenges and Future Directions**

# About the Editors

**Thiru Vanniasinkam** holds a BSc from Flinders University, Australia, and an Applied Science/Graduate Diploma (Honours) in Medical Laboratory Science and PhD in Microbiology/Immunology from the University of South Australia. She is a Fellow of the Australian Society for Microbiology (ASM), Secretary of the ASM Education Special Interest Group and Committee Member of the ASM New South Wales branch. She is a member of the American Society for Microbiology and is a Microbiology Society (UK) Champion. Dr. Vanniasinkam is Head of the Medical Science Discipline, School of Biomedical Sciences, Charles Sturt University, Australia. Her research interests include viral vectors, vaccinology and epidemiology of equine and poultry pathogens.

**Suresh K. Tikoo** received his BVSc & AH and MVSc from Govind Ballabh Pant University of Agriculture and Technology, India, and his PhD from the University of Saskatchewan, Canada. He is a Fellow of the National Academy of Veterinary Sciences, India. Prof. Tikoo has served as Head of the Viral Vectored Program at VIDO-InterVaC and Interim Executive Director of the School of Public Health, University of Saskatchewan. Currently, he is Professor and Director of the Vaccinology and Immunotherapeutics Graduate Program at the School of Public Health and a Research Fellow at VIDO-InterVac, University of Saskatchewan. His research interests include molecular virology, viral vectors, DNA viruses, vaccinology, mucosal immunization and virus–host interactions.

**Siba K. Samal** received his BVSc & AH from Orissa University of Agriculture and Technology, India, MVSc from the Indian Veterinary Research Institute, and MS and PhD from Texas A&M University. He was a Postdoctoral Fellow at Baylor College of Medicine and at Plum Island Animal Disease Research Center. Prof. Samal is a Diplomate of the American College of Veterinary Microbiologists and a Fellow of the American Academy of Microbiology and the National Academy of Veterinary Science, India. He is currently a Professor at the University of Maryland and is former Chair and Associate Dean of Virginia-Maryland College of Veterinary Medicine. His research focuses on the molecular biology of paramyxoviruses, viral pathogenesis and development of paramyxoviruses as vaccine vectors.

**Part I**

**Fundamentals of Viral Vector Vaccine Development**

# Introduction to Veterinary Vaccines

Teshome Mebatsion

## Abstract

Vaccination of animals has been carried out for centuries, and it is the most cost-effective and sustainable method of controlling infectious diseases. Veterinary vaccines not only are important to animal health but also play a vital role in reducing transmission of zoonotic diseases to humans and in securing food supply for humans. Conventional inactivated (killed) or live-attenuated vaccines constitute the majority of licensed veterinary vaccines that are currently in use. The widespread use of these vaccines not only substantially contributed to animal welfare and public health but also led to a successful global eradication of rinderpest, one of the animal diseases with major economic consequences in many parts of the world. Despite these successes, there are some limitations associated with conventional vaccines, and there are still several diseases that have yet to be successfully treated, demonstrating the need for better and safer vaccines. Recombinant vaccines represent an attractive strategy by which some of the limitations of conventional vaccines can be overcome. In the recent past, the veterinary field has witnessed the most successful applications of recombinant vaccines where more than a dozen viral-vectored vaccines, subunit, DNA, and virus-like particles-based vaccines were licensed for veterinary use, and many more are under development. There is a wave of rationally designed vaccine innovations ahead of us to benefit animals, animal owners, and ultimately humans.

## Keywords

Antigens · Conventional vaccines · DNA vaccines · Rational design · Recombinant vaccines · Subunit vaccines · Vector vaccines · Veterinary vaccines · VLP-based vaccines

## Learning Objectives

After reading this chapter, you should be able to:

- Describe how veterinary vaccines were developed starting with conventional vaccines to genetically engineered ones with the emphasis on vaccines for viral and bacterial diseases
- Explain the progress made from traditional to technology-based modern vaccines using examples of licensed vaccines
- Discuss remaining challenges and future research directions in developing improved veterinary vaccines, where appropriate

Note: Veterinary vaccines for parasitic and noninfectious diseases (allergy, cancer, fertility, etc.) are not covered here, and the reader is advised to refer to recent excellent reviews on these subjects [1–3].

T. Mebatsion (✉)
Boehringer Ingelheim Animal Health, Athens, GA, USA
e-mail: teshome.mebatsion@boehringer-ingelheim.com

© Springer Nature Switzerland AG 2021
T. Vanniasinkam et al. (eds.), *Viral Vectors in Veterinary Vaccine Development*,
https://doi.org/10.1007/978-3-030-51927-8_1

# 1    Introduction

Vaccination aims to mimic the development of naturally acquired immunity. The terms "vaccine" and "vaccination" are derived from Variolae vaccinae, first coined by Edward Jenner in 1796 to describe the inoculation of materials collected from a lesion on a milkmaid suffering from cowpox to confer protection against the related human smallpox virus [4, 5]. Louis Pasteur is another pioneer who discovered how to make vaccines from attenuated microbes in the mid-nineteenth century. He developed the earliest vaccines against fowl cholera, anthrax, and rabies [6, 7]. Subsequent breakthroughs in vaccine development came out in the 1950s by adaption of in vitro preparation using chicken embryos and tissue culture cells for large-scale production [8, 9].

Vaccination is the most widely used tool in veterinary medicine to prevent and control animal diseases, and it represents the most cost-effective and sustainable intervention. Protection of livestock from infectious diseases contributes to better welfare, helps to enhance their productivity and profitability for livestock producers, and also ensures the provision of healthy and nutritious food such as eggs, milk, and meat products for consumers. Further benefit of immunization of livestock animals is the reduction in antibiotic use and a subsequent reduction in their residue in animal products contributing to public safety. On the other hand, the purpose of vaccination in companion animals is primarily aimed at their welfare by preventing particular infectious diseases. The close association between people and their pets would not be as carefree without vaccination. Regardless of the animal types, vaccination serves as a primary defense to prevent diseases in animals and also further transmission from animals to humans. Rabies is an excellent example of vaccine-preventable viral zoonotic disease which occurs in more than 150 countries and territories. Dogs are the main source of human rabies deaths, contributing up to 99% of all rabies transmissions to humans [10]. Through mandatory dog vaccination, dog-mediated rabies

has been eliminated from Western Europe, Canada, the United States, and Japan, demonstrating that rabies elimination is feasible through vaccination of dogs and prevention of dog bites. In the developed world, the fight against rabies is brought to another level by introducing vaccines that can be administered orally into wildlife to control rabies in wild animals and also prevent the spillover to domestic animals. An outstanding example of such a vaccine is RABORAL V-RG®, the first licensed vaccinia virus-based vector vaccine expressing the glycoprotein of rabies virus. RABORAL V-RG® allowed oral vaccination on a large scale using baits containing vaccine [11]. The introductory chapter is divided into conventional and genetically engineered veterinary vaccines sections, where a broader coverage is devoted to genetically engineered veterinary vaccines in general, and viral vector vaccines in particular, as it is the main subject of this textbook.

# 2    Conventional Live and Inactivated Veterinary Vaccines

The majority of the licensed veterinary vaccines that are currently in use are inactivated (killed) or live-attenuated/modified live vaccines (MLV). Vaccine strains for inactivated vaccine preparation can be pathogenic wild-type isolates that are subsequently killed to avoid risk of causing disease. Thus, inactivated or killed vaccines preparations are generally safe and do not pose a risk of reversion to virulence. However, they are unable to infect host cells and activate cytotoxic T cells, and consequently, they are less protective and generally require strong adjuvants to induce the required level of immunity. Because of a higher production cost and the need for adjuvants, inactivated vaccine preparations are relatively more expensive and generally require multiple administrations by parenteral route to induce optimal immunity [1, 12]. Recent research toward improving the effectiveness and duration of immunity of inactivated vaccines has focused on the development of improved adjuvants [13].

Strains for MLV preparations, in contrast, need to be sufficiently attenuated by serial passaging of viruses or bacteria in an unnatural host or cell with the hope that random mutations result in a non-virulent, but replicative competent infectious agent [1]. Such vaccine preparations can replicate in target cells and induce both cellular and humoral immunity and generally do not require an adjuvant to be effective. As opposed to inactivated vaccine preparations, live vaccines can be administered by mass administration routes such as spray, drinking water, in ovo, etc. The major drawback of live vaccine preparations is the potential risk of reversion to pathogenic wild type as well as being a potential source of environmental contamination. Despite such disadvantage, MLV products have played a major role in successful disease control and eradication. An excellent example is the eradication of rinderpest from the globe that was declared on 25 May 2011 [14]. This achievement is critically dependent on the use of the "Plowright" vaccine [15], which is an attenuated vaccine produced from the Kabete O strain passaged 90 times in tissue culture [16]. In summary, conventional live and inactivated vaccines for a wide range of infectious diseases have been available for several decades and are still being developed for some emerging diseases. They are widely used in veterinary medicine and contribute considerably to the improvement of animal and public health [1, 2, 17].

## 3 Genetically Engineered Veterinary Vaccines

The availability of rich information on viral and bacterial genomes along with the advancement of available genetic tools allowed specific modifications or deletions to be introduced into the genome, with the aim of producing well-defined and stably attenuated live or inactivated viral or bacterial vaccines. Because of their relatively smaller size, viral genomes were used to be more amenable to genetic manipulations than bacterial genomes, and as a result, most of the genetically engineered licensed vaccines are based on viral vaccines. However, recent advances in genome editing opened up new avenues for multiple applications by allowing genetic material to be added, removed, or altered at particular locations in the genome of organisms much larger than viruses [18, 19].

Despite the method used, the ability to identify and selectively delete genes from a pathogen has allowed the development of "DIVA vaccines" which, combined with suitable diagnostic assays, allow differentiating infected from vaccinated animals (DIVA). The DIVA principle not only extends to vaccines with deleted genes but also includes subunit and other types of vaccines that induce an antibody response that is different from the antibody response produced by the wild-type organism. An accompanying diagnostic method, tailored for optimal sensitivity and specificity toward the epitopes distinguishing the vaccine from the wild-type, accounts for the DIVA diagnostics test. The first DIVA vaccine was used to differentiate between infected and vaccinated pigs for pseudorabies [20]. Such DIVA vaccines and their companion diagnostic tests are now available for several diseases including infectious bovine rhinotracheitis, classical swine fever, foot-and-mouth disease, etc. [17].

## 4 Subunit and Virus Like Particles (VLP)-Based Vaccines

Subunit-based vaccines contain only part of the virus or bacteria that is capable of inducing protective immune response against that component only. Thus, subunit vaccines can be compatible with the DIVA principle as long as the pathogen consists of another protein capable of consistently inducing antibodies in wild-type infected animals. The protective antigen or multiple of antigens of pathogens can be generated as recombinant proteins in various expression systems such as E. coli, yeast, baculovirus-insect cell, etc. to be used as subunit vaccines. A large quantity of recombinant proteins can be expressed, purified, and formulated in most of the cases with a potent adjuvant to be used as a safe and nonreplicating subunit or VLP vaccines. A typical subunit veterinary vaccine composed of a single protein is a recombinant OspA vaccine for the purpose of preventing Lyme disease in dogs caused by the spirochete, *Borrelia burgdorferi*. Interestingly, even a non-adjuvanted recombinant OspA (Recombitek® Lyme) was demonstrated to completely prevent *Borrelia burgdorferi* infection in vaccinated dogs [21]. Diagnostic assessment of Lyme "positive" or "negative" is made using a popular test kit (SNAP®), which is a patient-side, lateral flow ELISA technology that can detect antibody or antigen to a variety of vector-borne parasites, including C6 antibody to *Borrelia burgdorferi* [22]. An example of a more complex subunit vaccine was also developed against porcine contagious pleuropneumonia, a severe disease of pigs with hemorrhagic necrotizing pneumonia and high mortality in the acute form. The disease is caused by Actinobacillus pleuropneumoniae, and prevention by vaccination has been severely restricted by the prevalence of 15 different serotypes. The two available subunit vaccines are composed of either acellular A. pleuropneumoniae four extracted proteins (Porcilis® APP) or five recombinant proteins (Pleurostar APP™), which confer some degree of cross-protection against all tested serotypes [1]. There is currently a large amount of scientific interest in the identification of immunogenic and protective antigens for animal pathogens and expressing them in a safe heterologous expression system; thereby, handling of virulent or partially virulent microbes can be eliminated by the manufacturer as well as the end user.

As shown above, the end product of a subunit vaccine is recombinant protein(s), whereas in VLP-based vaccines, the protein(s) further spontaneously assemble into supramolecular structures resembling infectious viruses or, in some cases, subviral particles. Thus, VLPs are structurally similar to infectious viruses and thus are highly immunogenic but, because they lack viral nucleic acid, are noninfectious and safe [23]. In addition, VLPs do not induce antibody responses to internal or nonstructural viral proteins, thereby allowing distinction between vaccinated and infected animals [24]. These advantages have made VLPs attractive vaccine candidates for many viral diseases [25–29]. In the veterinary field, two baculovirus-expressed VLP-based vaccines were approved and widely used (Ingelvac® CircoFLEX and Porcilis PCV®) to provide protection against disease caused by porcine circovirus type 2 (PCV2) [30, 31]. PCV2, a member of the Circoviridae family, is associated with postweaning multisystemic wasting syndrome, a swine disease characterized by wasting, weight loss, respiratory distress, and diarrhea, that has a severe economic impact on production [32]. The immunogens in both vaccines are VLPs formed by the assembly of the single capsid protein encoded by the open reading frame 2 of PCV2. The assembly of these VLPs from single proteins allows a cost-effective vaccine production at large scale in the baculovirus-insect cells expression system. Although VLPs have been produced for a wide range of viruses under experimental condition, clearly not all are equally suitable for the development of commercial vaccines, due to several challenges ranging from scalability and cost-effectiveness to the need of co-expressing multiple proteins, including viral envelope proteins [30, 33, 34].

## 5 DNA Vaccines

Similar to subunit or VLP-based vaccines, DNA vaccines are encoding only part(s) of the pathogen, but antigen is produced intracellularly mimicking antigen expression by replication of live pathogen, thereby leading to the development of both humoral and cellular immune response. In addition, innate immune responses are also stimulated as plasmids contain molecular elements such as unmethylated CpG motifs that are not prevalent in mammal, avian, and fish cells [35, 36]. DNA vaccines are safe as production of plasmids does not involve manipulation of infective antigens. As immune responses are developed only against those coded antigens in the plasmid, DNA vaccines are also compatible with the DIVA principle. The DNA vaccines licensed so far are veterinary vaccines that include APEX-IHN® to prevent infectious hematopoietic necrosis (IHN) in farm-raised salmon and West

Nile-Innovator® DNA to prevent horses from West Nile disease. The first USDA-licensed therapeutic vaccine, ONCEPT® Canine Melanoma, is also a DNA vaccine that is proven to extend the lives of dogs treated for oral melanoma [37]. One of the main drawbacks of DNA vaccination in large animals and poultry has been their relative low efficacy and relatively high cost of goods mainly due to the large amounts of DNA needed to be injected in order to achieve a strong response. Thus, several strategies are under development in order to achieve better responses with less amounts of DNA.

# 6    Viral Vector-Based Vaccines

Viruses have evolved sophisticated structures and mechanisms to infect cells, and hence, they serve as efficient delivery vectors of various antigens. The concept of viral vector was introduced in 1972 when Jackson et al. created recombinant DNA from the SV40 virus by genetic engineering [38]. A decade later, the use of vaccinia virus as a transient gene expression vector was described [39, 40]. Thereafter, nonpathogenic or host-restricted viruses carrying foreign genes have been used as delivery vehicles that can be administered into a host, creating protective immunity to the inserted proteins. Although host-restricted vector viruses will not be productively replicating within the tissues of the vaccinated animals, they are able to express the foreign gene [41]. Viral vector-based vaccines are mostly used without an adjuvant, and like subunit and DNA vaccines, they allow for differentiating infected from vaccinated animals (DIVA). The first licensed viral vector-based vaccine (RABORAL V-RG®) is an oral rabies vaccine bait that contains an attenuated ("modified-live") recombinant vaccinia virus vector expressing the rabies virus glycoprotein gene of Evelyn-Rokitnicki-Abelseth rabies virus [42, 43]. RABORAL V-RG® has been in continuous use since 1987 when it was first field-tested in foxes in Belgium [44]. Thereafter, approximately 250 million doses have been distributed globally [11]. Before its introduction, rabies control in wildlife relied mostly on depopulation and the vaccination of individual animals. RABORAL V-RG® allowed oral vaccination on a large scale using vaccine-containing baits, and several countries have used RABORAL V-RG® safely without any adverse effects and have achieved complete rabies control [45, 46].

Subsequent poxvirus-based licensed products include TROVAC™-AIV H5, which is a bivalent recombinant fowlpox virus expressing the H5 antigen of avian influenza virus. This product has had a conditional license for emergency use in the United States since 1998 and has been widely used in Central America, with over four billion doses administered [47]. As the vaccinated birds will not develop antibodies against matrix protein/nucleoprotein, this vaccine can also be used with a DIVA approach. Additional fowlpox-based bivalent vaccines include Vectormune® FP-LT and Vectormune® FP-MG, which are indicated as aids in the prevention of fowlpox and infectious laryngotracheitis and fowlpox and Mycoplasma gallisepticum, respectively [48].

The most extensively used poxvirus vector platform in animal health is the ALVAC platform based on canarypox virus. Canarypox virus has the advantage of being more host-restricted than vaccinia virus. Because of the host restriction, canarypox virus recombinants produce abortive infections in mammalian cells, but they still effectively express inserted foreign genes. Currently, the canarypox virus or ALVAC vector platform has been used as a vaccine vector for a range of veterinary diseases of companion animals, including canine distemper, feline leukemia, rabies, West Nile, and equine influenza [49–54]. With the exception of equine influenza vaccine (PROTEQFLU®), all ALVAC-based vaccines are non-adjuvanted. Interestingly, the PROTEQFLU® contains a polymer adjuvant, and through the induction of both cell-mediated and humoral immunity, it is claimed to produce sterile immunity 2 weeks after the second of two doses [55].

Outside of poxviruses, the most successful and widely used vaccine vector in veterinary medicine is herpesvirus of turkeys (HVT). First and foremost, HVT has been widely accepted as a safe and effective vaccine against Marek's disease in chickens for almost 50 years [56, 57], providing an excellent vector backbone for the induction of protection against two or more poultry diseases. HVT has also unique features including the ability to cause persistent infection. As a result, a single dose of vaccine delivered in ovo to 18-day-old embryo or subcutaneously to 1-day-old chicks at the hatchery [58] induces a lifelong immunity. Like poxviruses, herpesvirus genomes can accommodate long fragments of foreign DNA without compromising their ability for normal replication [59]. These promising properties of HVT as a vector have led to the development and commercialization of HVT-based recombinant vaccines against important poultry diseases, including IBD, ND, and ILT [VAXXITEK® HVT + IBD, Innovax®-ILT, VECTORMUNE® ND, etc.]. One of the limitations of this extremely effective vector platform is the interference when more than one recombinant HVT vaccines are simultaneously administered. Hence, most of the recent research in this area has focused on either expressing more than one foreign genes or developing compatible viral vectors that could be combined with HVT in inducing protection against multiple avian diseases in multivalent vaccines [60].

## 7    Future Perspectives and Research Directions

**Combination Vaccines** Prevention rather than treatment is the most effective means of controlling infectious diseases in animal health. With the number of vaccines growing, combination vaccines are becoming more important. Protection against several diseases with fewer injections while maintaining the efficacy and safety of single-component vaccines helps not only to reduce costs but also to simplify overloaded immunization schedules [61]. In using combination vaccines, each component of the vaccine must be assessed individually and in combination to avoid an inappropriate immune response as a result of unwanted interaction of

the various components with each other [62]. In general, one observes higher interference in live multivalent vaccines than in killed multivalent vaccines. Thus, the challenge in this respect is to be able to combine multiple antigens in one vaccine without substantially reducing the efficacy and safety of the individual components.

The enormous potential of expressing multiple protective antigens from two or more pathogens in a single vector also opens the door for a new generation of multivalent viral vector vaccines. This is especially true of larger viral vectors, like herpesviruses and poxviruses, where there are few restrictions imposed by gene packaging limits. However, as we attempt to push the biological boundary by inserting multiple genes, issues like genetic instability of the vector and/or insert or less efficient replication of the vector backbone becomes more of a problem. Genetic instability is one of the major challenges in making viral vector-based vaccine candidates as some vectors are simply incompatible with certain sequences, length, or configuration of the insert. Consequently, the virus acting as the vector may introduce mutations or deletions into the inserted gene as well as into part of the vector backbone to abrogate proper expression of the transgene(s). It is, therefore, necessary to analyze the stability of these vectors during the early stages of vaccine research by serial rounds of replication in cell culture as well as in target animal species as required by the regulatory agencies (reversion to virulence or back passage studies). Obviously, the efficacy of a multiantigen vector should be satisfactory to meet the regulatory requirements as well as customer's need under field conditions.

**Overcoming Maternal/Preexisting Immunity** Maternal immunity, a form of passive immunity, plays a vital role in protecting young animals against severe disease upon infection with virulent field virus. However, maternal immunity is not without its negative effects. It often interferes with active immunization and is the most common cause of vaccine failure in animals. As vaccinations to most animals are preferably given at a young age, overcoming the interference by maternal immunity after vaccination is one of the greatest challenges in animal health. Although maternally derived antibody (MDA) interference is complex as many factors can influence the outcome after vaccination, one possibility that may be pursued to achieve this goal is to implement a heterologous prime-boost strategy. An excellent example that demonstrated the usefulness of this strategy in circumventing MDA interference used a vectored fowlpox virus expressing the hemagglutinin gene of avian influenza H5. The fowlpox-AI recombinant was administered at 1 day of age for early priming of chickens with MDA against fowlpox or avian influenza followed by a booster vaccination with whole virus-inactivated vaccine [63].

**More and Improved Bacterial Vaccines** Apart from improving animal health and productivity, veterinary vaccines have a significant impact on public health through reductions in the nontherapeutic use of antibiotics and hormones and their residues in the human food chain. The emergence of antibiotic-resistant bacteria in food-producing animals as well as in humans and the increasing restriction of the use of nontherapeutic antibiotic in livestock evidently promote the use of vaccine rather than antibiotics. Unfortunately, for a number of bacterial pathogens, vaccines are either still missing or some of the existing vaccines are not completely protective. One of the reasons for the poor protection is the existence of many different serotypes for a given disease and the poor level of cross-protection between serotypes. The growing power of combining sequencing, structural, and computational approaches may help the design of novel cross-protective immunogens suitable for future improved bacterial vaccine development.

**Rational Antigen Design** Prospects for new vaccines stem primarily from advances in genetic engineering and the ability to define the antigens responsible for inducing protective immunity. Recent advances in genomics, structural, and computational biology are opening the door to new technologies such as reverse vaccinology that allow designing novel vaccines against diseases unamenable to traditional vaccine development strategies [64, 65]. These approaches allow identification of a broader spectrum of vaccine candidates, including proteins that had not been identified and/or not abundant. The ability to rationally design candidate antigens can provide more cross-protective antigen candidates against antigenically variable pathogens [66]. In this regard, the parallel discovery and development of new and improved adjuvants for recombinant targets is essential to obtain the desired level and duration of protective immune response against different target pathogens. To speed up the implementation of these new technologies in animal health, closer interaction between research groups developing human and veterinary vaccines should be encouraged as common technologies can be applied in each areas. Consequently, animal vaccine research scientists can lean on medical research for some breakthrough technologies, but may also contribute to human vaccine development, as they are able to bridge the gap in translating results obtained in small-rodent models to large animal (human) application.

# 8 Summary

Much progress has been made in expanding the range of existing veterinary vaccines and introducing new technology-based vaccines with increased efficacy and reduced side effects. Due to the ability to directly test new vaccine candidates in target animal species, veterinary vaccines have a relatively quicker route to the market and thus are at the forefront of testing and commercializing innovative technologies, as exemplified by the successful licensing of vectored, subunit, DNA, and VLP-based veterinary vaccines. However, there are still many problems that remain to be addressed, and there is ample scope to incorporate new knowledge and technologies such as mRNA, replicon, and particle-scaffold-based platforms into vaccine design to fill the gaps. A better understanding of the molecular and immunological disease processes as well as the interaction between pathogens and the host immune system is likely to be required to improve the effectiveness of new and improved veterinary vaccines. Moreover, the interchange between scientists working on animal and human disease will remain essential to be prepared for the ever-present threat of new and reemerging diseases. The increased recognition of the "One Health" concept undoubtedly helps to foster this research collaboration. Based on the past significant advances in vaccinology and future opportunities for new innovations, it is beyond doubt that vaccines will continue to play a crucial role in maintaining and improving the health of animals and ultimately humans.

# References

1. Meeusen ENT, Walker J, Peters A, Pastoret PP, Jungersen G. Current status of veterinary vaccines. Clin Microbiol Rev. 2007;20:489–510.
2. Jorge S, Dellagostin OA. The development of veterinary vaccines: a review of traditional methods and modern biotechnology approaches. Biotechnol Res Innov. 2017;1:6–13.
3. Jivani HM, Mathapati BS, Javia BB, Padodara RJ, Nimavat VR, Barad DB, et al. Veterinary vaccines: past, present and future. Int J Sci Environ Technol. 2016;5:3473–85.
4. Willis NJ. Edward Jenner and the eradication of smallpox. Scott Med J. 1997;42:118–21.
5. Winkelstein W Jr. Not just a country doctor: Edward Jenner, scientist. Epidemiol Rev. 1992;14:1–15.
6. Pasteur L. Sur les maladies virulentes, et en particulier sur la maladie appelee vulgairement cholera des poules. C R Acad Sci. 1880;90:249–8.
7. Pasteur L. Methode pour prevenir la rage apres morsure. C R Acad Sci. 1885;101:765–74.
8. Weller TH, Enders JF, Robbins FC, Stoddard MB. Studies on the cultivation of poliomyelitis viruses in tissue culture. I. the propagation of poliomyelitis viruses in suspended cell cultures of various human tissues. J Immunol. 1952;69:645–71.
9. Syverton JT, Scherer WF. Studies on the propagation in vitro of poliomyelitis viruses. I. Viral multiplications in tissue cultures employing monkey and human testicular cells. J Exp Med. 1952;96:355–67.
10. World Health Organization. Rabies: Key facts. 2018. https://www.who.int/newsroom/fact-sheets/detail/rabies
11. Maki J, Guiot AL, Aubert M, Brochier B, Cliquet F, Hanlon CA, King R, Oertli EH, Rupprecht CE,

Schumacher C, Slate D, Yakobson B, Wohlers A, Lankau EW. Oral vaccination of wildlife using a vaccinia–rabies-glycoprotein recombinant virus vaccine (RABORAL V-RG®): a global review. Vet Res. 2017;48:57.
12. Delany I, Rappuoli DR, Gregorio ED. Vaccines for the 21st century. EMBO Mol Med. 2014;6:708–20.
13. Bergman JG, Muniz M, Sutton D, Fensome R, Ling F, Paul G. Comparative trial of the canine parvovirus, canine distemper virus and canine adenovirus type 2 fractions of two commercially available modified live vaccines. Vet Rec. 2006;159:733–6.
14. Organisation mondiale de la santé (Office international des épizooties [OIE]. No more deaths from rinderpest. OIE's recognition pathway paved way for global declaration of eradication by FAO member countries in June. 25 May 2011. http://www.oie.int/forthe-media/press-releases/detail/article/no-more-deaths-from-rinderpest/
15. Roeder P. Rinderpest eradication—is it feasible? In: Olsen I, Gjøen T, editors. Proceedings of the international veterinary vaccine and diagnostics conference. Oslo: Reprosentralen, University; 2006. p. 61–2.
16. Plowright W. The production and use of rinderpest cell culture vaccine in developing countries. World Anim Rev. 1972;1:14–8.
17. van Oirschot JT. Vaccinology: present and future of veterinary viral vaccinology: a review. Vet Q. 2001;23:100–8.
18. Doerflinger M, Forsyth W, Ebert G, Pellegrini M, Herold MJ. CRISPR/Cas9-the ultimate weapon to battle infectious diseases? Cell Microbiol. 2017;19:2. https://doi.org/10.1111/cmi.12693.
19. Loureiro A, da Silva GJ. CRISPR-Cas: converting a bacterial defence mechanism into a state-of-the-art genetic manipulation tool. Antibiotics (Basel). 2019;8(1):E18. https://doi.org/10.3390/antibiotics8010018.
20. van Oirschot JT, Rziha HJ, Moonen PJ, Pol JM, van Zaane D. Differentiation of serum antibodies from pigs vaccinated or infected with Aujeszky's disease virus by a competitive enzyme immunoassay. J Gen Virol. 1986;67:1179–82.
21. Wikle RE, Fretwell B, Jarecki M, Jarecki-Black JC. Canine lyme disease: one-year duration of immunity elicited with a Canine OspA Monovalent lyme vaccine. Int J Appl Res Vet Med. 2006;4:23–8.
22. Eschner AK, Mugnai K. Immunization with a recombinant subunit OspA vaccine markedly impacts the rate of newly acquired Borrelia burgdorferi infections in clientowned dogs living in a coastal community in Maine, USA. Parasit Vectors. 2015;8:92.
23. Brun A, Bárcena J, Blanco E, Borrego B, Dory D, Escribano JM, Le Gall-Reculé G, Ortego J, Dixon LK. Current strategies for subunit and genetic viral veterinary vaccine development. Virus Res. 2011;157:1–12.
24. Capua I, Terregino C, Cattoli G, Mutinelli F, Rodriguez JF. Development of a DIVA (Differentiating Infected from Vaccinated Animals)

strategy using a vaccine containing a heterologous neuraminidase for the control of avian influenza. Avian Pathol. 2003;32:47–55.

25. Buonaguro L, Tornesello ML, Buonaguro FM. Virus-like particles as particulate vaccines. Curr HIV Res. 2010;8:299–309.

26. Jennings GT, Bachmann MF. The coming of age of virus-like particle vaccines. Biol Chem. 2008;389:521–36.

27. Ramqvist T, Andreasson K, Dalianis T. Vaccination, immune and gene therapy based on virus-like particles against viral infections and cancer. Expert Opin Biol Ther. 2007;7:9971007.. Review

28. Roy P, Noad R. Virus-like particles as a vaccine delivery system: myths and facts. Adv Exp Med Biol. 2009;655:145–58.. Review

29. Spohn G, Bachmann MF. Exploiting viral properties for the rational design of modern vaccines. Expert Rev Vaccines. 2008;7:43–54.. Review

30. Mena JA, Kamen AA. Insect cell technology is a versatile and robust vaccine manufacturing platform. Expert Rev Vaccines. 2011;10:1063–81.. Review

31. Fachinger V, Bischoff R, Jedidia SB, Saalmüller A, Elbers K. The effect of vaccination against porcine circovirus type 2 in pigs suffering from porcine respiratory disease complex. Vaccine. 2008;26:1488–99.

32. Segalés J, Domingo M. Postweaning multisystemic wasting syndrome (PMWS) in pigs. A review. Vet Q. 2002;24:109–24.

33. Roldão A, Vicente T, Peixoto C, Carrondo MJ, Alves PM. Quality control and analytical methods for baculovirus-based products. J Invertebr Pathol. 2011;107(Suppl):94–105.. Review

34. Vicente T, Roldão A, Peixoto C, Carrondo MJ, Alves PM. Large-scale production and purification of VLP-based vaccines. J Invertebr Pathol. 2011;107 (Suppl):42–8.. Review

35. Bauer S, Pigisch S, Hangel D, Kaufmann A, Hamm S. Recognition of nucleic acid and nucleic acid analogs by toll-like receptors 7, 8 and 9. Immunobiology. 2008;213:315–28.. Review

36. Mutwiri G, Pontarollo R, Babiuk S, Griebel P, van Drunen Littel-van den Hurk S, Mena A, Tsang C, Alcon V, Nichani A, Ioannou X, Gomis S, Townsend H, Hecker R, Potter A, Babiuk LA. Biological activity of immunostimulatory CpG DNA motifs in domestic animals. Vet Immunol Immunopathol. 2003;91:89–103.. Review

37. Bergman PJ, Camps-Palau MA, McKnight JA, Leibman NF, Craft DM, Leung C, et al. Development of a xenogeneic DNA vaccine program for canine malignant melanoma at the animal medical center. Vaccine. 2006;24:4582–5.

38. Jackson DA, Symons RH, Berg P. Biochemical method for inserting new genetic information into DNA of simian virus 40: circular SV40 DNA molecules containing lambda phage genes and the galactose operon of Escherichia coli. Proc Natl Acad Sci U S A. 1972;69:2904–9.

39. Mackett M, Smith GL, Moss B. Vaccinia virus: a selectable eukaryotic cloning and expression vector. Proc Natl Acad Sci U S A. 1982;79:7415–9.

40. Panicali D, Paoletti E. Construction of poxviruses as cloning vectors: insertion of the thymidine kinase gene from herpes simplex virus into the DNA of infectious vaccinia virus. Proc Natl Acad Sci U S A. 1982;79:4927–31.

41. McFadden G. Poxvirus tropism. Nat Rev Microbiol. 2005;3:201–13.

42. Kieny MP, Lathe R, Drillien R, Spehner D, Skory S, Schmitt D, Wiktor T, Koprowski H, Lecocq JP. Expression of rabies virus glycoprotein from a recombinant vaccinia virus. Nature. 1984;312:163–6.

43. Blancou J, Kieny MP, Lathe R, Lecocq JP, Pastoret PP, Soulebot JP, Desmettre P. Oral vaccination of the fox against rabies using a live recombinant vaccinia virus. Nature. 1986;322:373–5.

44. Pastoret PP, Brochier B, Languet B, Thomas I, Paquot A, Bauduin B, Kieny MP, Lecocq JP, De Bruyn J, Costy F, et al. First field trial of fox vaccination against rabies using a vaccinia–rabies recombinant virus. Vet Rec. 1988;123:481–3.

45. Wandeler AI, Capt S, Kappeler A, Hauser R. Oral immunization of wildlife against rabies: concept and first field experiments. Rev Infect Dis. 1988;10(Suppl 4):649–53.

46. Rupprecht CE, Charlton KM, Artois M, Casey GA, Webster WA, Campbell JB, Lawson KF, Schneider LG. Ineffectiveness and comparative pathogenicity of attenuated rabies virus vaccines for the striped skunk (Mephitis mephitis). J Wildl Dis. 1990;26:99–102.

47. Bublot M, Pritchard N, Swayne DE, Selleck P, Karaca K, Suarez DL, Audonnet JC, Mickle TR. Development and use of fowlpox vectored vaccines for avian influenza. Ann N Y Acad Sci. 2006;1081:193–201.

48. Zhang GZ, Zhang R, Zhao HL, Wang XT, Zhang SP, Li XJ, Qin CZ, Lv CM, Zhao JX, Zhou JF. A safety assessment of a fowlpox-vectored Mycoplasma gallisepticum vaccine in chickens. Poult Sci. 2010;89:1301–6.

49. Taylor J, Meignier B, Tartaglia J, Languet B, VanderHoeven J, Franchini G, Trimarchi C, Paoletti E. Biological and immunogenic properties of a canarypox-rabies recombinant, ALVAC-RG (vCP65) in non-avian species. Vaccine. 1995;13:539–49.

50. Stephensen CB, Welter J, Thaker SR, Taylor J, Tartaglia J, Paoletti E. Canine distemper virus (CDV) infection of ferrets as a model for testing morbillivirus vaccine strategies: NYVAC- and ALVAC-based CDV recombinants protect against symptomatic infection. J Virol. 1997;71:1506–13.

51. Tartaglia J, Jarrett O, Neil JC, Desmettre P, Paoletti E. Protection of cats against feline leukemia virus by vaccination with a canarypox virus recombinant, ALVAC-FL. J Virol. 1993;67:2370–5.

52. Schlecht-Louf G, Mangeney M, El-Garch H, Lacombe V, Poulet H, Heidmann T. A targeted

mutation within the feline leukemia virus (FeLV) envelope protein immunosuppressive domain to improve a canarypox virus-vectored FeLV vaccine. J Virol. 2014;88:992–1001.

53. Minke JM, Siger L, Cupillard L, Powers B, Bakonyi T, Boyum S, Nowotny N, Bowen R. Protection provided by a recombinant ALVAC(®)-WNV vaccine expressing the prM/E genes of a lineage 1 strain of WNV against a virulent challenge with a lineage 2 strain. Vaccine. 2011;29:4608–12.

54. Edlund Toulemonde C, Daly J, Sindle T, Guigal PM, Audonnet JC, Minke JM. Efficacy of a recombinant equine influenza vaccine against challenge with an American lineage H3N8 influenza virus responsible for the 2003 outbreak in the United Kingdom. Vet Rec. 2005;156:367–71.

55. Minke JM, Audonnet JC, Fischer L. Equine viral vaccines: the past, present and future. Vet Res. 2004;35:425–43.

56. Swayne DE. Diseases of poultry. 13th ed. Ames: Wiley; 2013.

57. Okazaki W, Purchase HG, Burmester BR. Protection against Marek's disease by vaccination with a herpesvirus of turkeys. Avian Dis. 1970;14:413–29.

58. Morgan RW, Gelb J Jr, Schreurs CS, Lutticken D, Rosenberger JK, Sondermeijer PJ. Protection of chickens from Newcastle and Marek's diseases with a recombinant herpesvirus of turkeys vaccine expressing the Newcastle disease virus fusion protein. Avian Dis. 1992;36:858–70.

59. Afonso CL, Tulman ER, Lu Z, Zsak L, Rock DL, Kutish GF. The genome of Turkey herpesvirus. J Virol. 2001;75:971–8.

60. Baron MD, Iqbal M, Nair V. Recent advances in viral vectors in veterinary vaccinology. Curr Opin Virol. 2018;29:1–7.

61. Halsey NA. Safety of combination vaccines: perception versus reality. Pediatr Infect Dis J. 2001;20 (Suppl):S40–4.

62. Sanyal G, Shi L. A review of multiple approaches towards an improved hepatitis B vaccine. Expert Opin Ther Pat. 2009;19:59–72.

63. Richard-Mazet A, Goutebroze S, Le Gros FX, Swayne DE, Bublot M. Immunogenicity and efficacy of fowlpox-vectored and inactivated avian influenza vaccines alone or in a prime-boost schedule in chickens with maternal antibodies. Vet Res. 2014;45:107.

64. Dellagostin OA, Grassmann AA, Hartwig DD, Félix SR, da Silva ÉF, McBride AJ. Recombinant vaccines against leptospirosis. Hum Vaccin. 2011;7:1215–24.

65. Rappuoli R, Pizza M, Del Giudice G, De Gregorio E. Vaccines, new opportunities for a new society. Proc Natl Acad Sci U S A. 2014;111:12288–93.

66. Seib KL, Zhao X, Rappuoli R. Developing vaccines in the era of genomics: a decade of reverse vaccinology. Clin Microbiol Infect. 2012;18(Suppl 5):109–16.. Review

# What Is Required to Develop a Viral Vector Vaccine: Key Components of Vaccine-Induced Immune Responses

Philip J. Griebel

**Abstract**

Major challenges in vectored vaccine design include identification of protective antigens or epitopes and induction of protective immune responses in the appropriate target population. This chapter provides an overview of the many methods developed to identify vaccine antigens for which genes can then be cloned or synthesized and inserted into vaccine vectors. The types of protective immune responses required, whether humoral or cell-mediated and either systemic or mucosal, are then be discussed. The vaccine vector selected must be able to induce a protective immune response that effectively controls or prevents infection and prevents clinical disease. Achieving these objectives may dictate the route of vaccine delivery and necessitate combining two or more prime-boost strategies. Immune responses to the vaccine vector itself may also restrict vector usage and dictate vaccination protocols. These considerations may be of particular importance when considering different target populations for vaccination, such as either very young or elderly individuals. Finally, screening vaccine vector candidates for efficacy is dependent upon the availability of an appropriate disease model. This is not always feasible, especially with highly virulent emerging diseases. It may be possible to screen vaccines vectors only for immunogenicity and safety prior to testing in a clinical trial. Knowledge of disease pathogenesis and host-pathogen interactions must then be used to inform the selection, design, and delivery of an appropriate vaccine vector.

**Keywords**

Antibody · Epitopes · Mucosal immunity · Protective antigen · Systemic immunity · T cells · Vaccine interference

**Learning Objectives**

After reading this chapter, you should be able to:

- Explain the importance of identification of a protective antigen or epitope as a key step in selecting the appropriate pathogen gene for delivery in a vaccine vector.
- Recognize that protective immunity may be mediated by antibody or T cells or require a combination of both types of effector responses
- Explain the role of either mucosal or systemic immunity in the prevention of infection or clinical disease and that preventing pathogen shedding will inform the selection of the type of vaccine vector to use and the route of vector delivery

P. J. Griebel (✉)
University of Saskatchewan, Saskatoon, Canada
e-mail: philip.griebel@usask.ca

© Springer Nature Switzerland AG 2021
T. Vanniasinkam et al. (eds.), *Viral Vectors in Veterinary Vaccine Development*,
https://doi.org/10.1007/978-3-030-51927-8_2

- Recognize the importance of appropriate animal models in evaluating vaccine vector immunogenicity, safety, and efficacy

## 1    Introduction

Molecular biology has made it possible to engineer vaccine vectors expressing transgenes-encoding proteins or polypeptides from a wide variety of pathogens. It is also possible to modify the biological properties of vaccine vectors to alter their interaction with the host immune system or alter vector tropism. This chapter will focus on interactions between the vaccine vector and the host immune system and how knowledge of this interaction can be used to optimize selection and delivery of vaccine vectors. Major issues addressed will include first the need to identify a protective antigen or epitope. Secondly, it is important to consider the type of immune response required to either prevent or control infection, to significantly reduce clinical disease, or to prevent shedding and transmission of a pathogen. Finally, the population being targeted for vaccination may have a significant impact on vector selection when considering key aspects of vaccine immunogenicity, safety, and efficacy.

## 2    Knowledge of Target Antigens

Vaccines are an effective tool for the prevention or reduction of pathogen infection, the control of clinical disease, or reducing the shedding and transmission of infectious agents. One or more of these outcomes may be sufficient for a vaccine to achieve what is termed "protective immunity." Protective immunity is achieved by an interaction between the vaccine and the adaptive immune system that results in the induction of antigen-specific effector and memory immune responses. Therefore, the design of an effective vaccine is predicated on the identification of protein(s) or polypeptides that can induce immune responses and confer some form of protection. Identification

of a protective protein or polypeptide is of particular importance when designing vaccine vectors. The vaccine vector will deliver a gene which can then be expressed and translated into a single-pathogen protein or polypeptide. Therefore, appropriate selection of the protein or polypeptide to be expressed by the vaccine vector is a critical first step in designing a vector to induce a protective immune response.

Many different approaches have been used to select pathogen proteins that may be antigenic and induce a protective immune response (Table 1). For many pathogens, antibody responses may be sufficient to either reduce the level of infection, control clinical disease, or reduce pathogen shedding and transmission. In these situations, vaccine proteins were traditionally identified by determining which pathogen protein(s) reacted with antibodies present in convalescent serum [1]. This approach, when combined with isolation and sequencing of individual proteins, can identify multiple antigenic proteins but does not confirm which proteins induced a protective antibody response. It is possible to further refine screening of B-cell antigens to identify specific peptide epitopes. A variety of peptide formats have been used to screen peptides for reactivity with relevant antibodies, including random peptide arrays [2] and phage display libraries [3]. Further information regarding which protein or peptide induced a protective immune response is essential, however, before selecting the gene to be delivered in a vaccine vector. An appropriate animal disease model is usually required to further screening vaccine candidates for induction of protective antibody responses, but in vitro correlates of immune protection, such as neutralization of viral or bacterial infectivity and attachment, may also be used.

For many intracellular pathogens, including viruses and bacteria, identification of protective T-cell antigens or epitopes is also challenging. Immunogenic proteins and peptides have been identified through in vitro reactivation of T cells isolated from blood or other lymphoid tissues following recovery of an individual from infection [4]. The production of MHC tetramers made it possible to identify peptides recognized by T

**Table 1** Screening approaches used to identify potential vaccine antigens

| Methodology | Outcome | References |
|---|---|---|
| 1D and 2D Western blots with convalescent serum | B-cell antigens | Vytvytska et al. [1] |
| B-cell epitopes | Epitopes reacting with antibodies | Whittemore et al. [2] |
| | | Ellis et al. [3] |
| T-cell epitopes | Immunogenic peptides | Laing et al. [4] |
| | | Sidobre and Kronenberg [5] |
| DNA immunization and disease challenge | Immunogenic and protective antigens | Manovich et al. [6] |
| Genomic sequencing and bioinformatics | Predicted proteins, with function and localization in the pathogen | Green and Baker [7] |
| (Reverse vaccinology) | | Sanchez-Trincado et al. [8] |

cells when presented within the context of specific MHC molecules [5], but this methodology is not readily available for use in species of veterinary interest. More recently, the use of DNA vaccines has facilitated screening pools of potential vaccine candidates for either T- or B-cell responses. When this approach is combined with a disease challenge model, it is then possible to identify individual proteins that induce a protective level of immunity [6].

The prediction of potential vaccine antigens was further enhanced when whole genome sequences became available for a wide variety of pathogens. Genome sequence data could be used to identify genes-encoding proteins conserved among members of a heterogeneous pathogen population or identify proteins unique to a pathogen and not closely related but non-pathogenic species. Bioinformatic criteria, including protein function and location within a pathogen, can be applied during this selection process to further focus on the list of potentially relevant vaccine antigens [7]. Bioinformatic criteria have also been developed and applied to predict specific T- or B-cell epitopes within selected antigens [8]. This approach has been useful when applied to proteins known to be involved in the induction of a protective immune response. Information generated through these approaches can be used to synthesize genes encoding polypeptides containing multiple immunogenic epitopes, including both T- and B-cell epitopes. Thus, while antigen discovery has been greatly accelerated, there is still a need to screen proteins or peptides for their capacity to induce a protective immune response.

## 3 Knowledge of Optimal Route for Vector Delivery

Over 90% of pathogens enter through mucosal surfaces of the body, which include the eyes, respiratory, gastrointestinal tract, and reproductive tract (Fig. 1). Further, distinct immune effector mechanisms are induced within mucosa-associated lymphoid tissue (MALT; Fig. 1) versus systemic lymphoid tissues, which include blood, spleen, and lymph nodes draining skin, muscle, and some internal organs (Fig. 1). The induction of B cells producing IgA antibody in MALT is a defining difference between the mucosal and systemic immune systems (Fig. 1). IgA secretion is a key defense required for maintaining the integrity of mucosal epithelial surfaces and preventing both the invasion and shedding of pathogens. Therefore, an understanding of disease pathogenesis is critical for determining whether a vaccine vector should induce primarily a mucosal versus systemic immune response.

**Fig. 1** Comparison of mucosal and systemic immune system structure and function and vaccine delivery routes available for induction of adaptive immune responses within each immune system

| Mucosal ImmuneSystem |
| --- |
| **Location** <br> Eyes, Nose, Mouth <br> Gastrointestinal tract <br> Reproductive tract |
| **Organized Lymphoid Tissue** <br> Tonsils <br> Peyer's patches <br> Solitary lymphoid follicles <br> Lymph nodes draining mucosa |
| **Effector Mechanisms** <br> Antibody - secretory IgA <br> T cells – Cytotoxic, helper, regulatory <br> Innate Lymphoid Cells |
| **Vaccine Delivery Route** <br> Ocular <br> Intranasal <br> Oral <br> Vaginal |

| SystemicImmune System |
| --- |
| **Location** <br> Skin <br> Muscle <br> CNS and Internal Organs |
| **Organized Lymphoid Tissue** <br> Blood <br> Spleen <br> Lymph nodes draining skin, muscle, CNS, and internal organs |
| **Effector Mechanisms** <br> Antibody – IgM, IgG, IgE <br> T cells – Cytotoxic, helper, regulatory <br> Innate Lymphoid Cells |
| **Vaccine Delivery Route** <br> Intradermal <br> Subcutaneous <br> Intramuscular <br> Intravenous |

## 4    Knowledge of Immune Responses Induced by Vaccine Vectors

Vaccine vectors offer numerous advantages and opportunities to tailor effector responses and optimize immune protection. The selection of a vaccine vector that targets epithelial cells at mucosal surfaces provides an opportunity to induce secretory (S) IgA responses (Fig. 1). The capacity of SIgA to bind a specific protein provides a key defense in preventing pathogen or toxin interactions with mucosal epithelial cells. This is extremely beneficial when the pathogen or toxin has direct cytotoxic or metabolic effects that compromise mucosal barrier function or integrity.

Intracellular pathogens, such as viruses, may invade mucosal surfaces and subsequently target tissues or organs throughout the body. Cellular immune response may then be required to control and eliminate these infections. Vaccine vectors that express recombinant proteins provide an ideal delivery vehicle for the induction of both antibody and cellular immune responses. If vaccine vectors can be designed to specifically target dendritic cells (DCs), this provides an opportunity for recombinant proteins to be processed by the endogenous antigen-processing pathway. This route of antigen processing optimizes induction of both CD4 T-helper cells and CD8 cytotoxic T cells, through antigen presentation on major histocompatibility complex II and I, respectively. Thus, targeting a vaccine vector to the appropriate antigen-presenting cell can provide a broad spectrum of immune effector functions, including both antibody and cell-mediate immune effector functions.

Using vaccine vectors to target mucosal surfaces is also very attractive from the perspective of vaccine safety. This route of vaccine delivery eliminates the use of needles and associated risks of iatrogenic disease transmission. There are, however, a number of potential risks that must be considered when delivering vaccine vectors to a mucosal site. Vaccine delivery may be less efficient if targeting mucosal sites where natural barriers, such as mucopolysaccharides and mucociliary clearance, limit vector interactions with mucosal epithelial cells. Further, replication-competent vaccine vectors may induce vector-specific antibody and cell-mediated immune responses. Vector-specific immune responses, especially SIgA, may limit subsequent use of the same vaccine vector by blocking vector interactions with mucosal epithelium. Booster vaccinations are frequently required to ensure induction of protective immune responses in a high proportion of the vaccinated population and adequate duration of immunity. Therefore, alternative vaccine boost strategies may be required if vector-specific antibody or cell-mediated immune responses are induced during the primary vaccination. The induction of T regulatory cells (Tregs) is also a possibility when a vaccine vector targets mucosal surfaces, where tolerogenic DCs frequently reside. Therefore, when targeting mucosal surfaces with a vaccine vectors, it is critical to evaluate whether the vector induces effector immune responses or Tregs (Fig. 1). Tregs may subsequently limit immune responses when the host is infected with a pathogen.

Knowledge of disease pathogenesis may also be critical for selecting the most effective vaccine vector to induce protective immunity at specific sites within the body. A pathogen may have a specific tropism for epithelial cells present at one or more of the many mucosal surfaces (Fig. 1), resulting in localized infection. The concept of the "common mucosal immune system" is based on numerous observations that immunization at one mucosal site can induce an effector response at other unrelated mucosal surface [9]. These vaccine studies have clearly demonstrated, however, that shared effector responses at mucosal surfaces are not always reciprocal, with some mucosal induction sites more effective in generating a common mucosal immune response than other mucosal sites. Therefore, vaccine vector selection and delivery may need to be tailored to induce a protective immune response at one specific mucosal site. Vector transgene expression at the appropriate mucosal site may, however, be influenced by a variety of factors. These factors include vector tropism for a specific mucosal epithelial surface, possible pre-existing viral vector immunity at a mucosal site, and innate and acquired immune responses induced by the vaccine vector.

Much of the discussion regarding interactions between vaccine vectors and the immune system focused on immune responses following mucosal delivery of a vaccine vector. Similar concerns exist, however, when considering vaccine vectors delivered parenterally to induce systemic immune responses (Fig. 1). Effector responses induced at mucosal versus systemic immune induction sites may be different (Fig. 1), but similar concerns exist regarding induction of vector-specific immune responses, targeting antigen-presenting cells, and interactions with the innate immune system. Therefore, when selecting a vector for vaccine delivery, it is important to first determine whether mucosal or systemic effector responses are required to protect against clinical disease and control infection and disease transmission.

There has been an increased focus on the analysis of innate immune responses induced by specific vaccine vectors [10]. Characterizing innate immune responses induced by vaccine vectors is important for understanding host interactions with vaccine vectors, immunogenicity of vaccine vectors, and possible implications for vaccine safety. Innate immunity plays an important role in determining the magnitude, duration, and specific type of adaptive immune response induced by a vaccine. Therefore, analyzing and comparing innate immune responses induced by vaccine vectors at both mucosal and systemic immune induction sites (Fig. 1) may provide information critical for optimizing vector selection. It may be necessary to either alter the type of vaccine vector used and the route of vector delivery (Fig. 1) or specifically engineer the vector to ensure induction of protective immune responses by the protein encoded by the transgene. Inflammation is a key component of innate immune responses, and the level of inflammation induced by a vaccine vector may be important when considering adverse local or systemic effects associated with the use of a vaccine vector.

## 5 Knowledge of Target Population for Immunization with Vaccine Vectors

Newborns of all species are the most susceptible to infectious disease due to a naïve and immature immune system. Passive transfer of maternal antibody and T cells may provide some protection from infection, but the level of protection may vary greatly, depending on the specificity and amount of maternal antibody transferred. Rapid decay of maternal antibody in the newborn can also result in susceptibility to infection. Therefore, interest has increased in the use of vaccines during the neonatal period to accelerate onset of protective immunity as maternal immunity wanes [11]. The use of vaccine vectors in neonates provides an opportunity to induce either mucosal or systemic immunity. The selection of vectors will, however, be critical to ensure there is no vaccine interference by maternal antibody and to ensure vaccine safety when there is an immature immune system.

An increasing population of elderly people has created an awareness that immune senescence in the elderly may compromise immune responses to vaccines [12]. Situations where immune competence may also be compromised exist for species of veterinary interest. Many of these situations are associated with stress responses as animals go through transitions periods, such as transportation, parturition, or the separation of young animals from their mothers [13]. These stress responses may compromise responses to vaccines, but a variety of strategies have been suggested to augment vaccine efficacy, including increasing the dose of vaccine antigen, increasing the frequency of vaccination, and using more potent adjuvants to enhance activation of the immune system. Tailored design of vaccine vectors may be able to address a number of these issues and augment immune responses induced by a vaccine. Delivery of an increased antigen dose may be addressed by engineering vaccine vectors to either increase transgene expression or prolong the duration of antigen expression following vector delivery. The immunogenicity of vaccine vectors may be enhanced by incorporating specific immune stimulatory cytokines [14], designing vectors that stimulate greater innate immune responses [15], or strategic use of vaccine vectors for prime-boost vaccination.

## 6 Knowledge of Vaccine Use Prophylactically or Therapeutically

A discussion of immune responses induced by vaccine vectors must also consider whether the vaccine is being used for prophylactic or therapeutic treatment. Prophylactic vaccines aim to reduce the risk of infection, prevent clinical disease, or reduce the risk of disease transmission through herd immunity. To achieve these objectives, a prophylactic vaccine should establish protective immunity prior to the period when there is a risk of infection. As discussed with neonates, this is a challenge since immunization must begin early in life. This highlights the need to understand the time required for onset of immunity following vaccination and whether protective immunity is achieved following a single or multiple vaccinations. Furthermore, it is important to determine the duration of protective immunity and immune memory following vaccination since this will define the number of vaccinations required to protect individuals during the period when they are at risk of infection. These factors need to be considered when evaluating the efficacy of a vaccine vector and determining how a vector may be most effectively used in a vaccine program.

Therapeutic vaccines are used following infection, usually with the objective of modulating immune responses to either enhance protective immune responses or suppress responses that cause immune pathology. The design of effective therapeutic vaccines requires detailed knowledge of the mechanism by which a pathogen is able to persist and evade the immune system [16]. Designing therapeutic vaccines will require knowledge of pathogen antigens able to induce protective immune responses, but also the exact nature of protective immune responses, whether B cell or T cell, must be known. These correlates of protective immunity can then be used to guide the design and selection of vaccine vectors that induce immune responses of the correct specificity required to control or clear a persistent infection. The ability to engineer vaccine vectors to express both antigens and immune-modulating molecules [14] may be essential to achieve these objectives.

# 7 Summary

Many factors must be considered when developing a vaccine vector that induces a protective immune response. Immune responses must be focused on the appropriate antigen to induce a protective immune response. Further, these immune responses must be present at the appropriate site in the body to prevent either infection or dissemination of infection throughout the body and reduce pathogen shedding or disease transmission. Further, the onset and duration of immunity induced by the vaccine vector must be appropriate to protect the target population throughout the risk period of infection. A broad array of vaccine vectors are currently available or in development, and this provides the opportunity to match the requirements of an effective vaccination program with the most appropriate vaccine vector. Molecular biology is also making it possible to further tailor the attributes of vaccine vectors to optimize their tropism, level of transgene expression, interactions with the host immune system, and safety. The full potential of vaccine vectors to meet the ever-changing vaccine challenges has yet to be determined.

# References

1. Vytvytska O, Nagy E, Blüggel M, Meyer HE, Kurzbauer R, Huber LA, Klade CS. Identification of vaccine candidate antigens of Staphylococcus aureus by serological proteome analysis. Proteomics. 2002;2 (5):580–90.
2. Whittemore K, Johnston SA, Sykes K, Shen L. A general method to discover Epitopes from Sera. PLoS One. 2016;11(6):e0157462.
3. Ellis SE, Newlands GF, Nisbet AJ, Matthews JB. Phage-display library biopanning as a novel approach to identifying nematode vaccine antigens. Parasite Immunol. 2012;34(5):285–95.
4. Laing KJ, Magaret AS, Mueller DE, Zhao L, Johnston C, De Rosa SC, Koelle DM, Wald A, Corey L. Diversity in CD8(+) T cell function and epitope breadth among persons with genital herpes. J Clin Immunol. 2010;30(5):703–22.
5. Sidobre S, Kronenberg M. CD1 tetramers: a powerful tool for the analysis of glycolipid-reactive T cells. J Immunol Methods. 2002;268(1):107–21.
6. Manovich JK, Chapman D, Hansen DT, Robida MD, Loskutov A, Craciunescu F, Borovkov A, Kibler K, Goatley L, King K, Netherton CL, Taylor G, Jacobs B, Sykes K, Dixon LK. Immunization of pigs by DNA prime and recombinant Vaccinia virus boost to identify and rank African swine fever virus immunogenic and protective proteins. J Virol. 2018;92(8):e02219-17.
7. Green BA, Baker SM. Recent advances and novel strategies in vaccine development. Curr Opin Microbiol. 2002;5(5):483–8.
8. Sanchez-Trincado JL, Gomez-Perosanz M, Reche PA. Fundamentals and methods for T- and B-cell epitope prediction. J Immunol Res. 2017;2017:2680160. https://doi.org/10.1155/2017/2680160.
9. McGhee JR, Xu-Amano J, Miller CJ, Jackson RJ, Fujihashi K, Staats HF, Kiyono H. The common mucosal immune system: from basic principles to enteric vaccines with relevance for the female reproductive tract. Reprod Fertil Dev. 1994;6(3):369–79.
10. Teigler JE, Phogat S, Franchini G, Hirsch VM, Michael NL, Baroucha DH. The canarypox virus vector ALVAC induces distinct cytokine responses compared to the vaccinia virus-based vectors MVA and NYVAC in rhesus monkeys. J Virol. 2014;88 (3):1809–14.
11. Griebel PJ. Mucosal vaccination of the newborn: an unrealized potential. Expert Rev Vaccines. 2009;9 (1):1–3.
12. Boraschi D, Italiani P. Immunosenescence and vaccine failure in the elderly: strategies for improving response. Immunol Lett. 2014;162(1 Pt B):346–53.
13. Chen Y, Arsenault R, Napper S, Griebel P. Models and methods to investigate acute stress responses in cattle. Animals. 2015;5:1268–95.
14. Raggo C, Habermehl M, Babiuk LA, Griebel PJ. In vivo effects of a recombinant bovine herpesvirus-1 vector expressing bovine interferon-gamma. J Gen Virol. 2000;81:2665–73.
15. Hendrickx R, Stichling N, Koelen J, Kuryk L, Lipiec A, Greber UF. Innate immunity to adenovirus. Hum Gene Ther. 2014;25(4):265–84.
16. Spohn G, Bachmann MF. Exploiting viral properties for the rational design of modern vaccines. Expert Rev Vaccines. 2008;7(1):43–54.

# Viruses and the Evolution of Viral Vectors

Carla Giles and Thiru Vanniasinkam

**Abstract**

Since the first documented, widespread use of vaccines in the eighteenth century, vaccines have become a common health preventative for humans and animals against a range of pathogens. This chapter covers the key developments in the use of viruses in vaccine development that have occurred since viruses were first identified as potential vaccine and gene therapy vectors.

**Keywords**

Adenovirus · DNA viruses · Poxvirus · RNA viruses · Veterinary vaccines · Viral vector

## Learning Objectives

After reading this chapter, you should be able to:

- Explain how viruses were first used as vaccine vectors
- State key events in the timeline of viral vector development
- List examples of viruses used as vaccine vectors
- Compare various viral vector platforms using examples of vaccines developed using that technology

## 1 Introduction

The concept of a non-bacteriological agent causing disease had been present for many years. However, in 1892, Dmitri Ivanovsky described the first strong evidence that tobacco mosaic virus was not bacteriological after sap filtrate remained infectious after being passed through Chamberland filters which can filter bacteria. This work was replicated in 1898 by Martinus Beijerinck, who coined the term virus to indicate the pathogen causing the infection was non-bacteriological.

In 1796, 100 years prior to the identification of viruses, Edward Jenner developed the first vaccine for the human disease smallpox caused by the variola virus. He used the distinct but antigenically similar cowpox virus as an antigen and, in doing so, created the world's first vaccine. Over the centuries since his discovery, vaccines have become a common health preventative for humans and animals with a plethora of viral, bacterial and parasitic vaccines developed and routinely administered.

The limited availability of broad-spectrum antivirals available to veterinarians has made the need for vaccines against viruses imperative.

C. Giles (✉)
Centre for Aquatic Animal Health & Vaccines, Department of Primary Industries Parks Water & Environment, Launceston, TAS, Australia
e-mail: carla.giles@dpipwe.tas.gov.au

T. Vanniasinkam
School of Biomedical Sciences, Charles Sturt University, Wagga wagga, NSW, Australia
e-mail: tvanniasinkam@csu.edu.au

© Springer Nature Switzerland AG 2021
T. Vanniasinkam et al. (eds.), *Viral Vectors in Veterinary Vaccine Development*,
https://doi.org/10.1007/978-3-030-51927-8_3

With the increasing antimicrobial and antiparasitic resistance seen worldwide, the need for effective vaccines to prevent infections has increased, particularly in intensive agricultural industries including poultry, pigs and aquaculture.

Over the years, many types of vaccine technologies ranging from attenuated to subunit have been utilised in the development of various veterinary vaccines. Some have been more successful than others. In the 1990s, technologies such as DNA vaccines were a key focus for vaccinologists; however, due to the lack of efficacy of DNA vaccine candidates, this approach soon dropped out of favour. However, one DNA vaccine was licensed in Canada for use in Atlantic salmon for infectious haematopoietic necrosis virus [21].

More recently, since the dawn of the 'molecular biology and genomics' era, researchers have been manipulating genomes of various microorganisms. This has allowed for the study of various organisms as potential vectors, whether this is bacterial, viral or parasitic, in the design of vaccines. Vectors have been evaluated for their potential use to deliver a gene or peptide. Since the early 1980s, the use of viral vector vaccines has been explored as an alternative mechanism of gene/peptide delivery [45, 51, 62], particularly in circumstances where empirical vaccines have not been effective [25, 86].

Even where vaccines are available to prevent disease caused by a particular pathogen, there may be a need for more efficacious vaccines against some pathogens, and generally, viral vector-based vaccines are one vaccine modality that researchers consider when looking at approaches to make a more immunogenic vaccine. Figure 1 describes the evolution timeline of viruses and vaccines.

One particular area where the viral vector platform is thought to be particularly relevant is in vaccinating neonates. Neonates have often been challenging to immunise using traditional empirical vaccines due to their naïve immune system and the possible interference by maternal antibodies. However, viral vectors have had success in stimulating the naïve immune system of neonates to produce a protective immune response [31, 95].

The early success of some viral vector-based vaccines has led to a plethora of viral vector-based vaccines being developed and evaluated in studies across the globe. One key factor in creating a successful viral vector-based vaccine is a virus with a genome that is relatively easy to manipulate. Several viruses have now been intensively studied for their ability to be developed as vaccine vectors. These include adenovirus; poxviruses, particularly vaccinia virus; flaviviruses; and lentiviruses; all have been used as vaccine candidates in both humans and animals. Apart from the ease of their genome manipulation, these viruses are chosen for their stability, relatively safe profile, large transgene capacity, high expression of the transgene, low pathogenicity and suitability for engineering to provide prophylactic and therapeutic protection [25, 86].

In the veterinary field, viral vectors have been successfully licensed, and vaccines based upon this technology are commercially available. Vectors include the attenuated canarypox virus (ALVAC) [69], engineered to express antigens from a range of pathogens including equine influenza, West Nile virus and canine distemper virus [53]. Fowlpox virus ( TROVAC ) has been engineered as a vaccine against avian influenza virus and Newcastle disease virus in poultry [13, 84].

Vaccines against rabies have also been successfully engineered using different viral vectors for wildlife and domestic animal vaccination [11, 45, 83, 85, 87]. Many more viral vector-based vaccines are in clinical trials for veterinary applications [5]. Viral vectors are also being explored for immunisation of salmon [7, 98] and other aquaculture species [15]. It must be noted that similar to vaccines directed at wildlife, vaccines for fish grown in ocean net cages need to be based on a safe viral vector that cannot revert to its wild type or replicate as the viral

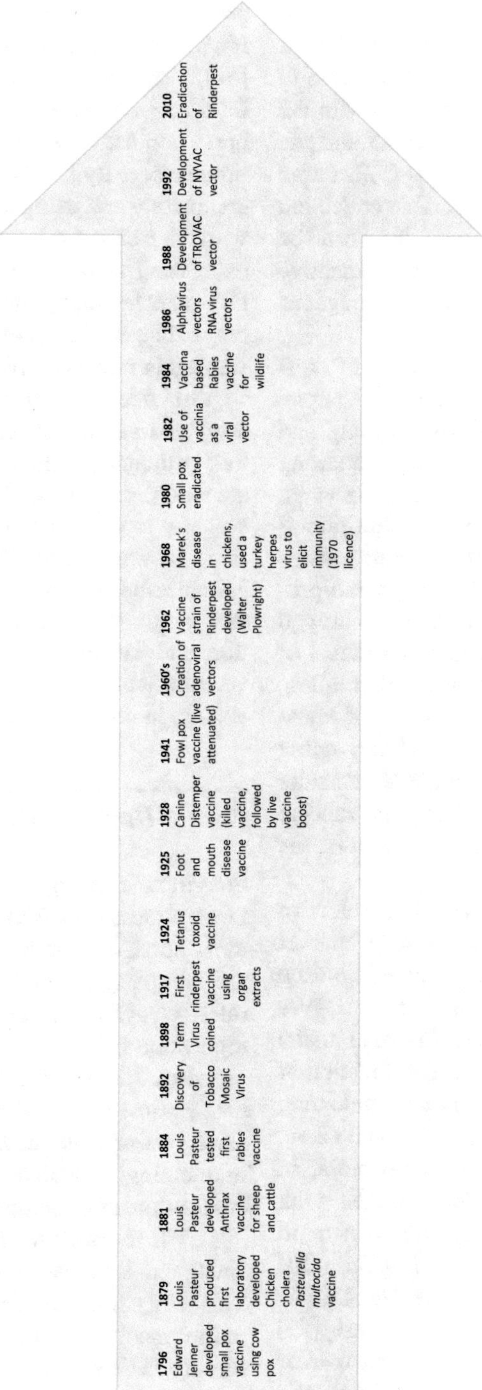

**Fig. 1** Timeline of the evolution of viruses, vaccine and viral vector development

The timeline contains the following entries:

|  |  |
|---|---|
| 1796 | Edward Jenner developed small pox vaccine using cow pox |
| 1879 | Louis Pasteur produced first laboratory developed Chicken cholera *Pasteurella multocida* vaccine |
| 1881 | Louis Pasteur developed Anthrax vaccine for sheep and cattle |
| 1884 | Louis Pasteur tested first rabies vaccine |
| 1892 | Discovery of Tobacco Mosaic Virus |
| 1898 | Term Virus coined |
| 1917 | First rinderpest vaccine using organ extracts |
| 1924 | Tetanus toxoid vaccine |
| 1925 | Foot and mouth disease vaccine |
| 1928 | Canine Distemper vaccine (killed vaccine, followed by live vaccine boost) |
| 1941 | Fowl pox vaccine (live attenuated) |
| 1960's | Creation of adenoviral vectors |
| 1962 | Vaccine strain of Rinderpest developed (Walter Plowright) |
| 1968 | Marek's disease in chickens, used a turkey herpes virus to elicit immunity (1970 licence) |
| 1980 | Small pox eradicated |
| 1982 | Use of vaccinia as a viral vector |
| 1984 | Vaccinia based Rabies vaccine for wildlife |
| 1986 | Alphavirus vectors RNA virus vectors |
| 1988 | Development of TROVAC vector |
| 1992 | Development of NVAC vector |
| 2010 | Eradication of Rinderpest |

vector will be used in the presence of other aquatic species, and there is potential to contaminate the environment [76].

A replication-deficient virus can have similar or increased immunogenicity when compared to its replicating counterpart, as is the case with the modified vaccinia virus Ankara (MVA) and the New York vaccinia (NYVAC) virus [52]. Large DNA viruses often have a considerable collection of immune evasion mechanisms that can be directed against both the innate and adaptive immune pathways, which can be advantageous in a vaccine vector.

Different viral vectors are often suited to different vaccine regimens. For example, adenoviruses are proficient at both priming and boosting the immune system, producing a strong T- and B-cell response. However, fowlpox virus, influenza and DNA (plasmid) vectors are principally suited as priming vectors. Vaccinia viruses are highly effective as boosting vectors and produce a strong T-cell response. It seems as though the adenoviruses have an extended high level of antigen expression that is ideal for B-cell priming, whereas vaccinia virus has short bursts of transgene expression suiting B-cell boosting rather than priming. However, it is essential to consider the route of delivery when determining a vaccine regimen as this can affect the expression of the transgene [25, 32].

Early modified viral vaccines using selective deletion generated the ability to produce 'marker vaccines' coupled with appropriate diagnostic assays which allowed for the use of the DIVA (differentiating infected from vaccinated animals) principles to be followed and the identification of vaccinated and unvaccinated animals, allowing for subsequent appropriate measures to be taken. These DIVA vaccines have been influential in eliminating viruses from national herds, with many nations willing to utilise this system to eliminate disease outbreaks and gain OIE disease-free status. The most notable DIVA-based vaccines are foot-and-mouth disease [35] and classical swine fever, where subunit marker vaccines are coupled with appropriate diagnostic assays [9, 56].

Diagnostic tests can effectively differentiate between vaccinates and wild-type immunity providing for effective control strategies to be implemented as described in Fig. 2 (Image from [61]). This is an example of the 2007 equine influenza outbreak in Australia, where a naïve horse population was exposed to imported equine influenza. By stopping horse movement completely and using a ring vaccination program with a canarypox viral vector (ALVAC) expressing equine influenza genes coupled with DIVA technology, equine influenza was able to be eradicated from Australia, and EI free status was again reinstated [61].

Viral vector vaccines will continue to be pursued as a viable option for the difficult-to-vaccinate pathogens as their safety margin and efficacy are well documented. DIVA viral vector vaccines are easy to engineer, offer a viable way to determine between vaccinates and naturally infected animals and will continue to be useful in veterinary fields. This chapter will outline the generation and evolution of some of the major viral vectors with further details on specific viral vectors in other chapters.

## 2    Poxviruses

Poxviruses are large, enveloped, double-stranded DNA viruses from 200 to 300 nm in size, with the most notable virus being smallpox. Pox virions are complex, contain enzymes associated with mRNA synthesis and have a cytoplasmic replication [66].

Edward Jenner first used a poxvirus to vaccinate against variola virus, the causative agent of small pox in 1798, and, in doing so, founded the immunology sciences. This discovery led to the eradication of smallpox from existence in 1980 [42, 66]. In modern times, in order to eradicate smallpox, scientists used a live vaccinia virus. This virus was attenuated through cell passage; unfortunately, the origin of this virus is a myth. Phylogenetic studies suggest this virus was originally a horse poxvirus; however, the true history of vaccinia has been lost to time. The eradication

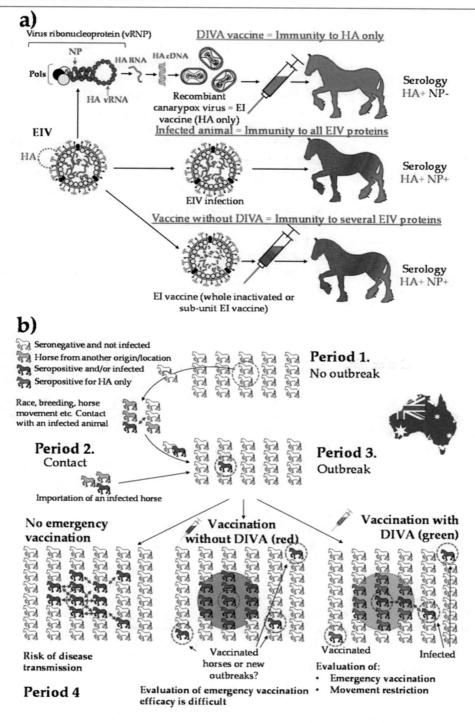

**Fig. 2** The importance of differentiating infected from vaccinated animals (DIVA). The use of an EI vaccine with DIVA capability is an asset to maintain an effective surveillance during an outbreak while emergency vaccination is implemented [3]. (a) The canarypox-based EI vaccine is a live-attenuated canarypox virus with the EIV HA gene (green) inserted in its genome (one EIV HA per canarypox vector). The canarypox-based EI vaccine induces a seroconversion limited to the EIV HA antigen after immunisation (green horse). Infection with EIV or

of smallpox was aided by the restriction of small-pox to the human host as poxviruses vary dramat-ically with their host infectivity [66, 69].

In the 1980s, there was a flood of interest in poxviruses as viral vectors primarily due to the rise of successful cloning techniques and molecu-lar genetics. One early example is the use of vaccinia virus to express hepatitis B antigens [75]. This finding provided the initial platform for expressing foreign genes in viral vectors [51, 62]. Primarily two approaches have been successfully used to increase the safety of poxviruses. One approach has been the deletion of known genes involved in viral metabolism, extracellular virus formation and host interaction. The other involves successive viral passage in cell culture or a non-host species and the subsequent isolation of less virulent variants [66].

The Copenhagen vaccinia virus strain was used to develop the NYVAC strain of vaccinia. This virus contains multiple gene deletions and 18 open reading frames, each with targeted deletions that prevent the synthesis of unwanted novel gene products, are highly attenuated, are confirmed safety and maintained proficiency to induce a robust immune response to foreign inserted antigens [82]. The NYVAC platform has been used to vaccinate pigs against pseudorabies [12]; however, it has not been commercialised. Instead, an attenuated (gene-deleted) pseudorabies virus has been commercialised [30]. Considered a 4th-generation vaccinia vaccine, the NYVAC

vector has been genetically engineered to have lost targeted genes [42].

Another commonly used vaccinia strain is the MVA, which was isolated from over 500 passages in chicken embryo cells [52]. MVA is a third-generation vaccinia vaccine as the virus was attenuated by cell passage [42]. This virus has lost the ability to replicate in human cells, is not pathogenic in immunocompromised animals and was used as a smallpox vaccine in the late 1970s [39]. This vaccinia strain has been tested as a vaccine vector or for delivering cancer therapeu-tics in both humans and animals [26, 58].

Avipox viruses (avian origin poxviruses) are restricted to replicate in avian species. In non-avian species, avipox viruses are proven to be safe and efficacious vectors, where they are suicidal vectors and are unable to replicate in non-avian species. Some avipox viruses are attenuated in other avian species; for example, pigeon pox is innately attenuated in chickens and thus is an ideal vector in this species [66]. The attenuation of the avipox viruses in other species opens up many benefits in terms of using these vectors in vaccine development. Other examples of attenuated viral vectors include fowlpox and canarypox virus (TROVAC is the attenuated fowlpox, and ALVAC the attenuated canarypox virus) [63].

Live-attenuated fowlpox has been used in poultry to control disease since the 1940s. In the 1980s, it was considered as a potential viral vec-tor due to its extreme stability (no cold-chain

**Fig. 2** (continued) immunisation with whole inactivated or subunit EI vaccines induces a seroconversion to several EIV antigens, including the EIV HA (green) and the nucleoprotein (NP, red) [12]. (b) An equid population naïve for EI (Period 1). Due to horse movement and/or importation of an infected animal (Period 2), an EI out-break is detected (green + red horse, Period 3). Prevention and control measures are implemented (Period 4). In the absence of emergency EI vaccination, disease control relies primarily on movement restriction, active surveil-lance and biosecurity measures and is heavily dependent on the horse population density. A virus such as EIV is likely to spread quickly, especially in a naïve population such as those in Australia in 2007. Emergency vaccination

is implemented to support these measures. If an EI vaccine without DIVA capability is used (green + red vaccine), any seroconversion (green + red horse) detected outside the vaccination buffer zone should be considered as a potential EI case (i.e., it is not possible to discriminate between a vaccinated horse that moved from the vaccina-tion buffer zone and a new infected horse). The use of an EI vaccine with DIVA capability (green vaccine) allows scientists to follow the spread of EIV infection inside the vaccination buffer zone, to identify real EI outbreak and infected horses (green + red horses) outside the vaccina-tion buffer zone and to control the implementation of specific measures such as movement restriction. Image available via licence: CC BY 4.0 [61] storage required), the low cost of production and the ability to insert multiple genes for expression [84]. The TROVAC-based viral vector has most notably been used and licensed in Central and South America as an avian influenza vaccine [13], and the DIVA approach can also be used with this vaccine.

Canarypox was first isolated from a single-pox lesion on a canary. The virus was passaged 200 times through chicken embryo fibroblasts and purified using a plaque purification technique. The attenuated version of this virus, ALVAC, has been proven to be safe, cannot replicate and is safe to use in a wide variety of species including canaries, mice, horses, cats and dogs. The inserted transgene does not appear to affect the stability or host tropism of the vector [69].

Due to ALVAC's rigorous evaluation prior to registration, its approval for use as a veterinary vaccine vector in Europe has been a relatively straightforward process [69]. Currently, ALVAC is used as a vector platform for the development of vaccines against a range of veterinary diseases including equine influenza and West Nile virus in horses, canine distemper virus in dogs and feline leukaemia virus in cats [53, 69].

A vaccinia-vectored rabies vaccine, expressing the rabies G protein [45], was developed with the purpose of wildlife vaccination in rabies-endemic areas due to the re-emergence of rabies in wildlife reservoirs, particularly the red fox. Vaccinia was chosen due to its stability at a wide range of temperatures, efficacy and safety. The introduction of this vaccine saw the dramatic reduction in rabies cases in both wildlife and domestic carnivores in both Europe and North America [10, 11, 65].

# 3    Adenoviruses

Adenoviruses are medium-sized, 90–100 nm, non-enveloped double-stranded DNA viruses, with a linear genome of 26–48 Kbp, encoding 22–40 genes. This virus infects a broad range of host species, including mammal, bird, amphibian, reptile and fish [19]. Adenoviruses are capable of

multiplicity reactivation whereby at least two lethally damaged viral genomes interact within a host cell and recombine to form a viable cell [99].

Adenoviruses are a popular vector candidate due to their enhanced ability to produce:

- Antigen-specific immune responses to transgenes
- Strong CD8$^+$ T-cell and B-cell responses
- Low pathogenicity
- Readily infect mucosal surfaces
- Induce systemic infections
- Stability once engineered

One limitation of the adenovirus vector is the presence of pre-existing immunity to this virus in many hosts. This is due to adenoviruses being a common pathogen of humans and animals. In humans, rare adenovirus serotypes or non-human serotype adenoviruses have been evaluated in order to circumvent this issue. Currently, work in re-engineering the virus capsid proteins has also been initiated [22, 25, 86].

Furthermore, the use of mastadenoviruses (mammalian adenoviruses) in the human context has been examined with both bovine and canine adenoviruses (BAdV-3 and CAdV-2, respectively) engineered to be E1 and E3 deleted and capable of replicating and growing in human cells [47, 54]. One of the first non-human adenoviruses to be utilised as a vector in human clinical trials was an ovine adenovirus used for the treatment of prostate cancer [94].

To use adenovirus as a vaccine vector, it is important to control the replication of the virus, in order to address the problem of virus shedding and ensure safety so the virus does not cause disease. Adenovirus has undergone various stages or 'generations' of gene deletions in the process of making better more efficacious vectors, which have been developed for gene therapy and vaccine vector purposes.

First-generation adenovirus vectors have been made by substituting an expression cassette for the E1 region and/or the E3 region. The E1 early promoter region encodes for proteins required for cellular replication and is located on the left end of the genome [2]. The E3 region encodes for

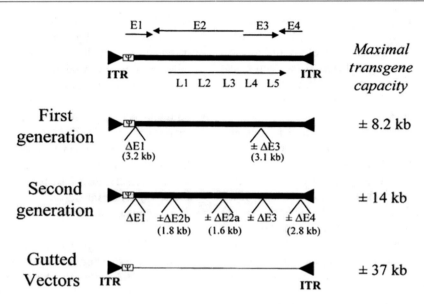

**Fig. 3** Genome organisation of first-generation, second-generation and gutted adenovirus vectors. The locations of the early and late transcription units (arrows) in the adenovirus genome (black bar) are shown on top of the figure. The ITRs are symbolised by triangles. The packaging signal Ψ is highlighted by an open box. Non-adenoviral sequences in gutless vectors are represented by a thin line. The sizes of the largest deletions are indicated for each region, and the maximal insert sizes are indicated for each type of vector [17]

proteins that reduce host defence mechanisms and is located on the right end of the genome. With the E1 region accepting inserts up to 5.1 kb, and when 3.1 kb of the E3 region was deleted, 8.2 kb of space was generated in these first-generation adenoviral vectors.

To assist with in vitro replication of the first generation E1-deleted viruses, specialised cell lines were developed to support adenovirus growth, especially when they are deficient in essential genes. These cell lines include HEK 293 (Human Embryonic Kidney) [34], 911 [29] and others. These cells were established by exposing them to sheared fragments of HAdV-5 DNA, which allowed for the expression of characteristics to support virus growth in the cells [34].

First-generation E1-deleted vectors are known for inducing a potent innate and adaptive immune response in vivo, particularly a Th1-type response [48]. In order to generate more genome space for larger inserts, second-generation vectors were constructed. This occurred by removing the E2b (terminal protein and DNA polymerase) [2], E2a

(DNA-binding protein) [101] and E4 regions of the genome [49], allowing for the insertion of 14 kb expression cassettes when combined with the first-generation deletions [17].

Third-generation or gutted or helper-dependent adenoviral vectors have been created whereby all of the viral genes are deleted except for the cis-acting sequences associated with viral DNA replication and packaging [78]. Theoretically, these gutted vectors can contain up to 37 kb of insertions. However, these vectors lack stability and can be difficult to produce. Figure 3 demonstrates the genome structures of each generation of adenovirus vector [17].

First-generation vectors are now easy to produce with numerous methods established [17]. Some first-generation adenovirus vectors, particularly HAdV-5, are documented for the induction of cytokine storms. That is releasing an overabundance of pro-inflammatory cytokines, including TNF-α and IL-10, and can cause toxicity. This is due to the ongoing expression of adenovirus genes 5–7 days post-administration, which can also aid in the development of a potent adaptive immune response [59]. Second-generation adenovirus vectors with the E2a region deleted have a longer gene expression time (20–40 days) [28], with E4-deleted adenovirus having transgene expression for 84 days [14, 93].

Studies have shown that neutralising antibody present before vaccination with an adenoviral vector or post prime vaccination can limit the effectiveness of the immune response of subsequent doses as was found in human clinical trials which utilised a homologous HAdV-5 vaccine regime [1]. Thus, a heterologous-based vaccine regimen that utilises multiple vectors may be more effective. In the past, some apprehension regarding the use of human adenoviral constructs in commercial, domestic animals has previously been discussed [33, 88]. Species tropism has also been problematic when utilising a non-host adenovirus [87, 100]; this highlights the importance of keeping the species specificity of the virus in mind when deciding on the most appropriate virus to use when developing a vector.

In order to avoid problems associated with pre-existing neutralising antibodies or to increase targeted sequestration of the vector, some researchers have employed capsid manipulation techniques or used chimeric adenoviruses [4, 16]. This mechanism changes the antigenic profile of the capsid or fibre proteins, preventing the naturally occurring antibodies from binding and the subsequent neutralisation of the vector.

An increase in the target receptor of a vector virus has also been achieved by utilising ovine adenovirus fibre on a human adenovirus capsid, altering receptor interaction and bio-distribution [60]. In prime/boost vaccine regimens, multiple different antigenic versions of a vector could be created to ensure optimal immunogenicity of the viral vector. These techniques may also potentially be used to create vectors that have specific host cell tropism.

In humans, the adenovirus hexon protein, which has a highly conserved C terminus, is an immunodominant antigen. Both $CD4^+$ and $CD8^+$ T cells develop adenovirus-specific memory that cross-react with different serotypes of the virus [81]. However, in various mammals, this immune cross-reactivity has not led to any problems concerning immunisation [31, 37]. It has been found that the adenovirus hexon capsid proteins can independently act as an adjuvant, and this has prompted further research into the role that these antigens may play when adenoviruses are used as vaccine vectors [25, 55].

The enhanced priming ability of adenoviruses is important when there is a requirement to elicit a robust immune response, for example, when priming a neonate with a relatively immature and naïve immune system. Adenoviral vectors have been successful in vaccinating canine pups against canine distemper virus (CDV), a highly contagious pathogen. A CDV transgene inserted into canine adenovirus 2 (CAdV-2) was replication-competent as it was only E3 deleted. Remarkably, this vaccine was able to elicit a strong immune response in puppies that were born to a CAdV-2- and CDV-positive dam and circumvent the maternal antibodies that were present in the puppies and provide protection against both pathogens. As this viral vector was replication-competent, it was shed in nasal secretions, and so the vaccine was not deemed suitable for commercial use [31].

As researchers work towards improving viral vector vaccines, heterologous prime boost protocols involving two different viral vectors have been frequently studied. One of the more widely used heterologous prime boost protocols involves the use of MVA and adenovirus vectors. Studies show that when these two vectors are used in heterologous prime boost regimens, each vector stimulates different branches of the immune system, making the vaccination regime more effective [25, 27].

Other mixed modality vaccination regimens have used adenoviral vectors with DNA or subunit vaccines include vaccines against *Mycobacterium tuberculosis* [40, 80] and malaria [24]. Generally, vaccinologists avoid using the same vector in prime boost protocols due to lack of efficacy when neutralising antibodies are generated during the prime. However, there are reports in the literature of approaches such as capsid manipulation, which may overcome issues associated with generating neutralising antibodies to the vaccine vector following the prime dose. This would make it possible to use the same vector in prime boost regimens [22].

Non-host adenoviruses are an important source of future vaccine vector development. For example, recently, vaccines based upon a chimpanzee adenovirus vector platform were shown to be efficacious in multiple mammalian species (cattle, sheep and goats) against Rift Valley fever virus. However, further studies are required [96], whereas avian herpesvirus vectors are also commonly used as vaccine vectors in poultry production, particularly herpesvirus of turkey [5]. These examples demonstrate a variety of applications for adenoviral vectors.

## 4 RNA Viruses

RNA viruses are known to cause numerous infections in humans, plants and animals, including diseases associated with epidemics and pandemics such as influenza and SARS. They contain an RNA genome that can be double- or single-stranded, positive or negative sense. Compared with DNA viruses, the development of RNA viruses as vaccine vectors is more recent and driven by the development of reverse genetics techniques for manipulating RNA virus genomes [79].

A range of viral vectors (based upon both double- and single-stranded RNA viruses) have been constructed, and several of these were licensed for use in vaccine development. These include a chimeric flavivirus targeting West Nile virus for use in horses and the alphavirus Venezuelan equine encephalitis virus (VEEV) vector targeting classical swine fever (CSFV).

The first veterinary alphavirus infectious clone for VEEV was described in 1989 [18]. Since then, alphavirus genome modification has been via cDNA manipulation of the clones to produce heterologous genes, with two types of clones developed. Replication-competent vectors (replicons) [70] and propagation-defective vectors both require helper gene assistance *in trans* for packaging [90].

VEEV vectors have been targeted for both veterinary and human settings. A VEEV vector vaccine was licensed, in the USA in 2012, for swine influenza expressing the HA gene against H3N2 [91]. In humans, this vector has been tested for infectious diseases and cancer therapy in clinical trials including HIV [97] and the seasonal flu [68, 72].

Salmonid alphavirus vectors have been developed [44, 57]. However, an inactivated vaccine was viable [43] and commercially licensed by Pharmaq (Zoetis). The only commercial DNA vaccine is targeted at the salmon and trout pathogen infectious haematopoietic necrosis virus [73]. To date, live-attenuated vector-based vaccine has not been approved for commercial use.

Other examples of RNA viral vectors include Flaviviridae, with the yellow fever vector vaccine targeting West Nile virus (WNV) and ChimeriVax-WN02® having participated in human clinical trials [8]. Various WNV vaccines are licensed in horses, including a single-dose live-attenuated yellow fever vector targeting WNV, PreveNile® by Intervet, which was reviewed [20].

Several pestiviral vectors have also been developed. Bovine viral diarrhoea virus vectors were developed to target classical swine fever virus (CSFV) in pigs and have been trialled extensively. These chimeric vaccines allow for DIVA differentiation [6, 41]; however, it requires multiple doses to be effective. Bovine parainfluenza 3 (bPIV-3) is antigenically similar and cross-reactive to human parainfluenza virus 3 and produces a protective immune response [89]. Due to this cross-protection, several vectors or chimeras have been developed for human use [36, 74].

The negative sense RNA viral genome of influenza virus was manipulated by reverse genetics [50]. In humans, the influenza vaccine FluMist® was licensed for seasonal influenza and is approved for people 2–49 years of age. This virus is cold-adapted and has the HA, and NA segments changed to be relevant to each season [38].

Another example of an RNA virus vector which has been widely researched is based upon the Newcastle disease virus (NDV). Several reverse genetic approaches to modify this virus have occurred [67, 71], and subsequent vectors developed [23, 46, 64, 92]. Many of these vectors target both Newcastle disease and avian influenza. In addition, more recently, the use of NDV as a potential cancer vaccine and gene therapy vector has been reported in the literature [3, 77].

# 5    Summary

While vaccines have been used for many decades, viral vector-based vaccines have only relatively recently been licensed for use in animals. It is likely that in the future, more viruses will be developed for use as vaccine vectors as researchers continue to improve on existing vaccine technologies. Advances may also influence viral vector-based veterinary vaccine development in viral vector-based human vaccines. While the focus of this chapter has been viral vectors used in vaccine development for infectious diseases, there will likely be viral vectors used in the development of vaccines for non-infectious conditions such as cancer and allergies in animals. If there is a demand for such vaccines, one potential market is companion animals such as dogs and cats.

# References

1. Abel B, Tameris M, Mansoor N, Gelderbloem S, Hughes J, Abrahams D, Makhethe L, Erasmus M, Md K, van der Merwe L. The novel tuberculosis vaccine, AERAS-402, induces robust and polyfunctional CD4+ and CD8+ T cells in adults. Am J Respir Crit Care Med. 2010;181(12):1407–17.
2. Amalfitano A, Hauser MA, Hu H, Serra D, Begy CR, Chamberlain JS. Production and characterization of improved adenovirus vectors with the E1, E2b, and E3 genes deleted. J Virol. 1998;72(2):926–33.
3. Amin ZM, Ani MAC, Tan SW, Yeap SK, Alitheen NB, Najmuddin SUFS, Kalyanasundram J, Chan SC, Veerakumarasivam A, Chia SL. Evaluation of a recombinant Newcastle disease virus expressing human IL12 against human breast cancer. Sci Rep. 2019;9(1):1–10.
4. Appaiahgari MB, Vrati S. Adenoviruses as gene/vaccine delivery vectors: promises and pitfalls. Expert Opin Biol Ther. 2015;15(3):337–51.
5. Baron MD, Iqbal M, Nair V. Recent advances in viral vectors in veterinary vaccinology. Curr Opin Virol. 2018;29:1–7. https://doi.org/10.1016/j.coviro.2018.02.002.
6. Beer M, Reimann I, Hoffmann B, Depner K. Novel marker vaccines against classical swine fever. Vaccine. 2007;25(30):5665–70.
7. Biacchesi S, Yu Y-X, Béarzotti M, Tafalla C, Fernandez-Alonso M, Brémont M. Rescue of synthetic salmonid rhabdovirus minigenomes. J Gen Virol. 2000;81(8):1941–5.
8. Biedenbender R, Bevilacqua J, Gregg AM, Watson M, Dayan G. Phase II, randomized, double-blind, placebo-controlled, multicenter study to investigate the immunogenicity and safety of a West Nile virus vaccine in healthy adults. J Infect Dis. 2011;203(1):75–84.
9. Blome S, Staubach C, Henke J, Carlson J, Beer M. Classical swine fever—an updated review. Viruses. 2017;9(4):86.
10. Brochier B, Aubert M, Pastoret P, Masson E, Schon J, Lombard M, Chappuis G, Languet B, Desmettre P. Field use of a vaccinia-rabies recombinant vaccine for the control of sylvatic rabies in Europe and North America. Revue Scientifique et Technique-Office International des Epizooties. 1996;15(3):947–80.
11. Brochier B, Kieny M, Costy F, Coppens P, Bauduin B, Lecocq J, Languet B, Chappuis G, Desmettre P, Afiademanyo K. Large-scale eradication of rabies using recombinant vaccinia-rabies vaccine. Nature. 1991;354(6354):520.
12. Brockmeier SL, Lager KM, Tartaglia J, Riviere M, Paoletti E, Mengeling WL. Vaccination of pigs against pseudorabies with highly attenuated vaccinia (NYVAC) recombinant viruses. Vet Microbiol. 1993;38(1–2):41–58.
13. Bublot M, Pritchard N, Swayne DE, Selleck P, Karaca K, Suarez DL, AUDONNET JC, Mickle TR. Development and use of fowlpox vectored vaccines for avian influenza. Ann N Y Acad Sci. 2006;1081(1):193–201.
14. Chen H-H, Mack LM, Kelly R, Ontell M, Kochanek S, Clemens PR. Persistence in muscle of an adenoviral vector that lacks all viral genes. Proc Natl Acad Sci. 1997;94(5):1645–50.
15. Chen Y, Guo M, Wang Y, Hua X, Gao S, Wang Y, Li D, Shi W, Tang L, Li Y. Immunity induced by recombinant attenuated IHNV (infectious hematopoietic necrosis virus)-GN438A expresses VP2 gene-encoded IPNV (infectious pancreatic necrosis virus) against both pathogens in rainbow trout. J Fish Dis. 2019;42:631–42.
16. Coughlan L, Alba R, Parker AL, Bradshaw AC, McNeish IA, Nicklin SA, Baker AH. Tropism-modification strategies for targeted gene delivery

using adenoviral vectors. Viruses. 2010;2 (10):2290–355.

17. Danthinne X, Imperiale M. Production of first generation adenovirus vectors: a review. Gene Ther. 2000;7(20):1707.

18. Davis NL, Willis LV, Smitht JF, Johnston RE. In vitro synthesis of infectious Venezuelan equine encephalitis virus RNA from a cDNA clone: analysis of a viable deletion mutant. Virology. 1989;171 (1):189–204.

19. Davison AJ, Benkő M, Harrach B. Genetic content and evolution of adenoviruses. J Gen Virol. 2003;84 (11):2895–908.

20. Dayan GH, Pugachev K, Bevilacqua J, Lang J, Monath TP. Preclinical and clinical development of a YFV 17 D-based chimeric vaccine against West Nile virus. Viruses. 2013;5(12):3048–70.

21. Dhar AK, Manna SK, Allnutt FT. Viral vaccines for farmed finfish. Virus. 2014;25(1):1–17.

22. Dharmapuri S, Peruzzi D, Aurisicchio L. Engineered adenovirus serotypes for overcoming anti-vector immunity. Expert Opin Biol Ther. 2009;9(10):1279–87.

23. DiNapoli JM, Kotelkin A, Yang L, Elankumaran S, Murphy BR, Samal SK, Collins PL, Bukreyev A. Newcastle disease virus, a host range-restricted virus, as a vaccine vector for intranasal immunization against emerging pathogens. Proc Natl Acad Sci. 2007;104(23):9788–93.

24. Douglas AD, de Cassan SC, Dicks MDJ, Gilbert SC, Hill AVS, Draper SJ. Tailoring subunit vaccine immunogenicity: maximizing antibody and T cell responses by using combinations of adenovirus, poxvirus and protein-adjuvant vaccines against *Plasmodium falciparum* MSP1. Vaccine. 2010;28 (44):7167–78.

25. Draper SJ, Heeney JL. Viruses as vaccine vectors for infectious diseases and cancer. Nat Rev Microbiol. 2010a;8(1):62–73.

26. Draper SJ, Heeney JL. Viruses as vaccine vectors for infectious diseases and cancer. Nat Rev Microbiol. 2010b;8(1):62.

27. Draper SJ, Moore AC, Goodman AL, Long CA, Holder AA, Gilbert SC, Hill F, Hill AVS. Effective induction of high-titer antibodies by viral vector vaccines. Nat Med. 2008;14(8):819–21.

28. Engelhardt JF, Litzky L, Wilson JM. Prolonged transgene expression in cotton rat lung with recombinant adenoviruses defective in E2a. Hum Gene Ther. 1994;5(10):1217–29.

29. Fallaux FJ, Kranenburg O, Cramer SJ, Houweling A, van Ormondt H, Hoeben RC, van der Eb AJ. Characterization of 911: a new helper cell line for the titration and propagation of early region 1-deleted adenoviral vectors. Hum Gene Ther. 1996;7(2):215–22.

30. Ferrari M, Brack A, Romanelli M, Mettenleiter TC, Corradi A, Mas ND, Losio M, Silini R, Pinoni C, Pratelli A. A study of the ability of a TK-negative and gI/gE-negative pseudorabies virus (PRV) mutant inoculated by different routes to protect pigs against PRV infection. J Veterinary Med Ser B. 2000;47 (10):753–62.

31. Fischer L, Tronel JP, Pardo-David C, Tanner P, Colombet G, Minke J, Audonnet J-C. Vaccination of puppies born to immune dams with a canine adenovirus-based vaccine protects against a canine distemper virus challenge. Vaccine. 2002;20 (29–30):3485–97.

32. Geiben-Lynn R, Greenland JR, Frimpong-Boateng K, Letvin NL. Kinetics of recombinant adenovirus type 5, vaccinia virus, modified vaccinia ankara virus, and DNA antigen expression in vivo and the induction of memory T-lymphocyte responses. Clin Vaccine Immunol. 2008;15(4):691–6.

33. Gogev S, Vanderheijden N, Lemaire M, Schynts F, D'Offay J, Deprez I, Adam M, Eloit M, Thiry E. Induction of protective immunity to bovine herpesvirus type 1 in cattle by intranasal administration of replication-defective human adenovirus type 5 expressing glycoprotein gC or gD. Vaccine. 2002;20(9–10):1451–65.

34. Graham F, Smiley J, Russell W, Nairn R. Characteristics of a human cell line transformed by DNA from human adenovirus type 5. J Gen Virol. 1977;36(1):59–77.

35. Grubman MJ. Development of novel strategies to control foot-and-mouth disease: marker vaccines and antivirals. Biologicals. 2005;33(4):227–34.

36. Haller AA, Miller T, Mitiku M, Coelingh K. Expression of the surface glycoproteins of human parainfluenza virus type 3 by bovine parainfluenza virus type 3, a novel attenuated virus vaccine vector. J Virol. 2000;74(24):11626–35.

37. Hammond JM, Johnson MA. Porcine adenovirus as a delivery system for swine vaccines and immunotherapeutics. Vet J. 2005;169(1):17–27.

38. Harper SA, Fukuda K, Cox NJ, Bridges CB. Using live, attenuated influenza vaccine for prevention and control of influenza. MMWR. 2003;52:1–8.

39. Hochstein-Mintzel V, Hänichen T, Huber H, Stickl H. An attenuated strain of vaccinia virus (MVA). Successful intramuscular immunization against vaccinia and variola (author's transl). Zentralblatt fur Bakteriologie, Parasitenkunde, Infektionskrankheiten und Hygiene Erste Abteilung Originale Reihe A: Medizinische Mikrobiologie und Parasitologie. 1975;230(3):283–97.

40. Hokey D, Tameris M, Henson D, Dheenadhayalan V, Blatner G, McClain B, Walker R, Hatherill M, Nduba V, Sacarlal J. Safety and immunogenicity of the Ad35/AERAS-402 tuberculosis vaccine in a phase 2 dose-finding clinical trial in BCG-vaccinated healthy infants (VAC5P. 1119). J Immunol. 2015;194(1 Suppl):73.74.

41. Hulst M, Westra D, Wensvoort G, Moormann R. Glycoprotein E1 of hog cholera virus expressed in insect cells protects swine from hog cholera. J Virol. 1993;67(9):5435–42.

42. Jacobs BL, Langland JO, Kibler KV, Denzler KL, White SD, Holechek SA, Wong S, Huynh T, Baskin CR. Vaccinia virus vaccines: past, present and future. Antivir Res. 2009;84(1):1–13.

43. Karlsen M, Tingbø T, Solbakk I-T, Evensen Ø, Furevik A, Aas-Eng A. Efficacy and safety of an inactivated vaccine against salmonid alphavirus (family Togaviridae). Vaccine. 2012;30(38):5688–94.

44. Karlsen M, Villoing S, Rimstad E, Nylund A. Characterization of untranslated regions of the salmonid alphavirus 3 (SAV3) genome and construction of a SAV3 based replicon. Virol J. 2009;6 (1):173.

45. Kieny M, Lathe R, Drillien R, Spehner D, Skory S, Schmitt D, Wiktor T, Koprowski H, Lecocq J. Expression of rabies virus glycoprotein from a recombinant vaccinia virus. Nature. 1984;312 (5990):163.

46. Kim S-H, Samal S. Newcastle disease virus as a vaccine vector for development of human and veterinary vaccines. Viruses. 2016;8(7):183.

47. Klonjkowski B, Gilardi-Hebenstreit P, Hadchouel J, Randrianarison V, Boutin S, Yeh P, Perricaudet M, Kremer EJ. A recombinant E1-deleted canine adenoviral vector capable of transduction and expression of a transgene in human-derived cells and in vivo. Hum Gene Ther. 1997;8(17):2103–15.

48. Liu Q, Muruve D. Molecular basis of the inflammatory response to adenovirus vectors. Gene Ther. 2003;10(11):935–40.

49. Lusky M, Christ M, Rittner K, Dieterle A, Dreyer D, Mourot B, Schultz H, Stoeckel F, Pavirani A, Mehtali M. In vitro and in vivo biology of recombinant adenovirus vectors with E1, E1/E2A, or E1/E4 deleted. J Virol. 1998;72(3):2022–32.

50. Luytjes W, Krystal M, Enami M, Parvin JD, Palese P. Amplification, expression, and packaging of a foreign gene by influenza virus. Cell. 1989;59 (6):1107–13.

51. Mackett M, Smith GL, Moss B. Vaccinia virus: a selectable eukaryotic cloning and expression vector. Proc Natl Acad Sci. 1982;79(23):7415–9.

52. Mayr A, Stickl H, Müller H, Danner K, Singer H. The smallpox vaccination strain MVA: marker, genetic structure, experience gained with the parenteral vaccination and behavior in organisms with a debilitated defence mechanism (author's transl). Zentralblatt fur Bakteriologie, Parasitenkunde, Infektionskrankheiten und Hygiene Erste Abteilung Originale Reihe B: Hygiene, Betriebshygiene, praventive Medizin. 1978;167(5–6):375–90.

53. Meeusen EN, Walker J, Peters A, Pastoret P-P, Jungersen G. Current status of veterinary vaccines. Clin Microbiol Rev. 2007;20(3):489–510.

54. Mittal SK, Prevec L, Graham FL, Babiuk LA. Development of a bovine adenovirus type 3-based expression vector. J Gen Virol. 1995;76 (1):93–102.

55. Molinier-Frenkel V, Lengagne R, Gaden F, Hong SS, Choppin J, Gahery-Segard H, Boulanger P, Guillet JG. Adenovirus hexon protein is a potent adjuvant for activation of a cellular immune response. J Virol. 2002;76(1):127–35.

56. Moormann RJ, Bouma A, Kramps JA, Terpstra C, De Smit HJ. Development of a classical swine fever subunit marker vaccine and companion diagnostic test. Vet Microbiol. 2000;73(2–3):209–19.

57. Moriette C, LeBerre M, Lamoureux A, Lai T-L, Brémont M. Recovery of a recombinant salmonid alphavirus fully attenuated and protective for rainbow trout. J Virol. 2006;80(8):4088–98.

58. Moss B, Carroll MW, Wyatt LS, Bennink JR, Hirsch VM, Goldstein S, Elkins WR, Fuerst TR, Lifson JD, Piatak M. Host range restricted, non-replicating vaccinia virus vectors as vaccine candidates. In: Novel strategies in the design and production of vaccines. Springer, New York; 1996. pp. 7–13.

59. Muruve DA, Cotter MJ, Zaiss AK, White LR, Liu Q, Chan T, Clark SA, Ross PJ, Meulenbroek RA, Maelandsmo GM. Helper-dependent adenovirus vectors elicit intact innate but attenuated adaptive host immune responses in vivo. J Virol. 2004;78 (11):5966–72.

60. Nakayama M, Both GW, Banizs B, Tsuruta Y, Yamamoto S, Kawakami Y, Douglas JT, Tani K, Curiel DT, Glasgow JN. An adenovirus serotype 5 vector with fibers derived from ovine atadenovirus demonstrates CAR-independent tropism and unique biodistribution in mice. Virology. 2006;350 (1):103–15. https://doi.org/10.1016/j.virol.2006.01. 037.

61. Paillot R, Marcillaud Pitel C, D'Ablon X, Pronost S. Equine vaccines: how, when and why? Report of the vaccinology session, French Equine Veterinarians Association, 2016. Reims: Multidisciplinary Digital Publishing Institute; 2017.

62. Panicali D, Paoletti E. Construction of poxviruses as cloning vectors: insertion of the thymidine kinase gene from herpes simplex virus into the DNA of infectious vaccinia virus. Proc Natl Acad Sci. 1982;79(16):4927–31.

63. Paoletti E, Taylor J, Meignier B, Meric C, Tartaglia J. Highly attenuated poxvirus vectors: NYVAC, ALVAC and TROVAC. Dev Biol Stand. 1995;84:159–63.

64. Park M-S, Steel J, García-Sastre A, Swayne D, Palese P. Engineered viral vaccine constructs with dual specificity: avian influenza and Newcastle disease. Proc Natl Acad Sci. 2006;103(21):8203–8.

65. Pastoret P-P, Brochier B. The development and use of a vaccinia-rabies recombinant oral vaccine for the control of wildlife rabies; a link between Jenner and Pasteur. Epidemiol Infect. 1996;116(3):235–40.

66. Pastoret P-P, Vanderplasschen A. Poxviruses as vaccine vectors. Comp Immunol Microbiol Infect Dis. 2003;26(5–6):343–55.

67. Peeters BP, de Leeuw OS, Koch G, Gielkens AL. Rescue of Newcastle disease virus from cloned cDNA: evidence that cleavability of the fusion protein is a major determinant for virulence. J Virol. 1999;73(6):5001–9.

68. Phase 1/2 clinical trial of an alphavirus replicon vaccine for Influenza. 2007. http://clinicaltrials.gov/show/NCT00440362

69. Poulet H, Minke J, Pardo MC, Juillard V, Nordgren B, Audonnet J-C. Development and registration of recombinant veterinary vaccines: the example of the canarypox vector platform. Vaccine. 2007;25(30):5606–12.

70. Rayner JO, Dryga SA, Kamrud KI. Alphavirus vectors and vaccination. Rev Med Virol. 2002;12(5):279–96.

71. Römer-Oberdörfer A, Mundt E, Mebatsion T, Buchholz UJ, Mettenleiter TC. Generation of recombinant lentogenic Newcastle disease virus from cDNA. J Gen Virol. 1999;80(11):2987–95.

72. A Safety and immunogenicity trial in adults 65 years of age or over to prevent influenza (AVX502-003). 2008. http://clinicaltrials.gov/show/NCT00706732

73. Salonius K, Simard N, Harland R, Ulmer JB. The road to licensure of a DNA vaccine. Curr Opin Investig Drugs. 2007;8(8):635.

74. Schmidt AC, McAuliffe JM, Huang A, Surman SR, Bailly JE, Elkins WR, Collins PL, Murphy BR, Skiadopoulos MH. Bovine parainfluenza virus type 3 (BPIV3) fusion and hemagglutinin-neuraminidase glycoproteins make an important contribution to the restricted replication of BPIV3 in primates. J Virol. 2000;74(19):8922–9.

75. Smith GL, Mackett M, Moss B. Infectious vaccinia virus recombinants that express hepatitis B virus surface antigen. Nature. 1983;302(5908):490–5.

76. Sommerset I, Krossøy B, Biering E, Frost P. Vaccines for fish in aquaculture. Expert Rev Vaccines. 2005;4(1):89–101.

77. Song H, Zhong L-P, He J, Huang Y, Zhao Y-X. Application of Newcastle disease virus in the treatment of colorectal cancer. World J Clin Cases. 2019;7(16):2143.

78. Steinwaerder DS, Carlson CA, Lieber A. Generation of adenovirus vectors devoid of all viral genes by recombination between inverted repeats. J Virol. 1999;73(11):9303–13.

79. Stobart CC, Moore ML. RNA virus reverse genetics and vaccine design. Viruses. 2014;6(7):2531–50.

80. Tameris M, Hokey D, Nduba V, Sacarlal J, Laher F, Kiringa G, Gondo K, Lazarus E, Gray G, Nachman S. A double-blind, randomised, placebo-controlled, dose-finding trial of the novel tuberculosis vaccine AERAS-402, an adenovirus-vectored fusion protein, in healthy, BCG-vaccinated infants. Vaccine. 2015;33(25):2944–54.

81. Tang J, Olive M, Pulmanausahakul R, Schnell M, Flomenberg N, Eisenlohr L, Flomenberg P. Human CD8+ cytotoxic T cell responses to adenovirus capsid proteins. Virology. 2006;350(2):312–22. https://doi.org/10.1016/j.virol.2006.01.024.

82. Tartaglia J, Perkus ME, Taylor J, Norton EK, Audonnet J-C, Cox WI, Davis SW, Van Der Hoeven J, Meignier B, Riviere M. NYVAC: a highly attenuated strain of vaccinia virus. Virology. 1992;188(1):217–32.

83. Taylor J, Meignier B, Tartaglia J, Languet B, VanderHoeven J, Franchini G, Trimarchi C, Paoletti E. Biological and immunogenic properties of a canarypox-rabies recombinant, ALVAC-RG (vCP65) in non-avian species. Vaccine. 1995;13(6):539–49.

84. Taylor J, Paoletti E. Fowlpox virus as a vector in non-avian species. Vaccine. 1988;6(6):466–8.

85. Taylor J, Trimarchi C, Weinberg R, Languet B, Guillermin F, Desmettre P, Paoletti E. Efficacy studies on a canarypox-rabies recombinant virus. Vaccine. 1991;9(3):190–3.

86. Thacker EE, Timares L, Matthews QL. Strategies to overcome host immunity to adenovirus vectors in vaccine development. Expert Rev Vaccines. 2009;8(6):761–77.

87. Tordo N, Foumier A, Jallet C, Szelechowski M, Klonjkowski B, Eloit M. Canine adenovirus based rabies vaccines. Dev Biol. 2007;131:467–76.

88. Torres JM, Alonso C, Ortega A, Mittal S, Graham F, Enjuanes L. Tropism of human adenovirus type 5-based vectors in swine and their ability to protect against transmissible gastroenteritis coronavirus. J Virol. 1996;70(6):3770–80.

89. van Wyke Coelingh KL, Winter CC, Tierney EL, London WT, Murphy BR. Attenuation of bovine parainfluenza virus type 3 in nonhuman primates and its ability to confer immunity to human parainfluenza virus type 3 challenge. J Infect Dis. 1988;157(4):655–62.

90. Vander Veen RL, Harris DH, Kamrud KI. Alphavirus replicon vaccines. Anim Health Res Rev. 2012a;13(1):1–9.

91. Vander Veen RL, Loynachan AT, Mogler MA, Russell BJ, Harris DH, Kamrud KI. Safety, immunogenicity, and efficacy of an alphavirus replicon-based swine influenza virus hemagglutinin vaccine. Vaccine. 2012b;30(11):1944–50.

92. Veits J, Wiesner D, Fuchs W, Hoffmann B, Granzow H, Starick E, Mundt E, Schirrmeier H, Mebatsion T, Mettenleiter TC. Newcastle disease virus expressing H5 hemagglutinin gene protects chickens against Newcastle disease and avian influenza. Proc Natl Acad Sci. 2006;103(21):8197–202.

93. Verma IM, Somia N. Gene therapy-promises, problems and prospects. Nature. 1997;389(6648):239–42.

94. Voeks D, Martiniello-Wilks R, Madden V, Smith K, Bennetts E, Both G, Russell P. Gene therapy for prostate cancer delivered by ovine adenovirus and mediated by purine nucleoside phosphorylase and fludarabine in mouse models. Gene Ther. 2002;9(12):759–68.

95. Wang Y, Xiang Z, Pasquini S, Ertl H. The use of an E1-deleted, replication-defective adenovirus recombinant expressing the rabies virus glycoprotein for

early vaccination of mice against rabies virus. J Virol. 1997;71(5):3677–83.

96. Warimwe GM, Gesharisha J, Carr BV, Otieno S, Otingah K, Wright D, Charleston B, Okoth E, Elena L-G, Lorenzo G. Chimpanzee adenovirus vaccine provides multispecies protection against Rift Valley fever. Sci Rep. 2016;6:20617.

97. Wecker M, Gilbert P, Russell N, Hural J, Allen M, Pensiero M, Chulay J, Chiu Y-L, Karim SA, Burke D. Phase I safety and immunogenicity evaluations of an alphavirus replicon HIV-1 subtype C gag vaccine in healthy HIV-1-uninfected adults. Clin Vaccine Immunol. 2012;19(10):1651–60.

98. Wolf A, Hodneland K, Frost P, Braaen S, Rimstad E. A hemagglutinin-esterase-expressing salmonid alphavirus replicon protects Atlantic salmon (Salmo salar) against infectious salmon anemia (ISA). Vaccine. 2013;31(4):661–9.

99. Yamamoto H, Shimojo H. Multiplicity reactivation of human adenovirus type 12 and simian virus 40 irradiated by ultraviolet light. Virology. 1971;45 (2):529–31.

100. Zakhartchouk A, Connors W, Van Kessel A, Tikoo SK. Bovine adenovirus type 3 containing heterologous protein in the C-terminus of minor capsid protein IX. Virology. 2004;320(2):291–300.

101. Zhou H, O'Neal W, Morral N, Beaudet AL. Development of a complementing cell line and a system for construction of adenovirus vectors with E1 and E2a deleted. J Virol. 1996;70(10):7030–8.

# The Role of Adjuvants in the Application of Viral Vector Vaccines

Timothy J. Mahony

## Abstract

Adjuvants are formulated in conventional vaccines to ensure immune recognition and the subsequent development of innate and/or adaptive immune responses. A principal advantage of using a viral vector is that no adjuvant is required as the vector alone can achieve the desired goals of eliciting the required immune responses. However, in some cases, the viral vector may be attenuated to achieve the desired level of in vivo safety, or it may not stimulate the most appropriate responses for the pathogen of interest. Consequently, the incorporation of an adjuvant may be required to augment the immune responses to the viral vector of interest including any heterologous antigen(s) it may encode. This chapter will describe with examples the two broad classes, convention and molecular, of adjuvants that have been utilized to improve the performance of live viral vectors. Potential unintended consequences of the use of adjuvants and the strategies used to minimize these effects while maintaining activities will also be discussed.

## Keywords

Adjuvant · Antigen expression · Cytokine expression · Immune response · Molecular adjuvant · Vaccine · Viral vector

**Learning Objectives**

After reading this chapter, you should be able to:

- Describe the function of an adjuvant
- Explain why an adjuvant might be incorporated into a viral vectored vaccine
- Describe the different types of adjuvants that might be used with a viral vaccine
- Describe the advantages of adjuvants with viral vectored vaccines
- Explain some of the potential unintended consequences of molecular adjuvants

## 1 Introduction

A conventional vaccine typically consists of two main components, the immunogen and the adjuvant. The role of the immunogen is to act as a template for the host's immune system to use the development of strong and specific responses to the pathogen of interest and to protect the host from infection and/or disease. Generally, the administration of an immunogen alone will fail to elicit an immune response. While the immunogen may be recognized and cleared by the

T. J. Mahony (✉)
The University of Queensland, Queensland Alliance for Agriculture and Food Innovation, St Lucia, Australia
e-mail: t.mahony@uq.edu.au

© Springer Nature Switzerland AG 2021
T. Vanniasinkam et al. (eds.), *Viral Vectors in Veterinary Vaccine Development*,
https://doi.org/10.1007/978-3-030-51927-8_4

immune system, the lack of an immune response is most likely due to there being insufficient stimulation of the immune sensing systems to trigger the development of an immune response. Of course, the lack of response of the immune system to potential antigens in isolation is a fail-safe control system to prevent the reaction to foreign materials, potential antigens, that are encountered during the lifetime of the organism.

Therefore, adjuvants are formulated with immunogens to prevent immediate clearance and stimulate immune recognition, ideally leading to the development of specific immune response. At the same time, the adjuvant should not overstimulate the immune system; while this might generate strong and potential protective responses, it can be deleterious to the host.

The classical model of the immune response to infection can be broadly placed into one of two categories. The first is where the response results in a dominant cell-mediated response referred to as Th1, while the second category is a dominant antibody response referred to as Th2. It is rare that a response to infection will be skewed into either of these categories; consequently, the outcome of infection will be referred to as predominantly Th1 (cell mediated, characterized by expression of IL-12, IFN-γ) or predominantly Th2 (antibody mediated, characterized by expression of IL-4, IL-10). The ideal vaccine will mimic the required balance between Th1 and Th2 responses, as might be observed following infection with the pathogen of interest.

How the immune response to vaccination ends up as either a Th1 or a Th2 response is a result of a complex interplay of the various components of the immune system which are described in detail elsewhere in this volume. This chapter will describe the application of different types of adjuvants used in conjunction with viral vectors in attempts to improve the immune responses to the antigen/pathogen of interest. Broadly speaking, these adjuvants will be grouped into two broad categories. While it is difficult to absolutely classify adjuvants strictly into each of these classes, for the purposes of this chapter, the following definitions will be applied. A formulation-based adjuvant is one which is added to the viral

vector postproduction to act as a broad immune system stimulant, while a molecular adjuvant will be defined as an adjuvant which is designed/selected to interact with specific pathway (s) within the immune system and is encoded within the vector of interest.

## 2  Adjuvants and Viral Vectored Vaccines

A fundamental advantage of using a live viral vector for vaccination is that during the course of the infection cycle, sufficient immunostimulatory or "danger" signals are created to facilitate the eliciting of strong and durable immune responses to the vector. As a consequence, if the host is exposed to the pathogen of interest, the immune system is able to respond to this exposure with rapid and specific responses based on the memory immune responses. Of course, most viruses cannot be used directly in the host of interest if it retains the capacity to cause disease. To reduce this risk, the first step in the development of a live viral vector is to reduce its capacity to cause disease, a process commonly referred to as attenuation.

However, the attenuated potential viral vector needs to strike a balance between infecting the host, undergoing replicating, and inducing the required immune responses to give the desired level of protection. At the same time, it is essential to ensure the vector does not retain the capacity to cause disease. A viral vector which is considered sufficiently attenuated may still retain the capacity to cause disease in a proportion of the population of interest, particularly those with compromised immune system. Moreover, an inadequately attenuated viral vector may have an increased likelihood of reversion to virulence whereby the wild-type genotype is restored through one of several mechanisms, leading to possible dissemination and disease in a susceptible population.

In contrast, if the viral vector is too attenuated, it may have insufficient in vivo replication capacity to stimulate the immune system of host and therefore fail to elicit protective immune responses against the pathogen of interest. The attenuation of viral vectors can be an inexact and poorly understood process. A common approach has been the sequential passage of the virus of interest in susceptible cultivated cells in the laboratory. After an empirically determined number of passages, the putative viral vector is then tested in the host of interest to assess the degree of attenuation. If the outcome of such testing is disease, then the virus of interest may be further passaged before further in vivo testing. While this can be a time-consuming process, many highly successful vaccines have been generated in this manner. Viral attenuation has also been successfully achieved using serial passage in model or alternate hosts with the resulting progeny viruses regularly tested in the host of interest (where possible) to determine the remaining capacity to cause disease.

Another strategy which has proven highly effect for attenuation is the deletion/inactivation of genes known to increase the virulence of the virus of interest. An example of this is the thymidine kinase (TK) homologue from the herpesviruses. The TK protein is involved in the nucleotide salvage pathways during the herpesvirus replication process and is known to be associated with virulence in several herpesviruses. Importantly, herpesviruses which lack the TK gene or encode a gene which expressed a nonfunctional TK protein are known to have a reduce capacity to cause disease in their natural host. Moreover, these TK-negative strains are typically able to replicate with similar efficiency in vitro to the parental wild-type and thus are highly amenable to exploitation as live viral vaccines and recombinant viral vectors.

Thymidine kinase-negative mutants for use as vaccine vectors have been developed in several ways. The first method is passaging the parent virus in the presence of nucleotide analogues. These drugs are a class of antivirals which are highly effective against the herpesviruses. The nucleotide analogues typically have higher affinities for viral TK enzymes compared to cellular TK enzymes and, when phosphorylated, are incorporated in the nascent viral genomes by the viral DNA polymerase, thus interrupting the virus replication cycle.

Consequently, from a safety point of view, viral vectors which are robustly attenuated and therefore less likely to revert to virulent forms of the parent virus are more desirable. However, an unintended consequence of robust attenuation can be a reduced capacity to stimulate effective immune responses.

While sequential passage or virulence gene inactivation/deletion can be effectively used as an attenuation strategy, the use of replication-limited viral vectors has also been utilized. One strategy which is utilized for this purpose is the deletion of one or more genes which are required for viral vector replication and/or dissemination. The deleted elements are subsequently provided *in trans* to facilitated growth and production of the viral vector in the laboratory. However, when the modified viral vector is used in vivo, it undergoes a limited replication cycle, producing the antigen of interest and ideally eliciting the desired immune responses. Due to the limited replication and potentially reduced amounts of antigen compared to a replication-competent viral vector, these types of systems may benefit from the use of an adjuvant to augment immune responses.

In applications where the viral vector is delivering a heterologous antigen (i.e., an antigen from another pathogen), the eliciting of specific responses to vector antigens may be of secondary concern, where the primary objective is to ensure optimal immune responses to the heterologous antigen. If the delivered antigen is poorly immunogenic, then additional measures may be required to maximize the immune response to it. The solution to this problem is the inclusion of an adjuvant in the viral vector formulation. This can be achieved in one of two ways. The first would be to use a conventional adjuvant which is formulated with the viral vector as part of the vaccine manufacturing process, while the second is a molecular adjuvant encoded by the viral vector.

# 3    Conventional Adjuvants

As is often the case, the addition and evaluation of conventional adjuvants to viral vectored vaccines has received far more attention in human medicine compared to veterinary medicine. The difficulties in identifying a suitable viral vector/ adjuvant combination are well illustrated by the following example. Milicic et al. [1] evaluated 13 adjuvants in various formulations to improve the immune response to an adenovirus vector expressing a malaria antigen in a murine challenge model. The vector utilized in the study was a live, nonreplicating chimpanzee adenovirus (ChAd63), expressing the ME-TRAP antigen which can induce protective immune responses in mice from the malaria parasite in a lethal challenge model. When delivered either intradermally (i.d.) or intramuscularly (i.m.), this vaccine requires three doses to protect the mice from malaria challenge [1]. A previous study reported that immunization with ChAd63 and modified vaccinia virus Ankara (MVA), both expressing the ME-TRAP antigen, was able to protect mice following a single dose, suggesting that if appropriate immune responses can be generated, then the antigen is protective [2]. Milicic et al. [1] formulated the ChAd63 vaccine with various combinations of conventional adjuvants, and after vaccination, the mice were challenged with lethal doses of malaria parasites. The results of this study clearly demonstrate the difficulties faced when trying to identify an appropriate adjuvant to augment the immune responses to viral vector-based immunization. The key results of the challenge experiment are summarized in Table 1. Surprisingly, the most consistent finding of the study was the addition of the conventional adjuvants resulting in either decreased or complete loss of protective efficacy of the viral vectored antigen. However, two of the adjuvants did improve the survival rates in the vaccinated mice.

This study also confirmed that identifying a suitable adjuvant for the application of interest can be a time-consuming and difficult process. This may be further exacerbated if a small animal model is not available to test multiple adjuvants in parallel in an efficient manner.

Examples of veterinary viral vaccine vector combined with conventional adjuvants have also been reported. One such study investigated the potential of formulating a double gene-deleted live BoHV-1 vaccine with the conventional adjuvants, Polygen and QuilA, to protect young cattle (4–9 months) from heterologous BoHV-1 challenge [3]. Overall, the study reported more robust protection when the Polygen adjuvant was included in the vaccine formulation after two administered doses. Interestingly, the study initially planned to evaluate a second adjuvant, the saponin QuilA; however, a formulation including this adjuvant was not tested in cattle due to unanticipated loss of vaccine titer after formulation. In contrast, Charerntantanakul and Pongjaroenkit [4] demonstrated the QuilA was a highly effective adjuvant for a modified live porcine reproductive and respiratory syndrome virus (PRRSV) in porcine vaccination studies. In this study, the authors evaluated the responses of pigs immunized with either the viral vaccine alone or the viral vaccine with QuilA, with the adjuvant injected separately, but proximal, to the vaccine infection site. Pigs which received the viral vector and adjuvant exhibit increased expression of inflammatory cytokines including IFN-α, IFN-β, IFN-γ, IL-2, IL-13, and TNF-α. The viral vaccine was also more efficacious in these animals in challenge experiments, with fewer adjuvanted pigs being viremic, and those that did come viremic excreted less virus. The adjuvant had no effect on the antibody levels in either treatment group.

Generally speaking, the effectiveness of a live viral vaccine is dependent on infection and either replication or the level of abortive replication of the vector in the host of interest; consequently, use of these vaccines in younger animals can be problematic due to interference by maternal antibody. It has been suggested that an important advantage of using adjuvanted live vaccine vectors is the capacity to elicit protective immune responses in the presence of maternal antibody.

**Table 1** Summary of the protective capacity of a replication-limited chimpanzee adenovirus 63 (AdV) viral vector expressing the malaria protective epitope ME-TRAP (AdV-MT) when formulated with selected conventional adjuvants

| Vaccine-adjuvant | Percentage survival (%) | |
| --- | --- | --- |
| | Intradermal[a] | Intramuscular[a] |
| AdV63-MT | 35 | 35 |
| Unvaccinated | 0 | 0 |
| AdV-MT/MVA-ME-TRAP | 100 | NT[b] |
| AdV-MT/Abisco™ (12 µg) | 18 | 35 |
| AdV-MT/Abisco™-100 (24 µg) | NT[b] | 50 |
| AdV-MT/Alhydrogel (Al(OH)$_3$) | 0 | NT[b] |
| AdV-MT/Glycolipid-A (GLA) | 0 | NT[b] |
| AdV-MT/TiterMax® Gold | 15 | NT[b] |
| AdV-MT/CoVaccineHT™ | 18–40 | 80 |
| AdV-MT/Al(OH)$_3$ + GLA | 0 | NT[b] |
| AdV-MT/Al(OH)$_3$-Poly-Pam$_3$Cys | 0 | NT[b] |
| AdV-MT/Liposomes-GLA | 0 | NT[b] |
| AdV-MT/SE-GLA[c] | 15 | NT[b] |
| AdV-MT/SE-GLA-CpG[c] | 15 | NT[b] |
| CoVaccineHT™ (Control) | NT[b] | 0 |
| Abisco™ (24 µg) (Control) | NT[b] | 0 |

After immunization, the effect of each adjuvant combination was evaluated in a lethal murine challenge model [adapted from [1]]
[a]Route of immunization
[b]Not tested (NT) or not reported
[c]SE-Stable emulsions of squalene-like oil in water

Zimmerman et al. [5] investigated the capacity of a commercial live multivalent vaccine (containing five modified live viruses) adjuvanted with Metastim® to elicit protective responses in calves. To mimic the effects of maternal antibody, the study supplemented colostrum taken from cows negative for bovine viral diarrhea virus 2 (BVDV-2) with milk containing antibodies to this virus. The groups of calves were fed either supplemented colostrum or unsupplemented colostrum, twice within 12 h of birth, and subsequently vaccinated at 4–5 weeks of age. The calves were challenged at 3.5 months (i.e., after the decay of material antibody) with virulent BVDV-2. The results demonstrated that the vaccine afforded similar levels of protection from severe disease in both groups, despite the presence of "maternal" antibody in supplemented colostrum group at the time of vaccination. While this study did not investigate the molecular mechanism(s) associated with the use of Metastim® adjuvant, it did suggest that the use of the adjuvant was able to overcome any negative effects of maternal antibody, allowing the

successful vaccination of the calves. This is a promising result as potential inference of maternal antibody with vaccination could lead to vaccination failure, leaving calves susceptible to disease once the maternal antibodies have decayed. Similarly, if vaccination must be delayed until the levels of maternal antibody have waned sufficiently for successful immunization, there is potential for calves to become susceptible to infection before protective responses in response to vaccination can develop.

A later study compared the immune responses of horses vaccinated with inactivated equine influenza A virus adjuvanted with either Metastim® or aluminium phosphate [6]. While both formulations stimulated expression of proinflammatory cytokines, higher expression of interferon-γ and IL-12 in horses receiving the Metastim® formulation suggests significantly stronger Th-1 immune response in these animals. As a Th1-type immune response is predominantly a cell-mediated response which considered desirable for protecting against viral infections, this adjuvant could be highly effective for augmenting the responses elicited by live viral vectors. However, specific testing of the Metastim® adjuvant within the viral vector of interest, including determining the type of immune response, is likely to be required on case-by-case basis to confirm this.

One of the most effective adjuvants known is complete/incomplete Freund's adjuvant system. Typically, the antigen of interest is formulated with complete Freund's adjuvant as a water-in-oil emulsion containing inactivated mycobacteria prior to delivery. Subsequently, booster immunizations are delivered with the antigen in incomplete Freund's adjuvant which lack the mycobacterial component. The use of the Freund's adjuvant system can be controversial as it can elicit very strong immune responses. This is evident through the use of the system in an autoimmune rodent model as it has the capacity to overcome tolerance [7]. While this is clearly a potent adjuvant, its use must be carefully evaluated prior to use to minimize the potential untended consequences.

A practical application of mycobacterial adjuvants was illustrated in an effort to increase the protective efficacy of modified live viral vaccine against PRRSV [8]. It was reported that the extracts of *Mycobacterium tuberculosis* were the most effective of the nine bacterial species tested in not only enhancing the immune response but also overcoming the potentially immunosuppressive effects of the vaccine. Intranasal immunization of pigs demonstrated the PRRSV vaccine adjuvanted with the *M. tuberculosis* extracts elicited primarily a Th1 immune response (increased expression of INF-γ and IL-12) and decreased immunosuppression (decreased expression of IL-10 and TGF-β) and was able to protect pigs from challenge [8, 9].

In summary, viral vectors have been evaluated in combination with conventional adjuvants with variable results. One common theme in these studies has been the need to empirically determine what is the most appropriate adjuvant for the application. While this may seem inefficient, there are potential benefits to identify a suitable adjuvant, such as successful use of a live viral vaccine in the presence of maternal antibody and/or reducing the number of doses needed to stimulate the required level of immune responses to protect the host from disease.

## 4    Molecular Adjuvants

For the purposes of this chapter, molecular adjuvants are defined as molecules that are encoded by the viral vector of interest and expressed with the specific aim of augmenting the immune response to either the vector or a heterologous antigen(s) from a pathogen which is also encoded by the vector. Clearly, molecular adjuvants differ from the adjuvants discussed in the previous section as they do not require specific formulation post-vaccine production. In general, a molecular adjuvant will be a specific effector molecule taken from a cellular pathway of the host with the aim of modifying the host's response to vaccination.

Broadly speaking, studies have explored the use of molecular adjuvants using two basic strategies. The first strategy is a process that could be referred to as molecular fusion. In this approach, the gene for the antigen of interest is fused to the gene encoding the adjuvant molecule. Typically, the goal of this type of adjuvant is to ensure the expressed antigen and adjuvant remain closely associated as this association will affect how the antigen interacts with the targeted molecular pathways of the immune system, thus influencing the type of immune response.

The second strategy involves the co-expression of the molecular adjuvant of choice in conjunction with the antigen of interest. In this approach, the antigen and adjuvant are likely to be heterologous to the viral vector, with the genetic elements of each inserted into the vector backbone as individual gene expression cassettes. In some cases, the adjuvant expression cassette may be added into an existing viral vector to augment the response to the vector. In this case, the viral vector could have been attenuated to act as a vaccine for the pathogen of interest; however, after preliminary efficacy testing, it may not have elicited the desirable levels of protection from disease. Consequently, an adjuvant(s) was evaluated to improve the immune responses and, ideally, levels of protection.

## 5    Antigen-Adjuvant Fusions

While molecular fusion can be highly effective in enhancing immune responses to the antigen of interest, there is a risk of inducing autoimmunity to the molecular adjuvant due to its close association with the antigen. Clearly, this is an undesirable outcome for the host as this could result in the development of autoimmune-associated diseases. Consequently, studies have evaluated strategies to address this issue while still maintaining adjuvant activity.

Using the live viral vector ChAd63-ME-TRAP described previously, Halbroth et al. [10] sort to improve vaccine efficacy by enhancing the CD8$^+$ T-cell responses to the ME-TRAP antigen through the use of a molecular adjuvant. At the same time, the study addressed the risk of inducing autoimmunity. In a previous study, the authors had demonstrated that fusion of the ME-TRAP antigen to the potential molecular adjuvant human MHC class II invariant chain (Ii chain or CD74) could enhance the required CD8$^+$ T-cell responses [11]. The Ii chain is a multifunctional protein involved in the MHC class II antigen presentation pathway and MHC class II assembly and prevents self-peptide from being presented [12]. The Ii chain is involved in the cellular trafficking of both MHC I and II molecules in immune cells [13].

While the Ii chain was a highly effective molecular adjuvant for the ME-TRAP, it was not progressed into human clinical trials due to concerns over autoimmunity. To address this issue, Halbroth et al. [10] sort to identify the critical component of the Ii chain that is required to improve vaccine efficacy and affect the stimulated immune responses by sequentially reducing the amount of the Ii chain fused to the ME-TRAP antigen. The study identified that the truncation which included only the transmembrane domain (TD-Ii, amino acids 1 to 73) of the Li chain was required to elicit protective immune responses in immunized mice.

Interestingly, the CD8+ T-cell responses were significantly higher in mice immunized with TD-Ii-ME-TRAP fusion protein expressed by the ChAd63 than those of mice immunized with ChAd63 expressing the complete li chain fused to the ME-TRAP antigen [10]. While investigations determined the properties of the Ii chain as a molecular adjuvant could be attributed to the transmembrane domain, even the use of this minimized human molecule as an adjuvant was considered to still retain the risk of inducing autoimmunity. To address this issue, the authors subsequently identified Ii chain molecules from other vertebrate species based on sequence similarities to the human molecule. The Ii chain of the shark was identified as striking the right balance between functional conservation and sequence divergence from the human molecule. Fusion of the transmembrane domain from shark Ii chain to the ME-TRAP antigen was shown to increase CD4$^+$ and CD8$^+$ T-cell responses compared to the ME-TRAP alone when delivered using the ChAd63 viral vector.

While this example is centered on an example of human disease, the framework used to identify effective molecular adjuvants could be readily applied to veterinary species. Indeed, given that the shark molecule was an effective adjuvant in murine studies, it would be of interest to determine if it could be utilized in veterinary species of economic importance.

## 6    Co-expressed Molecular Adjuvants

While live viral vectors were initially developed as a strategy to prevent homologous disease issues, the advent of recombinant DNA technology raised the prospect adding antigens from other pathogens into the viral vector of interest, potentially resulting the capacity to protect against multiple pathogens with one vaccination. More detail on the delivery of heterologous antigens using various veterinary viral vectors is described in other chapters of this volume. Following on from this innovation was the prospect of incorporating immunostimulatory molecules to augment the immune responses to the viral vector and/or the antigen(s) of interest. The possibility delivering molecules such as cytokines and other immune effector molecules to modulate the innate and/or adaptive immune responses to the viral vector and/or a heterologous antigen of interest has received much attention in subsequent years. In some cases, the use of a molecular adjuvant will be to skew the immune response toward what is known to be a protective immune response against the pathogen of interest. In this context, interferons, particularly interferon gamma (INF-γ), have been widely used to strengthen Th1-type immune responses where cell-mediated immunity is considered to be the desirable immune response against intracellular pathogens such as viruses, while the cytokine, IL-4, has been utilized to increase Th2-type responses where enhanced humoral immune responses are required to facilitate improved protection. In other instances, the molecular adjuvant can be used in an attempt to balance the immune response to the viral vector of interest. Moreover, molecular adjuvants have also been used to make the antigen of interest more likely to promote immune responses. Typically, this has involved the use of molecules with chaperoning properties which increase the likelihood of the antigen entering the relevant processing pathways. This section will describe examples of how molecular adjuvants have been used to augment the immune responses to viral vectors and where relevant the heterologous antigens encoded by them.

Early research to identify potential treatments for bovine respiratory disease explored the use of bovine INF-α, and recombinant bovine INF-α increases the resistance of cattle to secondary bacterial infections following bovine herpesvirus 1 challenge in a dual infection model [14–16]. The administration of INF-α prior to the primary viral challenge increases the survival of the animals more effectively than coadministration with the challenge virus. However, given that the mode of action of INF-α, along with INF-β, is to promote an antiviral state through mRNA degradation and inhibition of protein synthesis, this class of molecules may not be suitable for viral vector applications as

they may interfere with vector replication and prevent the development of effective immune responses. Molecular adjuvants that function to stimulate specific immune pathways without directly interfering with the replication of the vector are clearly more desirable.

Despite the potential negative drawbacks of interferons as molecular adjuvants, Raggo et al. [17] utilize a BoHV-1 vector (BoHV-1 ΔgC: rBgal) for the expression and delivery of functional bovine IFN-γ. Surprisingly, the expression IFN-γ did not affect the in vitro or in vivo replication of the BoHV-1 vector. Overall, there was limited evidence for the expression of IFN-γ having an effect on the immune responses of immunized cattle in respect to humoral and cell-mediated immunity. The responses in group immunized with the vector IFN-γ were not detectably different to those of the parent virus (BoHV-1ΔgC:rBgal). The study also demonstrated that the reactivated BoHV-1 ΔgC: rIFN-γ retained the capacity to express rIFN-γ interest in a biologically active form, demonstrating that the herpesviruses are potentially very good vectors for the delivery of functional biological molecules. Interestingly, it was reported that following reactivation, less IgA was detected in the nasal secretions of cattle immunized with the BoHV-1ΔgC:rIFN-γ compared to the parental BoHV-1ΔgC:rBgal viral vector. While this suggests immunomodulation by the viral vector, the mechanisms underlying this were not further investigated [17].

A similar study was later reported which sort to improve the immune responses to use a commercially registered BoHV-1 marker vaccine strain [18]. Marker vaccines are an excellent technology to improve the control of veterinary infectious diseases. Marker vaccines enable differentiation of infected and vaccinated animals (DIVA). Where the marker vaccine is a live viral vector, it will typically be derived from a virus strain which is considered safe for use as a vaccine. For use in DIVA applications, the vaccine strain will lack an immunological determinant, and consequently vaccinated animals will lack immune responses to the antigen. In contrast, animals infected by wild-type virus will have immune responses to the antigen. Using a specific test for the antigen of interest therefore can facilitate the differentiation of vaccinated-only animals. The capacity to apply DIVA concepts in eradication campaigns or to enable the use of vaccinating to control exotic disease incursions can prevent the slaughter of animals which are vaccinated only to achieve freedom from disease. As discussed previously, the deletion of one or more genes for attenuation or use as a marker vaccine can affect the capacity of the vector to elicit strong enough immune responses to protect from infection and/or development of disease. In some instances, viral strains with naturally occurring deletions of a major immunogenic gene have been exploited for use as marker vaccines. These viruses are excellent candidate viruses for exploitation as viral vectors as they are likely to have already adapted to the loss of the "marker gene", and therefore have excellent replication capacities.

Konig et al. [18] reported the construction of a series of recombinant BoHV-1 vectors expressing IL-2, IL-4, IL-6, IL-12, or IFN-γ as potential molecular adjuvants. The parental virus, BoHV-1 strain GK/D, was a BoHV-1 commercially registered live vaccine which lacked the glycoprotein E (gE) antigen and thus suitable for marker vaccine applications. When used to inoculate naïve calves, the recombinant vaccine viruses were shed in lower quantities and for less time compared to the unmodified parent virus. The reduction in the amount of vaccine virus shed was particularly noteworthy in the groups immunized with the BoHV-1:rIL-12 and BoHV-1:rIFN-γ. The results suggest that the expression of either IL-12 or IFN-γ was able to limit the replication of the vaccine vector. Despite the reduced amount of vaccine virus shedding, there was no effect on the antibody responses to the recombinant vaccines compared to the parent virus with the animals seroconverting within 14 days of immunization. The calves were subsequently challenged with virulent BoHV-1 to determine if the insertion of the incorporated cytokine gene was able to augment protection. The BoHV-1:rIL-2- and BoHV-1:rIL-4- vaccinated animals shed similar amounts of

challenge virus to the animals vaccinated with the parental vaccine virus [18]. In contrast, the calves vaccinated with the BoHV-1:rIL-6, BoHV-1:rIL-12, and BoHV-1:rIFN-γ vectors shed similar quantities of the challenge virus to the unvaccinated control group, albeit for less time. While the specific mechanism underpinning these results were not elucidated, the authors suggested that the vectored expression of IL-6, IL-12, and IFN-γ which resulted in significant reductions in the replication of the corresponding vectors could have interfered with the cell-mediated immune responses of the animals in these groups [18]. The results of this study clearly contrast with those described by Raggo et al. [17] who reported no effects of IFN-γ expression on BoHV-1 vector replication. When considered together, these studies suggest the selection of virus strain for exploitation as a vector can also affect the functionality of molecular adjuvants.

PRRSV is an important pathogen of swine with a worldwide distribution. While vaccines, including live viral vectors, have been developed, effective control of the virus has remained elusive. Key reasons for this lack of control include potential immunosuppression and strain-to-strain variation [19, 20]. Consequently, several studies have explored the use of molecular adjuvants to enhance the efficacy of PRRSV live vaccine vectors. Yu et al. [21] constructed a modified live PRRSV viral vaccine which expressed the granulocyte-macrophage colony-stimulating factor (GM-CSF). The GM-CSF molecule has been widely utilized as a vaccine adjuvant as it is implicated in the differentiation of monocytes to dendritic cells and the maturation/activation of dendritic cells which are important antigen-presenting cells [22]. The study did not test the capacity of the modified live PRRSV expressing GM-CSF to enhance immune responses in pigs or its efficacy in challenge studies. However, it was demonstrated that when the viral vector was used to infected immature dendritic cells, the cells had increased expression of MHC I and MHC II molecules on their surfaces. The viral vector-infected cells were also shown to be expressing higher levels of key inflammatory cytokines such as IL-1β, IL-4, and IFN-γ [21]. Despite these promising in vitro results, the viral vaccine was not evaluated in vivo in challenge studies. However, if the in vitro results could be replicated in vivo, the use of GM-CSF as a molecular adjuvant to improve the efficacy of PRRSV live viral vaccines is feasible.

A separate study sort to improve the efficacy of PRRSV-modified live vaccines by addressing the issue of vaccine-associated immunosuppression, by exploiting IL-15 and IL-18 as molecular adjuvants [23]. These interleukins were chosen as IL-15 is crucial for cytotoxic T-cell and natural killer cell functioning while IL-18 can be effective in improving Th1 responses, in addition to also contributing to natural killer cell activity. Interestingly, the authors were concerned about the potential negative effects of constitutive expression of the virally expressed IL-15 and/or IL-18 causing unintended effects in immunized pigs. This issue was addressed by expressing the cytokines in a modified form by fusing them to the glycosylphosphatidylinositol (GPI)-anchored protein, porcine CD59. Along with other surface displayed molecules, CD59 is clustered within membrane structures known as lipid rafts, and the CD59 molecule is a regulator of complement and is involved in T-cell activation [24]. In heterologous challenge studies with pigs vaccinated with one of the viral vector strain Suvaxyn (parental strain) or the IL-15 expressing vaccine (Suvaxyn:rIL-15) or the IL-18 expressing vaccine (Suvaxyn:rIL-18), all exhibited reduced lung lesions and reduced loads of the challenge virus. The loss of natural killer cell function in the Suvaxyn-vaccinated animals was prevented in animals vaccinated with Suvaxyn:rIL-15 or Suvaxyn:rIL-18, while vaccination with the Suvaxyn:rIL-15 results in a significantly increased numbers of IFN-γ-producing cells after 7 weeks, specifically CD4- CD8+ T cells and γδT cells, compared to the other vaccinated groups. The Suvaxyn:rIL-15-vaccinated group also had enhanced T-cell responses at 1 and 7 weeks after vaccination and 1 week after challenge. Importantly, the data were highly supportive of the Suvaxyn:rIL-15 as providing enhanced protection against heterologous challenge,

warranting further development to improve the control of PRRSV [23].

In an earlier study, Li et al. [25] employed an adenovirus vector to deliver a fused PRRSV antigen consisting of glycoproteins GP3 and GP5. To improve the anti-PRRSV immune responses to the GP3/GP5 poly-antigen, the authors sort to exploit the capacity of heat shock protein 70 (HSP70) as a molecular adjuvant. Eukaryotic and prokaryotic HSP70 are known to have the capacity to act as stimulators of the innate and adaptive immune responses [26]. The study utilized HSP70 from the porcine pathogenic bacterium *Haemophilus parasuis* as the molecular adjuvant. Three recombinant AdV were constructed, the first expressing the GP3-GP5 poly-antigen (AdVrecGP3-GP5), the second GP3-GP5 poly-antigen fused to the HSP70 via a polyglycine linker (AdVrecHSP70-GP3-GP5), and the third consisting of GP3-GP5 poly-antigen fused to the HSP70 via the 2A peptide from foot-and-mouth disease virus (AdVrecHSP70-2A-GP3-GP5). The foot-and-mouth disease virus 2A peptide is commonly used as a peptide linker as it can facilitate the expression of the two adjacent polypeptides as separate entities due to a poorly understood process referred to as bond skipping during the translation of this sequence in eukaryotic cells [27]. As a result of this process, the viral vectors encoding this protein would be expected to express the HSP70 adjuvant and GP3/GP5 antigen primarily as separate polypeptides, though some will also be expressed as a fusion polypeptide. The parental AdV (wtAdV) and the recombinant viral vectors were subsequently evaluated in piglet immunization studies. The AdV-GP3/GP5 adjuvanted with HSP70 elicited significantly higher levels of specific IgG to the GP3-GP5 antigen and PRRSV neutralizing antibody at 3 and 6 weeks post-immunization compared to AdVrecGP3-GP5 [25]. The levels of serum INF-γ and IL-4 determined with the AdVrecHSP70-GP3-GP5- and AdVrecHSP70-2A-GP3-GP5-immunized piglets have significantly higher levels of INF-γ at 6 weeks post-immunization compared to the AdVrecHSP70. While the levels of IL-4 were elevated in the groups with adjuvanted groups, only the AdVrecHSP70-2A-GP3-GP5 and AdVrecGP3-GP5 differed significantly [25]. Evaluation of the recombinant viral vectors in a challenge study with virulent PRRSV confirmed that animals immunized with HSP70-adjuvanted viral vectors had significantly reduced clinical scores (Day 14), reduced lung lesion scores (Day 14), and shed less virus (Day 7) after challenge [25]. Improved efficacy in the vectoring of norovirus antigens was also demonstrated with HSP70 as the molecular adjuvant using a vesicular stomatitis virus vector [28]. While the exact mechanisms of how HSP70 acts as molecular adjuvants remain to be determine, HSP are molecular chaperones, and similar to the Ii chain discussed previously, they may act to ensure the correct processing and continued association of the antigen of interest with other molecules involved in the immune response pathways.

One additional strategy which has been used to improve the efficacy of live viral vectors in the delivery of molecular adjuvants is the use of multiple vectors, where one expresses the antigen(s) of interest and the second vector expresses the adjuvant of interest. This approach has been utilized in the search for a universal human influenza vaccine [29]. The study evaluated the co-delivery of an adenoviral vector expressing the hemagglutinin and nucleoprotein from the influenza virus in combination with the same vector expressing either IL-1β or IL-18. These cytokines were chosen as previous studies had suggest they were essential for optimal antibody and T-cell responses to influenza A viruses [30, 31]. When testing in mouse intranasal immunization/challenge studies, co-immunization with the IL-1β expressing AdV was shown to increase the antibody responses to the hemagglutinin antigen. The study also demonstrated that the IL-1β adjuvant was able to increase the number of resident memory T cells in the lungs of vaccinated mice. An important outcome of this study was the IL-1β adjuvant facilitating the generation of immune responses against challenge with heterologous virus strains [29]. This suggests that the vector/adjuvant combination was able to elicit a robust immune response.

In summary, the examples discussed above suggest that identification of molecular adjuvants to increase the efficacy of viral vector vaccines requires careful consideration before selecting candidates for testing. In many cases, knowledge of the type of immune response required to protect from the pathogen of interest can help the selection process. One factor which complicates this selection process is the complex interplay between the host's immune system and the live viral vector.

## 7    Future Directions

The role of a vaccine adjuvant can be multifaceted. In some cases, an adjuvant is to ensure that the signaling threshold is reached so that the immune system reacts and produces an immune response, while more refine uses of adjuvants can include attempting to modulate the immune system to stimulate the desired type of responses to improve vaccine efficacy. The selection of the most appropriate adjuvants can be informed by understanding protective immune responses in the host/pathogen system of interest. However, as the examples provided in this chapter demonstrate, it can be difficult to accurately predict what the outcome of using the selected adjuvant might be.

One critical drawback of using molecular adjuvants which seek to utilize a specific component of the host immune system to augment the immune response to the antigen of interest is the risk of autoimmunity. It has been suggested that a percentage of the human population are genetically susceptible to vaccine-associated autoimmunity. While the mechanisms which underpin the development of autoimmunity are unclear, one pathway is through molecular mimicry. Molecular mimicry can occur if a component of the vaccine resembles a component from the vaccinated host. The developing immune response subsequently may then react to both the vaccine and host factors. That autoimmunity can result from antigenic similarity, it is clearly a concern where host factors are used as adjuvants. The example discussed previously, where Ii chain homologues from distantly related species were evaluated, offers a potential solution to this issue [10]. Though the successful identification of interspecies homologues will be dependent on conservation of functional pathways between species for this approach to have widespread application.

Another potential solution to the issue of autoimmunity would be to create synthetic homologues of the molecular adjuvant of interest. Bioinformatics combined with advances in DNA synthesis technology could facilitate the design and construction of novel molecular adjuvants which retain the adjuvant activity while at the same time are antigenically distinct from the host homologue. The development of such molecules could result in a new generation of adjuvants with considerably reduced risks of autoimmunity.

The mammalian immune system is a complex web of positive and negative feedback pathways which interact through multiple cell types. It is the balancing of all these signals that ultimately determines the type and intensity of immune responses that are generated following vaccination or infection. When a molecular adjuvant is expressed by the live viral vector in a constitutive manner, there is a risk that the constant stimulation (or inhibition) of the pathway(s) targeted for adjuvant modulation could lead to unintended consequences as demonstrated by the example of Jackson et al. [32] who aimed to development of a viral vectored immunocontraceptive vaccine for mice. The study utilized the poxvirus, ectromelia virus (ECTV), to deliver IL-4 as a molecular adjuvant. Some strains of laboratory mice are genetically resistant the development of mousepox which can develop following infection with ECTV. This resistance is mediated by a dominant Th1-type immune response characterized by expression of IL-2, IL-12, IFN-$\gamma$, and TNF-$\alpha$. The study elected to use IL-4 as a molecular adjuvant to increase the antibody responses (Th2) in vaccinated mice. Unexpectedly, the recombinant ECTV expressing IL-4 was highly virulent in the resistance mice, resulting in severe disease and high mortality rates [32]. The study concluded that the co-expression of IL-4 suppressed both the innate and adaptive immune responses to ETCV. This example reiterates the potential difficulty in utilizing adjuvants to manipulate immune responses to a live viral vector where there is no control over adjuvant expression.

One approach which may help to resolve issues around the use of molecular adjuvants is use modulated expression rather than constitutive expression. Modulation of adjuvant expression may allow more subtle manipulation of the immune system that may improve the subsequent responses while reducing undesirable effects. Inducible expression could be achieved in one of several ways. As an example, if a herpesvirus vector was being used, the gene encoding the molecular adjuvant could be placed under the control of a viral promoter. During the infection process, the expression of herpesvirus genes occurs in a highly regulated gene cascade [33]. The molecular adjuvant of interest could be cloned under the control of the viral promoter to facilitate expression when it is likely to be most effective. An alternative strategy would be to place the molecular adjuvant gene under the control of a promoter, where gene expression is dependent on the presence of a drug or some other small molecule. As a theoretical example, the viral vector of interest may have been engineered to express one or more pro-inflammatory cytokines to drive the recruitment of immune cells to the site of vaccination. However, long-term expression of pro-inflammatory cytokines may be undesirable due to potential inflammation-associated pathologies. If the pro-inflammatory cytokine genes were under the control of an inducible promoter, the promoter agonist could be included in the vaccination formulation. Immediately following vaccination, the presence of the agonist would facilitate the production of the pro-inflammatory cytokines. However, if the agonist is metabolized or diffuses from the site of vaccination, the expression of the pro-inflammatory cytokines could be downregulated, thus reducing any associated risks. Inducible transgene expression systems have been described for adenovirus and herpesvirus vector systems for use in gene therapy applications [34–37]. The adaption of these more controlled gene delivery strategies to viral vaccine vector applications may provide the improvements required to give more predictable outcomes when used in the host of interest.

# 8 Summary

There is a clear role for the use of adjuvants to augment the immune responses elicited in the host following the immunization with a live viral vector. A key driver of the need for adjuvants is ensuring the viral vector of interest is sufficiently attenuated to reduce its capacity to cause disease. Improved knowledge of host/pathogen interactions at the molecular level could help in the evaluation of the effectiveness of viral vector attenuation. Similarly, the level of understanding of the functional role of adjuvants and their interactions with the host immune system may aid in selecting the correct adjuvant and vector combinations.

# References

1. Milicic A, Rollier CS, Tang CK, Longley R, AVS H, Reyes-Sandoval A. Adjuvanting a viral vectored vaccine against pre-erythrocytic malaria. Sci Rep. 2017;7 (1):7284.

2. Reyes-Sandoval A, Rollier CS, Milicic A, Bauza K, Cottingham MG, Tang CK, et al. Mixed vector immunization with recombinant adenovirus and MVA can improve vaccine efficacy while decreasing antivector immunity. Mol Ther. 2012;20(8):1633–47.

3. Kalthoff D, Konig P, Trapp S, Beer M. Immunization and challenge experiments with a new modified live bovine herpesvirus type 1 marker vaccine prototype adjuvanted with a co-polymer. Vaccine. 2010;28 (36):5871–7.

4. Charerntantanakul W, Pongjaroenkit S. Co-administration of saponin quil A and PRRSV-1 modified-live virus vaccine up-regulates gene expression of type I interferon-regulated gene, type I and II interferon, and inflammatory cytokines and reduces viremia in response to PRRSV-2 challenge. Vet Immunol Immunopathol. 2018;205:24–34.

5. Zimmerman AD, Boots RE, Valli JL, Chase CC. Evaluation of protection against virulent bovine viral diarrhea virus type 2 in calves that had maternal antibodies and were vaccinated with a modified-live

vaccine. J Am Vet Med Assoc. 2006;228 (11):1757–61.

6. Horohov DW, Dunham J, Liu C, Betancourt A, Stewart JC, Page AE, et al. Characterization of the in situ immunological responses to vaccine adjuvants. Vet Immunol Immunopathol. 2015;164(1–2):24–9.

7. Billiau A, Matthys P. Modes of action of Freund's adjuvants in experimental models of autoimmune diseases. J Leukoc Biol. 2001;70(6):849–60.

8. Dwivedi V, Manickam C, Patterson R, Dodson K, Weeman M, Renukaradhya GJ. Intranasal delivery of whole cell lysate of Mycobacterium tuberculosis induces protective immune responses to a modified live porcine reproductive and respiratory syndrome virus vaccine in pigs. Vaccine. 2011;29(23):4067–76.

9. Dwivedi V, Manickam C, Patterson R, Dodson K, Murtaugh M, Torrelles JB, et al. Cross-protective immunity to porcine reproductive and respiratory syndrome virus by intranasal delivery of a live virus vaccine with a potent adjuvant. Vaccine. 2011;29 (23):4058–66.

10. Halbroth BR, Sebastian S, Poyntz HC, Bregu M, Cottingham MG, Hill AVS, et al. Development of a molecular adjuvant to enhance antigen-specific CD8 (+) T cell responses. Sci Rep. 2018;8(1):15020.

11. Spencer AJ, Cottingham MG, Jenks JA, Longley RJ, Capone S, Colloca S, et al. Enhanced vaccine-induced CD8+ T cell responses to malaria antigen ME-TRAP by fusion to MHC class ii invariant chain. PLoS One. 2014;9(6):e100538.

12. Villadangos JA, Driessen C, Shi GP, Chapman HA, Ploegh HL. Early endosomal maturation of MHC class II molecules independently of cysteine proteases and H-2DM. EMBO J. 2000;19(5):882–91.

13. Walchli S, Kumari S, Fallang LE, Sand KM, Yang W, Landsverk OJ, et al. Invariant chain as a vehicle to load antigenic peptides on human MHC class I for cytotoxic T-cell activation. Eur J Immunol. 2014;44 (3):774–84.

14. Babiuk LA, Lawman MJ, Gifford GA. Use of recombinant bovine alpha 1 interferon in reducing respiratory disease induced by bovine herpesvirus type 1. Antimicrob Agents Chemother. 1987;31(5):752–7.

15. Babiuk LA, Ohmann HB, Gifford G, Czarniecki CW, Scialli VT, Hamilton EB. Effect of bovine alpha 1 interferon on bovine herpesvirus type 1-induced respiratory disease. J Gen Virol. 1985;66 (Pt 11):2383–94.

16. Czarniecki CW, Anderson KP, Fennie EB, Bielefeldt Ohmann H, Babiuk LA. Bovine interferon-alphaI 1 is an effective inhibitor of bovine herpes virus-1 induced respiratory disease. Antiviral Res. 1985;(Suppl 1):209–15.

17. Raggo C, Habermehl M, Babiuk LA, Griebel P. The in vivo effects of recombinant bovine herpesvirus-1 expressing bovine interferon-gamma. J Gen Virol. 2000;81(Pt 11):2665–73.

18. Konig P, Beer M, Makoschey B, Teifke JP, Polster U, Giesow K, et al. Recombinant virus-expressed bovine

cytokines do not improve efficacy of a bovine herpes-virus 1 marker vaccine strain. Vaccine. 2003;22 (2):202–12.

19. Kimman TG, Cornelissen LA, Moormann RJ, Rebel JM, Stockhofe-Zurwieden N. Challenges for porcine reproductive and respiratory syndrome virus (PRRSV) vaccinology. Vaccine. 2009;27(28):3704–18.

20. Nan Y, Wu C, Gu G, Sun W, Zhang YJ, Zhou EM. Improved vaccine against PRRSV: current Progress and future perspective. Front Microbiol. 2017;8:1635.

21. Yu L, Zhou Y, Jiang Y, Tong W, Yang S, Gao F, et al. Construction and in vitro evaluation of a recombinant live attenuated PRRSV expressing GM-CSF. Virol J. 2014;11:201.

22. Yu TW, Chueh HY, Tsai CC, Lin CT, Qiu JT. Novel GM-CSF-based vaccines: one small step in GM-CSF gene optimization, one giant leap for human vaccines. Hum Vaccin Immunother. 2016;12(12):3020–8.

23. Cao QM, Ni YY, Cao D, Tian D, Yugo DM, Heffron CL, et al. Recombinant Porcine Reproductive and Respiratory Syndrome Virus Expressing Membrane-Bound Interleukin-15 as an Immunomodulatory Adjuvant Enhances NK and gammadelta T Cell Responses and Confers Heterologous Protection. J Virol. 2018;92 (13):e00007-18.

24. Kimberley FC, Sivasankar B, Paul MB. Alternative roles for CD59. Mol Immunol. 2007;44(1–3):73–81.

25. Li J, Jiang P, Li Y, Wang X, Cao J, Wang X, et al. HSP70 fused with GP3 and GP5 of porcine reproductive and respiratory syndrome virus enhanced the immune responses and protective efficacy against virulent PRRSV challenge in pigs. Vaccine. 2009;27 (6):825–32.

26. Srivastava P. Roles of heat-shock proteins in innate and adaptive immunity. Nat Rev Immunol. 2002;2 (3):185–94.

27. Daniels RW, Rossano AJ, Macleod GT, Ganetzky B. Expression of multiple transgenes from a single construct using viral 2A peptides in drosophila. PLoS One. 2014;9(6):e100637.

28. Ma Y, Duan Y, Wei Y, Liang X, Niewiesk S, Oglesbee M, et al. Heat shock protein 70 enhances mucosal immunity against human norovirus when coexpressed from a vesicular stomatitis virus vector. J Virol. 2014;88(9):5122–37.

29. Lapuente D. Storcksdieck Genannt Bonsmann M, Maaske A, Stab V, Heinecke V, Watzstedt K, et al. IL-1beta as mucosal vaccine adjuvant: the specific induction of tissue-resident memory T cells improves the heterosubtypic immunity against influenza a viruses. Mucosal Immunol. 2018;11(4):1265–78.

30. Ben-Sasson SZ, Hogg A, Hu-Li J, Wingfield P, Chen X, Crank M, et al. IL-1 enhances expansion, effector function, tissue localization, and memory response of antigen-specific CD8 T cells. J Exp Med. 2013;210(3):491–502.

31. Ben-Sasson SZ, Hu-Li J, Quiel J, Cauchetaux S, Ratner M, Shapira I, et al. IL-1 acts directly on CD4 T cells to enhance their antigen-driven expansion and differentiation. Proc Natl Acad Sci U S A. 2009;106 (17):7119–24.

32. Jackson RJ, Ramsay AJ, Christensen CD, Beaton S, Hall DF, Ramshaw IA. Expression of mouse interleukin-4 by a recombinant ectromelia virus suppresses cytolytic lymphocyte responses and overcomes genetic resistance to mousepox. J Virol. 2001;75(3):1205–10.

33. Roizman B. The function of herpes simplex virus genes: a primer for genetic engineering of novel vectors. Proc Natl Acad Sci U S A. 1996;93 (21):11307–12.

34. Dogbevia GK, Robetamanith M, Sprengel R, Hasan MT. Flexible, AAV-equipped genetic modules for inducible cof Gene expression in Mammalian brain. Mol Ther Nucleic Acids. 2016;5:e309.

35. Wieser C, Stumpf D, Grillhosl C, Lengenfelder D, Gay S, Fleckenstein B, et al. Regulated and constitutive expression of anti-inflammatory cytokines by nontransforming herpesvirus saimiri vectors. Gene Ther. 2005;12(5):395–406.

36. Jiang Y, Wei N, Zhu J, Zhai D, Wu L, Chen M, et al. A new approach with less damage: intranasal delivery of tetracycline-inducible replication-defective herpes simplex virus type-1 vector to brain. Neuroscience. 2012;201:96–104.

37. Chtarto A, Yang X, Bockstael O, Melas C, Blum D, Lehtonen E, et al. Controlled delivery of glial cell line-derived neurotrophic factor by a single tetracycline-inducible AAV vector. Exp Neurol. 2007;204 (1):387–99.

# Part II

# DNA Virus Vectors

# Adenovirus Vectors

Lisanework E. Ayalew, Amit Gaba, Wenxiu Wang, and Suresh K. Tikoo

## Abstract

Infectious diseases are the prime cause of morbidity and mortality in animals leading to heavy economic losses to the livestock/poultry industry. One of the most effective ways to control these losses is by use of recombinant vaccines. However, one of the requirements for the development of efficient recombinant veterinary vaccines is the availability of an efficient vaccine delivery system. The basic characteristic of the viruses to successfully deliver their genetic material into the host cells makes them a powerful tool for introducing foreign DNA into a variety of eukaryotic cells. While recombinant vectors from a number of viruses are available, vectors based on adenoviruses are being evaluated in the highest number of vaccination and gene therapy clinical trials globally. Interestingly, the last decade has witnessed a significant increase in the development and evaluation of species- or non-species-specific adenoviruses as vectors for veterinary vaccines. In fact, an adenovirus-vectored foot and mouth disease vaccine has been conditionally licensed for use in cattle in the USA. This chapter focuses on the current state of the research related to the development of adenovirus-vectored veterinary vaccines.

## Keywords

Adenovirus · DNA viruses · Veterinary vaccines · Viral vector · Vaccination · Replication-competent adenovirus · Replication-defective adenovirus · DISC adenovirus

## Learning Objectives

1. Adenoviruses are excellent vectors for delivery of veterinary vaccines.
2. Available knowledge of detailed molecular biology and easy methods of construction of recombinant adenoviruses.
3. A single immunization with recombinant adenovirus-expressing vaccine antigen can protect animals against disease.
4. The problem of pre-existing vector-specific immunity with use of species-specific adenovirus can be overcome.
5. More work is needed to prove safety and efficacy of replication-competent adenovirus vectors for veterinary vaccines.

L. E. Ayalew · A. Gaba
VIDO-InterVac, University of Saskatchewan, Saskatoon, SK, Canada

W. Wang
Shandong Binzhou Animal Science and Veterinary Medicine Academy, Binzhou, Shandong, China

S. K. Tikoo (✉)
VIDO-InterVac, University of Saskatchewan, Saskatoon, SK, Canada

Vaccinology & Immunotherapeutics Program, School of Public Health, University of Saskatchwean, Saskatoon, SK, Canada
e-mail: suresh.tik@usask.ca

© Springer Nature Switzerland AG 2021
T. Vanniasinkam et al. (eds.), *Viral Vectors in Veterinary Vaccine Development*,
https://doi.org/10.1007/978-3-030-51927-8_5

# 1    Introduction

Adenovirus was first isolated in the early 1950s from human adenoid by Wallace Rowe and colleagues [74]. Subsequently, numerous adenoviruses (more than 120 serotypes) have been isolated from a wide variety of species including mammals, reptiles, fishes, and birds [79]. Most of these adenoviruses are species specific and generally cause mild infection in their respective hosts. However, canine and avian adenoviruses are known to cause clinically important diseases. Over the last six decades, molecular biology of adenoviruses has been extensively studied, which has contributed significantly in understanding various biological processes like replication of DNA, gene expression, splicing mechanism, cell cycle, and cellular growth regulation [7]. The in-depth knowledge of biology of adenoviruses along with other characteristics like easy manipulation of the viral genome, ability to grow to high titers, efficient replication in both dividing and non-dividing cells, and large transgene carrying capacity has made adenoviruses as viral vectors of choice for gene and vaccine antigen delivery [1].

## 1.1    Adenovirus Classification

Adenovirus belongs to the family *Adenoviridae*, which is further divided into five genera, namely, *Mastadenovirus*, *Aviadenovirus*, *Atadenovirus*, *Siadenovirus*, and *Ichtadenovirus* [21]. A sixth genus, *Testadenovirus*, has been proposed recently to include adenoviruses from turtle [23]. Members of all genera have conserved 16 genes including DNA polymerase (Pol), DNA-binding protein (DBP), pre-terminal protein (pTP), IVa2, 52K, III, pIIIa, pX, pVII, pVI, 100K, pVIII, 33K, protease, hexon, and fiber.

Members of genus *Mastadenovirus* include (a) unique genes that encode genus-specific proteins (protein IX, core protein V, and some proteins encoded by E1, E3, and E4 regions) and

(b) longer and more complex inverted terminal repeats (ITRs) [21]. Members of genus *Aviadenovirus* include (a) larger genomes 43–45 kb [46] with short ITRs; (b) the presence of two fiber proteins per vertex; (c) absence of genes encoding protein V, protein IX, and E3 region proteins [34]; and (d) longer E4 region as compared to other genera [21]. Members of genus *Atadenovirus* (a) include genome shorter than *Mastadenovirus* but having high A + T content, (b) encode genus-specific unique structural proteins p32 and LH3, and (c) include absence of genes encoding protein IX and V [25, 33]. Members of genus *Siadenovirus* include (a) short genome and short ITRs, (b) presence of gene that encode sialidase enzyme, and (c) absence of E1, E3, and E4 region and genes encoding for proteins V and IX [21]. Members of *Ichtadenovirus* genus consist of only one species named white sturgeon adenovirus 1 (WSAdV-1). WSAdV-1 has the longest genome (48,395 bp) among all the known adenoviruses [12]. A sixth genus named *Testadenovirus* has been proposed based on characterization of a turtle adenovirus [23].

## 1.2    Adenovirus Structure

Adenovirus is a non-enveloped DNA virus with a typical icosahedral capsid (Fig. 1) that surrounds the double-stranded linear genome of 26 kb to 48 kb. The icosahedral capsid of adenovirus has a diameter of 65 nm to 90 nm and is composed of 13 proteins [7, 49]. Along with these major structural proteins (hexon, penton, and fiber), minor capsid proteins (proteins IIIa, VI, VIII, and IX) are also present in the virion. These minor proteins stabilize the virion structure by interacting among themselves and with the major capsid proteins [7]. The core of adenovirus is composed of viral DNA and core proteins V, VII, mu, TP, and IVa2 and cysteine protease. The terminal protein is covalently attached to 5' end of adenovirus DNA and is required for viral DNA replication [22].

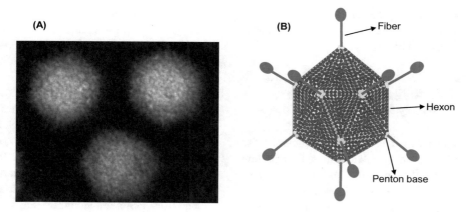

**Fig. 1** **Adenovirus structure.** (**a**) Electron micrograph of bovine adenovirus-3, a member of *Mastadenovirus* genus. (**b**) Schematic diagram of adenovirus virion depicting major structural proteins

## 1.3    Adenovirus Life Cycle

Adenovirus establishes the initial contact with host cell by interaction of the knob portion of the fiber protein with a cellular receptor. This initial attachment is followed by interaction of penton protein with receptors present on the cell surface, which activates rearrangement of host cell cytoskeleton leading to receptor-mediated endocytosis of the virion [57]. The increased acidification of endosome results in dismantling of capsid, exposing the lytic domain of protein VI leading to the disruption of endosomal membrane. The partially uncoated capsid released in the cytoplasm then makes use of microtubule motor protein dynein to reach to nuclear pore complex (NPC) [11]. The interaction of viral hexon protein with cytoplasmic nucleoporin Nup 214 [15] and protein IX with molecular motor kinesin results in further disassembly of virus capsid [83]. Finally, viral DNA in complex with protein VII is transported to the nucleus using cellular transport factors such as histone H1, transportin, and importins.

In the nucleus, host RNA polymerase II carries out transcription of adenovirus genome (reviewed in [75]), which is divided into early region (E), delayed early or intermediate region (I), and late region (L). The transcription of early region (E1, E3, and E4) genes produces non-structural proteins that are generally involved in cell cycle regulation and initiating viral transcription (E1),

evading host defense (E3), and regulating viral transcription and nuclear export (E4). The E2 early region encodes proteins that are essential for DNA replication. Next, the delayed early/ intermediate region encodes the structural proteins pIX and IVa2 involved in the activation of major late promoter (MLP), DNA packaging, and providing stability to virus capsid. This is followed by DNA replication utilizing protein priming model (reviewed in [22]). Finally, all late genes are produced from a single transcription unit known as major late transcription unit which is processed into different transcripts by splicing and usage of different polyA sites. The products of the late mRNAs are structural protein and scaffolding proteins, which regulate the late phase translation and assembly of virus particle [7].

The assembly of adenovirus particle occurs in the nucleus, where the hexon trimers and the penton capsomeres associate with each other as well as with other minor capsid proteins to form empty capsids. Adenovirus DNA is then packaged in these preformed capsids starting with the left end of the genome. This packaging requires (a) cis-acting AT-rich repeat sequences present at the left end of genome, (b) viral proteins, and (c) cellular proteins. However, some studies have reported that DNA packaging and capsid formation occur simultaneously [18]. Finally, activated adenovirus protease cleaves the structural proteins pIIIa, pVI, pTP, pVII, pVIII, and pμ making the progeny virion infectious (reviewed in Russell [75]), which is then released from the host cell by cell lysis.

## 2   Adenovirus as a Vaccine Delivery Vector

Though adenovirus has been used to identify the basic processes of a eukaryotic cell including RNA splicing [8], over the last three decades, adenovirus has been evaluated as a gene/vaccine delivery vehicle [81]. In fact, about 18.5% of the gene delivery trials in the world utilize adenovirus vectors [31]. A number of properties of adenovirus make them a vector of choice for vaccine delivery. These include:

(a)  Medium-sized stable genomes
(b)  Low pathogenicity
(c)  Ability to infect a broad spectrum of dividing and post-mitotic quiescent mammalian cells and not integrate into the host genome
(d)  Easy to manipulate, making cloning of transgenes very easy [63, 88]

Moreover, they can be delivered by different routes of administration and can be used in prime-boost protocols, and depending on the dose and route of administration, they can elicit mucosal, humoral, and cell-mediated immunity [63]. Multiple antigens from the same or different pathogens can be cloned and expressed in an adenovirus vector [13]. Paramount to clinical use of adenovirus vectors is the cost-effective, scalable, and reproducible production of the vectors in vitro using suspension- or anchorage-dependent cell lines [78].

### 2.1   Construction of Recombinant Adenovirus Vectors

A number of approaches have been used over the years for construction of recombinant adenovirus vectors. These approaches can be broadly categorized into three groups, namely, in vitro ligation, homologous recombination in

mammalian cells, and homologous recombination in bacteria. These methods allow insertion of foreign genes in any region of the adenovirus genome.

1. *In vitro ligation method.* This method takes advantage of the presence of unique restriction enzyme site (such as *Cla*I in human adenovirus-5) in the adenovirus genome [9]. Two major components of this system are (1) linear adenovirus DNA genome and (2) a plasmid that contains the left end of adenovirus genome, including the left ITR, the E1A enhancer sequence, and the packaging signal. The foreign gene is first cloned in the plasmid downstream of the viral sequences. The plasmid is then digested with the unique restriction enzyme, and the foreign gene-containing fragment is directly ligated to unique restriction enzyme-digested adenovirus genome. The resulting ligation product is then transfected in E1-complementing cells for the production of recombinant adenovirus (Fig. 2a).

Although the in vitro ligation method is simple, it has several disadvantages including low efficiency, generation of recombinant virus without gene of interest, generation of large number of wild-type virus, and limited insertion capacity. However, the in vitro ligation method has been improved further by introduction of more restriction sites in the E1 deletion region [59]. The improved system consists of two plasmids: (1) the vector plasmid that has an ampicillin-resistant gene and adenovirus genome with three unique restriction sites *I-Ceu*I, PI-*Sce*I, and *Swa*I and (2) the shuttle plasmid that has kanamycin-resistant gene and a multiple cloning site located between unique I-*Ceu*I and PI-*Sce*I sites. The gene of interest is first cloned in the shuttle vector. The vector and shuttle plasmids are then digested with I-*Ceu*I and PI-*Sce*I enzymes. The digested vector and shuttle plasmids are ligated together and further digested with *Swa*I to prevent the production of adenovirus genome without gene of interest. After *Swa*I digestion, the ligated plasmid is used to transform *E. coli*, and ampicillin is

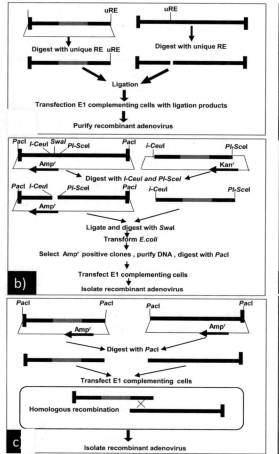
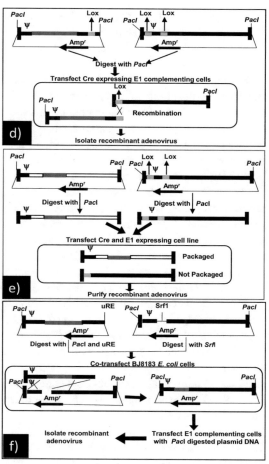

**Fig. 2 Construction of recombinant adenovirus**. Schematic diagram of adenovirus genome. (**a** and **b**) *In vitro ligation:* (**a**) one plasmid system; (**b**) two plasmid system; (**c**) *homologous recombination in E1-complementing cells*; (**d**) *homologous recombination in Cre- and E1-complementing cell lines using Lox sequence*; (**e**) *construction of gutless adenovirus vectors using Cre- and E1-complementing cell lines and Lox sequence*; (**f**) *homologous recombination in E. coli BJ5183*. BAdV-3 genome (■); plasmid DNA (——); transgene (■): unique restriction enzyme site (uRE). (Adapted from Refs. [59, 81])

used as selection marker to select positive clone. The vector DNA is isolated from positive clones, linearized with *Pac*I restriction enzyme, and transfected in E1-complementing cells to generate the recombinant virus (Fig. 2b).

2. *Homologous recombination in mammalian cells.* This method was first developed by Graham et al. in the mid-1990s using E1-complementing 293 cells [10]. This method uses two plasmids that contain overlapping fragments. One of the plasmid contains packaging signal, left end of adenovirus genome including left ITR, and gene of interest. The second plasmid has some overlapping adenovirus sequences compared to the first plasmid and the rest of the right end of viral genome but lacks DNA packaging signals. Both the plasmids are transfected in E1-complementing cells (such as 293) where recombination takes place and recombinant virus is generated. This system has a number of limitations including low recombination frequency and chance of generation of wild-

type virus, which require labor-intensive plaque purification step for isolation of a recombinant adenovirus (Fig. 2c).

However, this system has been significantly improved by development of Cre-lox site-specific system [40]. The major components of this system include (1) a Cre recombinase protein-expressing E1-complementing cell line (e.g. CRE8), a recombinant adenovirus with packaging signal flanked by two loxP sites (13 bp palindromic sequences recognized by Cre recombinase), and (2) a shuttle vector which has packaging signal, left ITR, the gene of interest, and loxP site. The viral DNA and shuttle vector are co-transfected in Cre-expressing E1-complementing cells. In Cre-expressing cells, the intramolecular recombination events lead to the generation of desired recombinant adenovirus (Fig. 2d, e). Later the flippase (FLP) recombinase/FLP recombinase target (FRT) system was developed for generation of recombinant adenovirus [64]. This system is similar to Cre-loxP system, but instead of Cre recombinase, it uses FLP recombinase which recognizes the FRT sequences.

3. *Homologous recombination in bacteria.* This method also uses a two-plasmid system, which makes use of a specialized recA-positive *E. coli* strain BJ5183 that has a highly efficient recombination capacity [17]. The vector plasmid contains full-length adenovirus genome flanked by two unique restriction sites (usually *Pac*I). The shuttle plasmid contains the homologous sequences to the viral genome on both sides of the region to be modified or to be used for insertion of foreign gene. The gene of interest is cloned in the shuttle vector. The shuttle plasmid is digested with restriction enzymes to isolate a DNA fragment containing gene of interest flanked by adenovirus genome sequence (homologous to sequence flanking the site/region of gene insertion in adenovirus genome in vector plasmid). Similarly, the vector plasmid containing full-length adenovirus genome is linearized by digestion with a suitable unique restriction enzyme located at or

near the site of insertion/deletion. The linearized vector plasmid along with the isolated DNA fragment containing gene of interest are co-transformed in BJ5183 where homologous recombination events occur leading to the generation of the desired recombinant adenovirus genome cloned in a plasmid. The recombinant adenovirus plasmid DNA is isolated from BJ5183 cells and is used to transform DH5α *E. coli*. The plasmid DNA isolated from DH5α cells is analyzed for positive clones by restriction enzyme digestion analysis. The desired plasmid DNA is then linearized by digestion with an appropriate restriction enzyme (e.g., *Pac*I) and is transfected in E1-complementing cells for the generation of recombinant adenovirus (Fig. 2f). One of the major advantages of this system is that the generation of wild-type adenovirus is extremely low, so no purification step is required. A further improvement in this system is the development of AdEasy system [42]. In AdEasy system, the vector plasmid containing full-length adenovirus genome is used as supercoiled DNA to transform *E. coli*, and the shuttle vector contains a kanamycin-resistant gene that is used to select the recombinant clones [55].

## 2.2 Type of Adenovirus Vectors

In order to use adenovirus as a vector for gene therapy and vaccination, adenovirus genomes are manipulated by deleting adenovirus-specific genes, both essential and non-essential for virus replication [20]. Moreover, adenovirus vectors are constructed by deleting the gene(s) required for capsid formation and virion assembly [5]. In addition to improving the foreign gene insertion capacity, these manipulations also make adenovirus vectors safe and efficient for the development of vaccines for use in human and animals (Fig. 3).

1. *First-generation adenovirus vectors.* The first-generation adenovirus vectors have deletions in the E1 and/or E3 regions. These vectors can

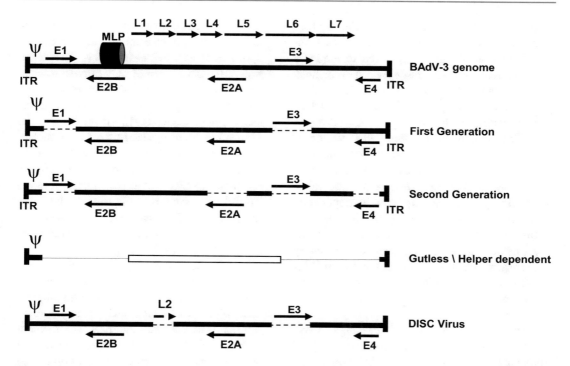

**Fig. 3 Type of adenovirus vectors.** Schematic diagram of BAdV-3 genome. BAdV-3 genome. (■); Transgene (▭); encapsidation signal (Ψ); early (E) region; L (late) region; major late promoter (MLP); deleted region (----); stuffer DNA (——). (Adapted from Refs. [1, 3, 98])

be replication-competent or replication-defective. While the replication-competent vectors retain intact E1 region with E3 region deleted, the replication-defective vectors lack partial- or full-length E1 region and E3 region. The replication-defective viruses are grown using E1-complementing cell lines (e.g., PERC 6). The E1- and E3-deleted regions serve for cloning transgene expression cassettes [20]. Both E3-deleted and E1–E3-deleted (leaky expression) adenovirus vectors produce some adenovirus proteins when introduced in animals, which results in cytotoxic T-cell responses leading to rapid clearance of the vector and elimination of transgene expression.

2. *Second-generation adenovirus vectors.* To reduce the toxicity and vector immunity associated with first-generation adenovirus vectors, second-generation adenovirus vectors have been developed. Second-generation adenovirus vectors are replication-defective, which lack E1, E3, and E2/E4 regions. Second-generation vectors exhibit long-term transgene expression than the first-generation adenovirus vectors in immunized animals with reduced ability to induce cell death [63, 81].

3. *Third-generation adenovirus vectors.* Third-generation adenovirus vectors were created to increase the safety by lowering vector immunogenicity, increase vector cloning capacity, and prolong duration of transgene expression. These vectors are also called helper-dependent or gutless vectors as nearly all viral genes are replaced by foreign sequences while leaving the packaging signal and the inverted terminal repeats (ITRs) intact. Production of gutless vectors involves intensive labor, which requires multiple passages in cell lines in the presence of a helper virus [63, 81].

4. *DISC (disabled infectious single cycle) adenovirus vector.* Replication-defective vectors are considered much safer and easy to get regulatory approval as progeny virus is not

produced and eventually shed in the environment. However, a high dose of replication-defective adenovirus is required to induce a protective immune response in the host, which does not make them cost-effective for use in the development of economical veterinary vaccines. In contrast, a low dose of replication-competent adenovirus vector is required to induce a protective immune response as it replicates in the host, thus making it cost-effective to develop economical veterinary vaccines. However, multiplication of replication-competent virus may lead to the shedding of progeny virions in the environment and thus may create regulatory problems, which may hinder the approval for use of replication-competent vectored vaccines in the field.

Recently, disabled infectious single cycle (DISC) adenovirus vectors have been isolated containing deletion of a protein involved in the formation of viral capsid [5]. DISC adenovirus is capable of abortive replication (viral DNA replication leading to the amplification of transgene expression) without producing infectious progeny virus (no capsid formation). Thus, use of DISC virus avoids shedding of progeny virus in the environment and reduces (economical) the amount of recombinant virus required to achieve efficient production of effective protective immune response. DISC adenoviruses are being developed and evaluated for the development of human [5] and veterinary vaccines [98].

## 2.3    Problems with Use of Adenovirus Vectors

1. *Induction of immune response.* Adenoviruses induce strong innate and adaptive immune responses (reviewed in [81]), which is attributed to their ability to infect immature dendritic cells, activating them to become mature antigen-presenting cells, thus promoting T-cell responses [63]. Immune cells including plasmacytoid dendritic cells

(pDCs), conventional DCs (cDCs), and macrophages produce interferon-α upon recognition of adenovirus infection. The detection of the interaction of the arginine-glycine-aspartic acid (RGD) motif of the penton base with macrophage β3 integrins results in the caspase-dependent IL-1β maturation. The viruses can be recognized by different molecular sensors inside the host such as nucleotide-binding oligomerization domain (NOD)-like receptors (NLRs), cytosolic DNA sensors, Toll-like receptors (TLRs), RIG-I-like receptors (RLRs), and effector molecules. Adenoviruses can be detected by macrophages in the cytosol by cDCs in a Toll-like receptor (TLR)-9-independent system and in the endosomes by pDCs via a TLR-9-dependent pathway. Moreover, adenovirus-ligand interactions with target cells can be inhibited by the classical and alternate complement pathways with factor B and C3 of the alternate complement pathway contributing to adenovirus-induced thrombocytopenia [53].

Adenoviruses also activate NFκ-β upon binding to factor C3 of the alternate complement pathway [53]. The interaction of the fiber protein of adenoviruses with their cellular receptors can also induce the expression of various proteins including PI3K kinase and junctional adhesion molecule-like protein that induce the production of various chemokines [89].

Adenovirus vectors, especially first-generation vectors, induce powerful anti-vector B- and T-cell immune response, which can lead to vector-mediated cytotoxicity and undesirable clinical side effects. In addition, the vectors induce strong innate immune responses by interacting with extracellular, intracellular, and membrane-bound innate immune sensing systems [41]. The innate immune response is characterized by high expression of neutrophils and macrophages. This results in tissue damage and rapid clearance of the vector from the body interfering with long-term transgene expression. To reduce immunogenicity of the vector backbone and prolong the period of transgene expression, third-generation vectors have been developed in which most or all of the adenovirus genomes are deleted (gutless/helper-dependent vectors) except the packaging signal and the ITRs [64].

2. *Pre-existing vector immunity.* The adaptive immunity against adenoviruses is elicited against both structural and non-structural proteins. The presence of pre-existing adaptive immunity and re-stimulation of long-lived memory $CD4^+$ T cells inhibit the efficacy of adenovirus vectors due to rapid clearance of the vectors from the body. Anti-adenovirus antibodies recognize capsid proteins including hexon, penton base, and fiber proteins. Like pre-existing neutralizing antibodies, pre-existing adenovirus-specific T cells can also reduce adenovirus vector transgene expression and immunity with greater consequences than humoral immunity due to their cross-reactive and multifunctional nature [81].

The neutralizing antibody-mediated effect is usually serotype specific; however, some cross-reactivity is also observed [14]. Hexon-specific antibodies appear to significantly contribute to the neutralization of adenovirus vectors in vivo. To counter pre-existing immunity, several strategies have been devised including replacing the exposed hypervariable loop region (HVR) of hexon with HVR of a different serotype to which antibodies are directed or using less prevalent adenovirus serotype as vector [56]. In addition, route of immunization can help in evading the effect of pre-existing adenovirus vector-specific neutralization antibodies. Mucosal immunization with recombinant adenovirus induces protective immune responses in animals containing adenovirus vector-specific pre-existing neutralizing antibodies, thus bypassing the effect of adenovirus vector-specific immune response [96]. Interestingly, subcutaneous immunization with recombinant adenovirus also induced protective immune responses in animals containing pre-existing adenovirus vector-specific neutralizing antibodies [27, 35].

## 2.4 Adenovirus as a Veterinary Vaccine Delivery Vector

Despite the problems indicated above, adenovirus vectors are still one of the best and widely evaluated viral vector platforms used for vaccine development. Although human adenovirus vectors have been used for vaccine development in animals/poultry, lately a number of non-human adenovirus-based vectors, both species-specific and non-species-specific animal adenoviruses, are being evaluated as vaccine delivery vehicles in animals/poultry [63]. Absence of pre-existing neutralizing antibodies and broad tropism of human and chimpanzee adenovirus make them attractive for use as vaccine delivery vector in animals/poultry (Table 1). However, the diverse tropism of human and chimpanzee adenovirus vectors may be a concern as use of these adenovirus vectors may increase the spread of the virus to other humans and mammals. In contrast, use of animal/poultry adenoviruses, including bovine adenovirus-3 (BAdV-3), porcine adenovirus-3 (PAdV-3), canine adenovirus-2 (CAdV-2), ovine adenovirus 287 (OAV287), fowl adenovirus-1 (FAdV-1), FAdV-8, or FAdV-10 (Table 1), appears advantageous as they are non-pathogenic and are species specific, thus reducing the threat of spread of the virus to humans and other animal species. Although the presence of pre-existing species-specific adenovirus vector-specific immunity in their natural hosts is a problem, this may be eliminated by choosing the appropriate route of immunization [27, 35, 96].

Usually, both replication-competent (e.g., E3-deleted) and replication-defective (E1–E3-deleted) vectors have been used as a vaccine delivery vehicle in animals/birds (Table 1). So far only high doses of replication-defective human adenovirus-5 [77] and chimpanzee adenovirus [90] vectors have been successfully used for developing and evaluating veterinary vaccines for the induction of efficacious protective immune responses in animals/birds (Table 1). In contrast, though species-specific replication-defective adenoviruses have been evaluated as veterinary

**Table 1** Adenovirus-vectored vaccines tested against diseases in natural hosts

| Vector[a] | Pathogen[b] (antigen) | Host | Route (dose) | Protection | References |
|---|---|---|---|---|---|
| HAdV-5 | PPR (H, F) | Sheep | I/M (2 doses) | Clinical protection | Rojas et al. [72] |
| | PPR (HA) | Goat | I/M (1 dose) | Complete protection | Herbert et al. [43] |
| | PRV (gD, gB gC) | Pigs | I/M (1 dose) | Complete protection | Monteil et al. [61] |
| | FMDV (poly P1) | Pigs | | Complete protection | Moraes et al. [62] |
| | SIV H3N2 (HA+ NP) | Pigs | I/M (1 dose) | Complete protection (HA + NP). Partial protection (HA) | Wesley et al. [92] |
| | FMDV (P1-2A-2B"-3B"-3C) | Calves | I/M (1 dose) | Complete protection against A24 serotype | Schutta et al. [77] |
| | FMDV | Calves | I/M (1 dose with or w/o adjuvant | Different doses varied protection | Barrera et al. [4] |
| | H5N1 (HA) | Chicken | S/C (1 dose) | Complete protection | Gao et al. [30] |
| BAdV-3 | BHV-1 (gD) | Calves | I/N (2 doses) | Partial protection | Zakhartchouk et al. [96] |
| | BHV-1 (gDt) + BoIL-6 | Calves | I/N (2 doses) | Partial protection | Kumar et al. [50] |
| PAdV-3 | TGEV (S) | Pigs | Oral | | |
| | CSFV (Gp55) (E2) | Pigs | S/C (1 dose) | Protection | Hammond et al. [36] |
| | CSFV (Gp55) (E2) | Pigs | S/C (2 doses) | | Hammond et al. [39] |
| | PRV (gD) | | S/C (2 doses) | Protection | Hammond et al. [37] |
| CAdV-2 | CDV (HA, F) | Puppies | S/C (2 doses) | Clinical protection | Fischer et al. [27] |
| ChAdOx1 | RVFV (N and C) | Cattle, Sheep and Goat | I/M (1 dose) | Solid protection | Warimwe et al. [90] |
| | RVFV (N and C) | Cattle | I/M (1 dose) (thermostabilized) | Not tested | Dulal et al. [24] |
| OAV287 | *Taenia ovis* (45 W antigen) | Sheep | I/M (1 dose after primary vaccination with DNA vaccine) | Good protection | Rothel et al. [73] |
| FAdV-1 (CELO) | IBDV (VP2) | Chicken | Oro-nasal (boost 14 days after prime) | Partial protection, complete protection (S/C or I/D) | Francois et al. [29] |
| FAdV-8 | IBV (S1) | Chicken | Oral (1 dose) | Protection (92.3% when vaccinated at day 6) | Johnson et al. [44] |
| FAdV-10 | IBDV (VP2) | Chicken | I/M, I/N, I/P, I/V | Complete protection | Sheppard et al. [80] |

(continued)

**Table 1** (continued)

| Vector[a] | Pathogen[b] (antigen) | Host | Route (dose) | Protection | References |
|---|---|---|---|---|---|
| HAdV-5/ MVA | ASFV (8 antigens) | Pigs | I/M | Protection against fatal disease (100%) | Goatley et al. [32] |

"partial coding sequence

[a]HAdV-5, human adenovirus-5; BAdV-3, bovine adenovirus-3; PAdV-3, porcine adenovirus-3; CAdV-2, canine adenovirus-2; ChAdOx1, chimpanzee adenovirus Ox1; OAV287, ovine adenovirus 287; FAdV-1, fowl adenovirus-1; FAdV-8, fowl adenovirus-8; FAdV-10, fowl adenovirus-10; CELO, chicken embryo lethal orphan; MVA, modified vaccinia Ankara

[b]PPR, peste des petits ruminants; PRV, pseudorabies virus; FMDV, foot and mouth disease virus; SIV, swine influenza virus; BHV-1, bovine herpes virus type 1; TGEV, transmissible gastroenteritis virus; CSFV, classical swine fever virus; CDV, canine distemper virus; RVFV, Rift Valley fever virus; IBDV, infectious bursal disease virus; IBV, infectious bronchitis virus; ASFV, African swine fever virus.


**Table 2** Examples of characteristics of species-specific adenovirus vectors under development

| Vector[a] | Genome size (bp) and GenBank acc. no. | Site of transgene insertion | Cargo space | Cell lines used to rescue virus[b] | References |
|---|---|---|---|---|---|
| ChAd38 | 36,521 AF394196 | E1A/E1B | 4.7 kb to 7.5 kb | 293 cells | Farina et al. [26] |
| CAdV-2 | 31,323 U77082 | E1A/E1B/E3 Helper dependent | ~30 kb | MDCK DKCre DKZeo | Soudais et al. [82] and Davison et al. [21] |
| CELO virus (FAdV-1) | 43,800 U46933 | ORF9/ORF10/ORF11 or promoter for ORF22? | 3.62 kb to 5 kb | LMH | Michou et al. [58] and Tutykhina et al. [85] |
| FAdV-9 | 45,063 NC000899 | ORF0, ORF1, and ORF2 (left end) and/or ORF19, TR2, ORF17, and ORF1 (right end) | 4 kb to 7 kb | CH-SAH | Ojkic et al. [65], Pei et al. [67, 69], Corredor and Nagy [19] |
| FAdV-4 | 45,667 GU188428 | ORF16, ORF17 | 3.21 kb | CH-SAH | Pei et al. [68] |
| BAdV-3 | 34,446 AF030154 | E1A/E3/E4 | 2.3 to 6.8 kb | VIDO R2 | Reddy et al. [70] and Baxi et al. [6] |
| PAdV-3 | 34,094 AF083132 | E1A/E3/E4 | 4.3 kb to 7.3 kb | VIDO R1 VR1BL | Zakhartchouk et al. [97] and Li et al. [51] |
| OAdV-7 | 29,576 U40839 | PVIII and fiber intergenic region/unique Sal-I site within ORF RH2 or between E4 and RHE transcription units | 4.32 kb to 7.32 kb | CSL503 HVO156 | Löser et al. [54] |

[a]ChAd38, chimpanzee adenovirus 38; CAdV-2, canine adenovirus-2; CELO, chicken embryo lethal orphan; FAdV-9, fowl adenovirus-9; FAdV-4, fowl adenovirus-4; BAdV-3, bovine adenovirus-3; PAdV-3, porcine adenovirus-3; OAV287, ovine adenovirus 287

[b]293 cells, human embryonic kidney cells transformed with HAdV-5 E1 proteins; MDCK, Madin-Darby canine kidney; DKCre, canine kidney cells transformed with CAdV-2 E1 proteins and expressing Cre recombinase; DKZeo, canine kidney cells transformed with CAdV-2 E1 proteins and resistant to zeocin; LMH, chicken hepatocellular carcinoma; DK, dog kidney; CH-SAH, chicken hepatocyte carcinoma cell, the same as LMH; ST, swine testicular; VIDO R2, fetal bovine retina cells transformed with HAdV-5 E1 proteins; VIDO R1, fetal porcine retina cells transformed with HAdV-5 E1 proteins; VR1BL, VIDO R1 cells expressing PAdV-3 E1B large protein; CSL503, fetal sheep lung fibroblast cell line; HVO156, fetal sheep skin fibroblast cell line

vaccine delivery vehicles, the results do not appear promising [71].

Earlier studies have indicated that replication-competent adenoviruses are safe for use as oncolytic virus or as vaccine delivery vehicle in humans [52]. As such, species-specific replication-competent adenovirus vectors including BAdV-3, PAdV-3, CAdV-2, OAV287, FAdV-1, FAdV-8, and FAdV-10 are being evaluated for the induction of protective immune responses against different pathogens in their natural hosts (Table 1). A number of characteristics of replication-competent adenovirus vectors, including acting as live-attenuated vaccine, inducing effective antibody and cellular immune response, and requirement of low dose of recombinant virus to produce an effective immune response [28], make them an ideal vector for developing adenovirus-vectored veterinary vaccines. Although shedding of replication-competent adenovirus in the environment may be a concern for regulatory agencies, protection-challenge experiments using replication-competent adenovirus vectors suggest that negligible amount of recombinant adenovirus may or may not be shed [35, 96].

Since species-specific adenovirus vectors currently being developed and evaluated as vaccine delivery vehicles for veterinary vaccines (Tables 1 and 2) are not pathogenic, the production and release of progeny particles in the environment may not prove to be a concern as animals are normally exposed to wild-type virus. Nevertheless, more animal testing of veterinary vaccines based on replication-competent adenoviruses should help to resolve the issue.

*Adenovirus-vectored vaccines in animals/poultry.* Though a number of viral vectored vaccines are licensed for use in animal/poultry [16, 87], so far no veterinary vaccine based on adenovirus vector platform has been licensed for use in animals/poultry. Exception to this is the successful licensing of human adenovirus-5-vectored rabies vaccine for use in wildlife [16]. Moreover, a number of characteristics of human adenovirus-vectored FMDV serotype A24 vaccine named AdtA24 including

induction of efficient protection against FMDV in cattle with no viremia by 7 days post vaccination, safe for use in production cattle, absence of detection of vaccine virus in milk of vaccinated cows, and no spread of vaccine virus in unvaccinated cattle or pigs led to the first USDA conditionally licensed FMDV vaccine in the USA in 2012 [4].

*Prime-boost.* Prime-boost protocols using adenovirus vectors have been effectively used [63, 99] in inducing effective cellular immune responses. This strategy involves the first delivery of a vaccine antigen (priming) using viral vector/naked DNA and second delivery of vaccine antigen (boosting) using the same or another distinct viral vector [2, 48]. Unlike use of homologous vector for priming-boosting [66], use of heterologous vectors for priming and boosting has been shown to generate high levels of both CD8+ and CD4+ T-cell responses leading to the induction of effective cellular immunity [76, 93]. A prime-boost strategy using naked DNA for priming and adenovirus vector for boosting immune response has proved an effective strategy of vaccination against different pathogens in animals ([38, 47, 84]: [99]). A prime-boost vaccination regimen based on priming with one serotype of human adenovirus vector and boosting with another serotype of human adenovirus vector expressing the same transgene has been developed [81]. Another strategy suggested the priming with non-human adenovirus (BAdV-3 or PAdV-3) and boosting with human adenovirus [60]. Recently, we demonstrated that priming with human adenovirus-5 expressing a vaccine antigen and boosted with PAdV-3 expressing the same vaccine antigen effectively augmented the antigen-specific immune response in mice (unpublished data). Prime-boost strategy using HAdV-5 and modified vaccinia Ankara (MVA) vectors expressing African swine fever virus antigens protect pigs against virulent challenge [32].

*Altered tropism.* In order to increase the utility of adenovirus vectors, various studies have focused to alter the tropism of adenovirus. The key role played by capsid proteins in the viral infectious pathway has suggested strategies to alter this process via modification of these proteins [95]. These approaches include chemical modification of capsid, bispecific complexes, and genetic modification of capsid proteins. Although the first two strategies have established the feasibility of adenovirus targeting, genetic modification of the capsid proteins for altering adenovirus tropism appears to also have practical utility. Genetic retargeting of adenovirus has been achieved by modification of knob region of fiber protein by exchanging knob, switching adenovirus fibers between serotypes, and insertion of epitopes in hexon, penton, and pIX capsid proteins [95]. For example, replacement of human adenovirus-5 fiber with sigma mucosal-targeting $\sigma 1$ protein of reovirus type 3 Dearing altered the tropism of human adenovirus-5 for mucosal surfaces of animals [91]. By choosing appropriate ligands, one can develop adenovirus vectors, which can deliver the vaccine antigens not only to different mucosal surfaces of animals but also to other animals (e.g., cats, dogs, poultry).

Several studies have demonstrated and provided proof of principle that the tropism of animal adenoviruses can be altered. Modification of knob region of OAV287 or BAdV-3 fiber protein altered the tropism of OAV287 [94] or BAdV-3 [3]. Incorporation of polylysine residues in the knob region of fiber protein of BAdV-3 altered the tropism of BAdV-3 for endothelial cells (unpublished data). Similarly, addition of "RGD" motifs at the C-terminus of BAdV-3 pIX increased the transduction efficiency of BAdV-3 for integrin-positive cells [97]. Since the largest accumulation of lymphoid tissue in the body is in the gastrointestinal tract [45], oral vaccination is considered to be the preferred route of immunization for inducing mucosal immune responses at the gastrointestinal tract (GIT). Although most of the animal adenoviruses currently being developed for animal vaccination do not have tropism for oral mucosa, it should be possible to develop an animal adenovirus which can be used as oral vaccine delivery vehicle in animals.

## 2.5 Future of Adenovirus-Vectored Veterinary Vaccines

A number of viral vectored veterinary vaccines currently licensed are based on use of canary pox virus, fowl pox virus, vaccinia virus, Marek's disease virus, and baculovirus vectors [87]. The last decade has seen a revolution in the development and evaluation of adenovirus-vectored vaccines for use in humans [81] and animals/ poultry (Tables 1 and 2). Since the production of the adenovirus-vectored vaccine can be easily scaled up, it is possible to produce cost-effective adenovirus vector-based veterinary vaccines without affecting the efficacy of veterinary vaccine. Earlier results suggest that even single immunization with adenovirus-vectored veterinary vaccine can induce protective immune responses in animals [24, 61, 62]. Moreover, one recombinant adenovirus-vectored vaccine has been shown to induce protection in different species of animals against Rift Valley fever [90]. These findings are encouraging for the industry to accept the adenovirus vector platform for large-scale production of adenovirus-vectored vaccine for use in animals/poultry [90].

Despite numerous advantages, not much success has been achieved in licensing adenovirus-vectored vaccines for use in humans and in animals/birds. Live human adenovirus-4 and human adenovirus-7 are licensed to be used as vaccines for military persons in the USA [86].

A live adenovirus-vectored vaccine against rabies has been licensed for use in wildlife (CFIA, Veterinary Biologics Licensed in Canada). Recently, recombinant adenovirus-based FMDV vaccine named AdtA24 has been conditionally licensed for use in the USA in 2012 ([4]). Future successful licensing and marketing of the adenovirus-vectored vaccine needs to develop collaborative engagement between governments, industry, and researchers. In addition, more evaluation of adenoviruses in animals and birds regarding cytotoxicity, stability of vectored viral genome, and presence of pre-existing neutralizing antibodies should help in clearing regulatory hurdles for successful licensing of adenovirus-vectored veterinary vaccines.

## 3 Summary

Currently, a number of human\non-human adenovirus vectored vaccines are being developed and evaluated for the induction of protective immune responses in animals. While successful use of human adenovirus as vaccine delivery vehicle requires production of human adenovirus-based veterinary vaccines at economical cost, the successful use of species-specific animal adenoviruses require the availability of efficient complementing cell lines for the production of animal adenovirus vectored veterinary vaccines and the elimination of the problem of viral vector-specific preexisting antibodies in animals. We believe that recent licensing of adenovirus vectored veterinary vaccines together with reports detailing important advances in adenovirus biology including in-depth studies detailing virus-host interactions should lead to the development, evaluation, and licensing of more efficient adenovirus vectored veterinary vaccines.

## References

1. Appaiahgari MB, Vrati S. Adenoviruses as gene/vaccine delivery vectors: promises and pitfalls. Expert Opin Biol Ther. 2015;15:337–51.
2. Aviles J, Bello A, Wong G, Fausther-Bovendo H, Qiu X, Kobinger G. Optimization of prime-boost vaccination strategies against mouse-adapted ebolavirus in a short-term protection study. J Infect Dis. 2015;212(Suppl 2):S389–97.
3. Ayalew LE, Kumar P, Gaba A, Makadiya N, Tikoo SK. Bovine adenovirus-3 as a vaccine delivery vehicle. Vaccine. 2015;33:493–9.
4. Barrera J, Brake DA, Schutta C, Ettyreddy D, Kamicker BJ, Rasmussen MV, Bravo de Rueda C, Zurita M, Pisano M, Hurtle W, Brough DE, Butman BT, Harper BG, Neilan JG. Versatility of the adenovirus-vectored foot-and-mouth disease vaccine platform across multiple foot-and-mouth disease virus serotypes and topotypes using a vaccine dose representative of the AdtA24 conditionally licensed vaccine. Vaccine. 2018;36:7345–52.
5. Barry M. Single-cycle adenovirus vectors in the current vaccine landscape. Expert Rev Vaccines. 2018;17:163–73.
6. Baxi MK, Deregt D, Robertson J, Babiuk LA, Schlapp T, Tikoo SK. Recombinant bovine adenovirus type 3 expressing bovine viral diarrhea virus glycoprotein E2 induces an immune response in cotton rats. Virology. 2000;278:234–43.
7. Berk A. Adenoviridae. Field's virology. 6th ed. Philadelphia: Lippincott Williams & Wilkins; 2013. p. 1704–31.
8. Berk A. Discovery of RNA splicing and genes in pieces. PNAS. 2016;113:801–5.
9. Berkner KL, Sharp PA. Generation of adenovirus by transfection of plasmids. Nucleic Acids Res. 1983;11:6003–20.
10. Bett AJ, Haddara W, Prevec L, Graham FL. An efficient and flexible system for construction of adenovirus vectors with insertions or deletions in early regions 1 and 3. Proc Natl Acad Sci USA. 1994;91:8802–6.
11. Bremner KH, Scherer J, Yi J, Vershinin M, Gross SP, Vallee RB. Adenovirus transport via direct interaction of cytoplasmic dynein with the viral capsid hexon subunit. Cell Host Microbe. 2009;6:523–35.
12. Brown J, Fauquet C, Briddon R, Zerbini M, Moriones E, Navas-Castillo J, King A, Adams M, Carstens E, Lefkowitz E. In: King AMQ, Lefkowitz E, Adams MJ, Carstens EB, editors. Virus taxonomy: Ninth report of the international committee on taxonomy of viruses. San Diego: Elsevier; 2012. p. 351–73.
13. Brownlie R, Kumnar P, Babiuk LA, Tikoo SK. Recombinant bovine adenovirus-3 co-expressing bovine respiratory syncytial virus glycoprotein G and truncated glycoprotein gD of bovine herpesvirus-1 induce immune response in cotton rats. Mol Biotechnol. 2015;57:58–64.
14. Bru T, Salinas S, Kremer EJ. An update on canine adenovirus type 2 and its vectors. Viruses. 2010;2:2134–53.
15. Cassany A, Ragues J, Guan T, Bégu D, Wodrich H, Kann M, Nemerow GR, Gerace L. Nuclear import of adenovirus DNA involves direct interaction of hexon with an N-terminal domain of the nucleoporin Nup214. J Virol. 2015;89:1719–30.
16. CFIA. Veterinary Biologics Licensed in Canada, 23, 01–2019.
17. Chartier C, Degryse E, Gantzer M, Dieterle A, Pavirani A, Mehtali M. Efficient generation of recombinant adenovirus vectors by homologous recombination in Escherichia coli. J Virol. 1996;70:4805–10.
18. Condezo GN, San Martín C. Localization of adenovirus morphogenesis players, together with visualization of assembly intermediates and failed products, favor a model where assembly and packaging occur

concurrently at the periphery of the replication center. PLoS Pathog. 2017;13:e1006320.

19. Corredor JC, Nagy É. The non-essential left end region of the fowl adenovirus 9 genome is suitable for foreign gene insertion/replacement. Virus Res. 2010;149 (2):167–74.

20. Danthinne X, Imperiale MJ. Production of first-generation adenovirus vectors: a review. Gene Ther. 2000;7:2707–14.

21. Davison AJ, Benkő M, Harrach B. Genetic content and evolution of adenoviruses. J Gen Virol. 2003;84:2895–908.

22. de Jong RN, van der Vliet PC, Brenkman AB. Adenovirus DNA replication: protein priming, jumping back and the role of the DNA binding protein DBP. Curr Top Microbiol Immunol. 2003;272:187–211.

23. Doszpoly A, Wellehan JF, Childress AL, Tarján ZL, Kovács ER, Harrach B, Benkő M. Partial characterization of a new adenovirus lineage discovered in testudinoid turtles. Infect Genet Evol. 2013;17:106–12.

24. Dulal P, Wright D, Ashfield R, Hill AVS, Charleston B, Warimwe GM. Potency of a thermostabilized chimpanzee adenovirus Rift Valley fever vaccine in cattle. Vaccine. 2016;34:2296–8.

25. Élő P, Farkas SL, Dán ÁL, Kovacs GM. The p32K structural protein of the Atadenovirus might have bacterial relatives. J Mol Evol. 2003;56:175–80.

26. Farina SF, Gao GP, Xiang ZQ, Rux JJ, Burnett RM, Alvira MR, Marsh J, Ertl HC, Wilson JM. Replication-defective vector based on a chimpanzee adenovirus. J Virol. 2001;75:11603–13.

27. Fischer L, Tronel JP, Pardo-David C, Tanner P, Colombet G, Minke J, Audonnet JC. Vaccination of puppies born to immune dams with a canine adenovirus-based vaccine protects against a canine distemper virus challenge. Vaccine. 2002;20:3485–97.

28. Fougeroux C, Holst PJ. Future prospects for the development of cost effective adenovirus vaccines. Int J Mol Sci. 2017;18:686.

29. Francois A, Chevalier C, Delmas B, Eterradossi N, Toquin D, Rivallan G, Langlois P. Avian adenovirus CELO recombinants expressing VP2 of infectious bursal disease virus induce protection against bursal disease in chickens. Vaccine. 2004;22:2351–60.

30. Gao W, Soloff AC, Lu X, Montecalvo A, Nguyen DC, Matsuoka Y, Robbins PD, Swayne DE, Donis RO, Katz JM, Barratt-Boyes SM, Gambotto A. Protection of mice and poultry from lethal H5N1 avian influenza virus through adenovirus-based immunization. J Virol. 2006;80:1959–64.

31. Gene Therapy Clinical Trials Worldwide database J. Gene Medicine Dec 2018.

32. Goatley LC, Reis A, Portugal R, Goldswain H, Shimmon GL, Hargreaves Z, Ho C-S, Montoya M, Sanchez-Cordon PJ, Taylor G, Dixon LK, Netherton CL. A pool of eight virally vectored African swine fever antigens protect pigs against fatal disease. Vaccine. 2020;8:234.

33. Gorman JJ, Wallis TP, Whelan DA, Shaw J, Both GW. LH3, a "homologue" of the mastadenoviral E1B 55-kDa protein is a structural protein of atadenoviruses. Virology. 2005;342:159–66.

34. Grgić H, Yang DH, Nagy E. Pathogenicity and complete genome sequence of a fowl adenovirus serotype 8 isolate. Virus Res. 2011;156:91–7.

35. Hammond JM, Johnson MA. Porcine adenovirus as a vaccine delivery system for swine vaccines and immunotherapeutics. Vet J. 2005;169:17–27.

36. Hammond JM, et al. Vaccination with a single dose of a recombinant porcine adenovirus expressing the classical swine fever virus gp55 (E2) gene protects pigs against classical swine fever. Vaccine. 2000;18:1040–50.

37. Hammond JM, et al. Vaccination of pigs with a recombinant porcine adenovirus expressing the gD gene from pseudorabies virus. Vaccine. 2001;19:3752–8.

38. Hammond JM, Jansen ES, Morrissy CJ, Goff WV, Meehan GC, Williamson MM, Lenghaus C, Sproat KW, Andrew ME, Coupar BE, Johnson MA. A prime-boost vaccination strategy using naked DNA followed by recombinant porcine adenovirus protects pigs from classical swine fever. Vet Microbiol. 2001;80:101–19.

39. Hammond JM, et al. Protection of pigs against 'in contact' challenge with classical swine fever following oral or subcutaneous vaccination with a recombinant porcine adenovirus. Virus Res. 2003;97:151–7.

40. Hardy S, Kitamura M, Harris-Stansil T, Dai Y, Phipps ML. Construction of adenovirus vectors through Cre-lox recombination. J Virol. 1997;71:1842–9.

41. Hartman ZC, Appledorn DM, Amalfitano A. Adenovirus vector induced innate immune responses: impact upon efficacy and toxicity in gene therapy and vaccine applications. Virus Res. 2008;132:1–14.

42. He TC, Zhou S, da Costa LT, Yu J, Kinzler KW, Vogelstein B. A simplified system for generating recombinant adenoviruses. Proc Natl Acad Sci. 1998;95(5):2509–14.

43. Herbert R, Baron J, Batten C, Baron M, Taylor G. Recombinant adenovirus expressing the haemagglutinin of peste des petits ruminants virus (PPRV) protects goats against challenge with pathogenic virus; a DIVA vaccine for PPR. Vet Res. 2014;45:24.

44. Johnson MA, Pooley C, Ignjatovic J, Tyack SG. A recombinant fowl adenovirus expressing the S1 gene of infectious bronchitis virus protects against challenge with infectious bronchitis virus. Vaccine. 2003;21:2730–6.

45. Jung C, Hugot J-P, Barrau F. Peyer's patches: the immune sensors of the intestine. Int J Inflamm. 2010: Article ID 823710.

46. Kaján GL, Davison AJ, Palya V, Harrach B, Benko M. Genome sequence of a waterfowl aviadenovirus, goose adenovirus 4. J Gen Virol. 2012;93:2457–65.

47. Kim SJ, Kim HK, Han YW, Aleyas AG, George JA, Yoon HA, Yoo DJ, Kim K, Eo SK. Multiple alternating immunizations with DNA vaccine and replication incompetent adenovirus expressing gB of pseudorabies virus protect animals against lethal virus challenge. J Microbiol Biotechnol. 2008;18:1326–34.

48. Kosinska AD, Johrden L, Zhang E, Fiedler M, Mayer A, Wildner O, Lu M, Roggendorf M. DNA prime-adenovirus boost immunization induces a vigorous and multifunctional T-cell response against hepadnaviral proteins in the mouse and woodchuck model. J Virol. 2012;86:9297–310.

49. Kovacs GM, LaPatra SE, D'Halluin JC, Benko M. Phylogenetic analysis of the hexon and protease genes of a fish adenovirus isolated from white sturgeon (*Acipenser transmontanus*) supports the proposal for a new adenovirus genus. Virus Res. 2003;98:27–34.

50. Kumar P, Ayalew LE, Godson DL, Gaba A, Babiuk LA, Tikoo SK. Mucosal immunization of calves with recombinant bovine adenovirus-3 co-expressing truncated form of bovine herpesvirus-1 gD and bovine IL-6. Vaccine. 2014;32:3300–6.

51. Li X, Babiuk LA, Tikoo SK. Analysis of early region 4 of porcine adenovirus type 3. Virus Res. 2004;104:181–90.

52. Lichtenstein DL, Wold WS. Experimental infections of humans with wild-type adenoviruses and replication-competent adenovirus vectors: replication, safety and transmission. Cancer Gene Ther. 2004;11:819–29.

53. Lopez-Gordo E, Podgorski II, Downes N, Alemany R. Circumventing antivector immunity: potential use of nonhuman adenoviral vectors. Hum Gene Ther. 2014;25:285–300.

54. Löser P, Hofmann C, Both GW, Uckert W, Hillgenberg M. Construction, rescue, and characterization of vectors derived from ovine atadenovirus. J Virol. 2003:11941–51.

55. Luo J, Deng Z-L, Luo X, Song W-X, Chen J, Sharff KA, Luu HH, Haydon RC, Kinzler KW, Vogelstein B, He T-C. A protocol for rapid generation of recombinant adenoviruses using the AdEasy system. Nat Protoc. 2007;2:1236–47.

56. Ma J, Duffy MR, Deng L, Dakin RS, Uil T, Custers J, Kelly SM, McVey JH, Nicllin SA, Baker AH. Manipulating adenovirus hexon hypervariable loops dictates immune neutralization and coagulation factor X- dependent in vitro and in vivo. PLoS Pathogens. 2015;11(2):e1004673.

57. Meier O, Greber UF. Adenovirus endocytosis. J Gene Med. 2004;6(Suppl. 1):S152–63.

58. Michou AI, Lehrmann H, Saltik M, Cohen M. Mutational analysis of the avian adenovirus CELO, which provides a basis for gene delivery vectors. J Virol. 1999;73:1399–410.

59. Mizuguchi H, Kay MA. Efficient construction of a recombinant adenovirus vector by an improved in vitro ligation method. Hum Gene Ther. 1998;9:2577–83.

60. Moffatt S, Hays J, HogenEsch H, Mittal SK. Circumvention of vector-specific neutralizing antibody response by alternating use of human and non-human adenoviruses: implications in gene therapy. Virology. 2000;272:159–67.

61. Monteil M, Le Pottier MF, Ristov AA, Cariolet R, L'Hospitalier R, Klonjkowski B, Eloit M. Single inoculation of replication-defective adenovirus-vectored vaccines at birth in piglets with maternal antibodies induces high level of antibodies and protection against pseudorabies. Vaccine. 2000;18:1738–42.

62. Moraes MP, Mayr GA, Mason PW, Grubman MJ. Early protection against homologous challenge after a single dose of replication-defective human adenovirus type 5 expressing capsid proteins of foot-and-mouth disease virus (FMDV) strain A24. Vaccine. 2002;20:1631–9.

63. Ndi OL, Barton MD, Vanniasinkam T. Adenoviral vectors in veterinary vaccine development: potential for further development. World J Vaccine. 2013;3:111–21.

64. Ng P, Beauchamp C, Evelegh C, Parks R, Graham FL. Development of a FLP/frt system for generating helper-dependent adenoviral vectors. Mol Ther. 2001;3:809–15.

65. Ojkic D, Krell PJ, Nagy E. Unique features of fowl adenovirus 9 gene transcription. Virology. 2002;302:274–85.

66. Palmowski MJ, Choi EM, Hermans IF, Gilbert SC, Chen JL, Gileadi U, Salio M, Van Pel A, Man S, Bonin E. Competition between CTL narrows the immune response induced by prime-boost vaccination protocols. J Immunol. 2002;168:4391–8.

67. Pei Y, Griffin B, de Jong J, Krell PJ, Nagy É. Rapid generation of fowl adenovirus 9 vectors. J Virol Methods. 2015;223:75–81.

68. Pei Y, Corredor JC, Griffin BD, Krell PJ, Nagy É. Fowl adenovirus 4 (FAdV-4)-based infectious clone for vaccine vector development and viral gene function studies. Viruses. 2018;10:E97.

69. Pei Y, Krell PJ, Nagy É. Generation and characterization of a fowl adenovirus 9 dual-site expression vector. J Biotechnol. 2018;266:102–10.

70. Reddy PS, Idamakanti N, Chen Y, et al. Replication-defective bovine adenovirus type 3 as an expression vector. J Virol. 1999;73:9137–44.

71. Reddy PS, Idamakanti N, Pyne C, Zakhartchouk AN, Godson DL, Papp Z, Baca-Estrada ME, Babiuk LA, Mutwiri GK, Tikoo SK. The immunogenicity and efficacy of replication-defective and replication-competent bovine adenovirus-3 expressing bovine herpesvirus-1 glycoprotein gD in cattle. Vet Immunol Immunopathol. 2000;76:257–68.

72. Rojas JM, Moreno H, Valcarcel F, Pena L, Sevilla N, Martin V. Vaccination with recombinant adenoviruses

expressing the peste des petits ruminants virus F or H proteins overcomes viral immunosuppression and induces protective immunity against PPRV challenge in sheep. PLoS One. 2014;9:e101226.

73. Rothel JS, et al. Sequential nucleic acid and recombinant adenovirus vaccination induces host-protective immune responses against Taenia ovis infection in sheep. Parasite Immunol. 1997;19:221–7.

74. Rowe WP, Huebner RJ, Gilmore LK, Parrott RH, Ward TG. Isolation of a cytopathogenic agent from human adenoids undergoing spontaneous degeneration in tissue culture. Proc Soc Exp Biol Med. 1953;84:570–3.

75. Russell WC. Adenoviruses: update on structure and function. J Gen Virol. 2009;90:1–20.

76. Schneider J, Gilbert SC, Hannan CM, Dégano P, Prieur E, Sheu EG, Plebanski M, Hill AV. Induction of CD8+ T cells using heterologous prime-boost immunisation strategies. Immunol Rev. 1999;170:29–38.

77. Schutta C, Barrera J, Pisano M, Zsak L, Grubman MJ, Mayr GA, Moraes MP, Kamicker BJ, Brake DA, Ettyreddy D, Brough DE, Butman BT, Neilan JG. Multiple efficacy studies of an adenovirus-vectored foot-and-mouth disease virus serotype A24 subunit vaccine in cattle using homologous challenge. Vaccine. 2016;34:3214–20.

78. Sharon D, Kamen A. Advancements in the design and scalable production of viral gene transfer vectors. Biotechnol Bioeng. 2018;115:25–40.

79. Shenk T. Adenoviridae: the viruses and their replication. In: Fields BN, Knipe DM, Howley PM, editors. Fields' virology. 4th ed. Philadelphia: Lippincott Williams & Wilkins; 2001.

80. Sheppard M, Werner W, McCoy RJ, Johnson MA. Fowl adenovirus recombinant expressing VP2 of infectious bursal disease virus induces protective immunity against bursal disease. Arch Virol. 1998;143:915–30.

81. Singh S, Kumar R, Babita A. Adenoviral vector based vaccines and gene therapies: current status and future prospects; 2018. https://doi.org/10.5772/intechopen.79697.

82. Soudais C, Skander N, Kremer EJ. Long-term in vivo transduction of neurons throughout the rat CNS using novel helper-dependent CAV-2 vectors. FASEB J. 2004;18:391–3.

83. Strunze S, Engelke MF, Wang IH, Puntener D, Boucke K, Schleich S, Way M, Schoenenberger P, Burckhardt CJ, Greber UF. Kinesin-1-mediated capsid disassembly and disruption of the nuclear pore complex promote virus infection. Cell Host Microbe. 2011;10:210–23.

84. Sun Y, Li N, Li HY, Li M, Qiu HJ. Enhanced immunity against classical swine fever in pigs induced by prime-boost immunization using an alphavirus replicon-vectored DNA vaccine and a recombinant adenovirus. Vet Immunol Immunopathol. 2010;137:20–7.

85. Tutykhina IL, Shul'pin I, Chvala IA, Shmarov MM, Logunov DY, Shcherbakova LO, Volkova MA, Mudrak NS, Borisov AV, Drygin VV, Naroditskii BS, Gintsburg AL. Construction of CELO recombinant adenoviruses expressing avian type an influenza virus hemagglutinin gene and its use as vaccine against avian influenza virus type a H5N1 and H7N1. Mol Genet Microbiol Virol. 2011;26:34–40.

86. United States Department of Food and drug Administration. Vaccines Licenced For use in United States of America

87. USDA. Vet Biological products. Licenses and Permittees. Jan 31st, 2019

88. Vellinga J, Rabelink MJWE, Cramer SJ, van den Wollenberg DJM, Van der Meulen H, Leppard KN, Fallaux FJ, Hoeben RC. Spacers increase the accessibility of peptide ligands linked to the C- terminus of adenovirus minor capsid protein IX. J Virol. 2004;78:3470–9.

89. Verdino P, Witherden DA, Havran WL, Wilson IA. The molecular interaction of CAR and JAML recruits the central cell signal transducer PI3K. Science. 2010;329:1210–4.

90. Warimwe GM, Gesharisha J, Carr BV, Otieno S, Otingah K, Wright D, Charleston B, Okoth E, Elena LG, Lorenzo G, et al. Chimpanzee adenovirus vaccine provides multispecies protection against Rift Valley fever. Sci Rep. 2016;6:20617.

91. Weaver EA, Camacho ZT, Hillestad ML, Crosby CM, Turner MA, Guenzel AJ, Fadel HJ, Mercier GT, Barry MA. Mucosal vaccination by adenoviruses displaying reovirus sigma 1. Virology. 2015;482:60–6.

92. Wesley RD, Tang M, Lager KM. Protection of weaned pigs by vaccination with human adenovirus 5 recombinant viruses expressing the hemagglutinin and the nucleoprotein of H3N2 swine influenza virus. Vaccine. 2004;22:3427–34.

93. Woodland DL. Jump-starting the immune system: prime-boosting comes of age. Trends Immunol. 2004;25:98–104.

94. Xu ZZ, Both GW. Altered tropism of an ovine adenovirus carrying fiber protein cell binding domain of human adenovirus-type 5. Virology. 1998;248:156–63.

95. Yoon AR, Hong J, Kim SW, Yun C-O. Redirecting adenovirus tropism by genetic, chemical and mechanical modification of the adenovirus surface for cancer gene delivery. Exp Opin Drug Deliv. 2016;13 (6):843–58.

96. Zakhartchouk AN, Pyne C, Mutwiri GK, Papp Z, Baca-Estrada ME, Griebel P, et al. Mucosal immunization of calves with recombinant bovine adenovirus-3: induction of protective immunity to bovine herpesvirus-1. J Gen Virol. 1999;80:1263–9.

97. Zakhartchouk A, Connors W, van Kersseg A, Tikoo SK. Bovine adenovirus type 3 containing heterologous protein in the C-terminus of minor capsid protein pIX. Virology. 2004;320:291–300.

98. Zhao X, Tikoo SK. Deletion of pV affects integrity of capsid causing defect in the infectivity of bovine adenovirus-3. J Gen Virol. 2016;97:2657–67.

99. Zhao HP, Sun JF, Li N, Sun Y, Wang Y, Qiu HJ. Prime-boost immunization using alphavirus replicon and adenovirus vectored vaccines induces enhanced immune responses against classical swine fever virus in mice. Vet Immunol Immunopathol. 2009;131:158–66.

# Poxvirus Vectors

Lok R. Joshi and Diego G. Diel

**Abstract**

The utility of poxviruses as expression vectors was first described in the early 1980s. Since then, poxviruses have been widely used as vaccine delivery platforms in human and veterinary medicine. The main features that make poxviruses excellent antigen delivery platforms and vaccine vectors are their large genome size with the presence of multiple immunomodulatory genes, the tolerance for large heterologous gene insertions, and their ability to induce cellular and humoral immunity. Initial attempts were focused on engineering vaccinia virus to express heterologous genes. Later, the potential of other poxviruses including avipoxvirus, parapoxvirus, and swinepox viruses as vectors was also explored with promising results. To address the safety concerns related to wild-type poxviruses, several highly attenuated, replication-defective strains have been developed mostly by serial passages in cell culture. Most of the recombinant poxviruses developed to date have targeted insertional inactivation of the thymidine kinase (TK) gene, in which the heterologous gene is inserted in the TK locus in the poxvirus genome. In recent years, other immunomodulatory genes have also been used to generate safer and multivalent poxvirus-vectored vaccine candidates. Poxvirus vectors have been shown to be very effective in heterologous prime-boost immunization regimes, where poxvirus vectors are used in combination with other killed or DNA vaccine formulations. To date multiple poxvirus-vectored vaccines have been licensed for use against a variety of animal pathogens including rabies virus (RabV), avian influenza virus (AIV), canine distemper virus (CDV), and West Nile virus (WNV).

**Keywords**

Poxvirus · Vector · Vaccine · Delivery platform · Immunity

**Learning Objectives**

After reading this chapter, you should be able to:

- Describe how poxviruses can be used in vaccine delivery platforms
- Explain how poxvirus vectors are constructed, selected, and utilized for vaccine delivery in domestic animal species
- Describe how poxvirus vectors are applied in veterinary medicine

L. R. Joshi · D. G. Diel (✉)
Department of Population Medicine and Diagnostic Sciences, College of Veterinary Medicine, Cornell University, Ithaca, NY, USA
e-mail: lrj36@cornell.edu; dgdiel@cornell.edu

© Springer Nature Switzerland AG 2021
T. Vanniasinkam et al. (eds.), *Viral Vectors in Veterinary Vaccine Development*,
https://doi.org/10.1007/978-3-030-51927-8_6

## 1    Introduction

Poxviruses are large complex viruses that belong to the family *Poxviridae*. The *Poxviridae* is divided into two subfamilies: the *Chordopoxvirinae* and *Entomopoxvirinae*. The subfamily *Chordopoxvirinae* comprises viruses that infect vertebrate animal species, whereas the *Entomopoxvirinae* contains viruses that mainly infect insects. Currently, there are 11 genera classified under the *Chordopoxvirinae* subfamily and 3 genera under the *Entomopoxvirinae* [1]. Poxviruses are classified in these genera on the basis of virus morphology, phylogeny, serological cross-reactivity, and host range [2]. The poxvirus species name usually refers to the host from which the virus was first isolated. While some poxviruses have restricted host range (i.e., variola virus known to infect only humans), there are many others, including cowpox virus (CPXV), buffalopox virus (BPXV), and monkeypox virus (MPXV), which have broad animal host range infecting multiple mammalian species including humans. Poxviral infections occur through different routes, including the skin (Orf virus), respiratory tract (variola virus), or oral route (ectromelia virus) [3] and are characterized by formation of skin lesions, usually evolving through the stages of papules, pustules, vesicles, nodules, and scabs [4].

Poxviruses are among the largest known viruses. Most poxviruses contain brick-shaped virions with a particle size ranging from 220 to 450 nm long × 140 to 260 nm wide × 140 to 260 nm thick. Parapoxviruses are oval-shaped with a particle size of 260 nm × 160 nm [4]. Virions are enveloped with the presence of surface tubules or surface filaments (Fig. 1a). Internally, the virions contain two lateral bodies and a dumbbell-shaped nucleoprotein core. The nucleoprotein core contains enzymes essential for virus replication and the nucleocapsid protein bound to the viral genome. The viral genome is a linear double-stranded DNA with the size ranging from 130 kbp to 375 kbp and encode between ~130 and 350 open reading frames (ORFs). The two strands of DNA are cross-linked at the termini due to presence of A+T-rich inverted terminal repeats (ITR) at the two ends of the genome. The central region of the genome is highly conserved across different poxviruses and encodes genes essential for viral transcription, replication, and virion assembly. Non-conserved genes that are involved in virus host range, immunomodulation, and virulence and pathogenesis are present at either end of the genome, flanking the conserved central genome core (Fig. 1b).

A unique feature of poxviruses is their replication site, which takes place in the cytoplasm of infected cells, making poxviruses an exception among DNA viruses. The replication mechanism of vaccinia virus (VACV) have been widely studied, and most of our understanding of poxvirus replication comes from VACV. Notably, transcription and expression of poxviral genes are temporally regulated, and the genes are classified as early, intermediate, or late genes based on the time of expression in relation to virus genome replication. In general, early genes are transcribed before replication, whereas intermediate and late genes are transcribed after the virus genome has been replicated. For VACV, for example, early, intermediate, and late genes are expressed in 20, 100, and 140 min after infection, respectively [5]. The poxvirus virion contains essential enzymes to initiate viral transcription upon infection. Therefore, early genes are transcribed within the virion core soon after the virus enters the cell, and mRNAs are extruded into the cytoplasm for translation. These early genes encode for transcription factors required for expression of intermediate genes. Some of the early proteins also play important roles in host immune evasion and modulation. Once early genes are expressed, uncoating of the virion core takes place, and DNA is released into the cytoplasm followed by viral DNA replication which occurs in discrete replication sites within the cytoplasm designated viral factories. The intermediate genes are expressed after DNA replication and encode for transcription factors required for the expression of late genes. The late genes encode proteins essential for virion assembly and early gene

A.

B.

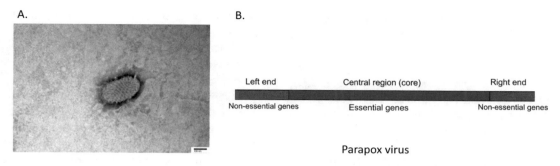

Parapox virus

**Fig. 1** Structure and genome of poxvirus. (**a**) Negative stain preparation of Orf virus (parapoxvirus). (**b**) Schematic representation of poxvirus genome. The essential genes are present in the central part of the genome. The non-essential genes which play role in immunomodulation are present in either end of the genome

transcription factors which will be packaged within the virion core. After virus assembly, enveloped virions are released by budding, whereas non-enveloped virions are released by cell lysis. The understading of these basic biological properties of poxviruses is important in designing poxvirus vectors.

## 2 Construction of Poxvirus Vectors

Poxviruses hold a unique place in the history of immunization. In 1796, Edward Jenner demonstrated that smallpox could be prevented by using CPXV as a vaccine [6]. Later, VACV was widely used to immunize people against smallpox, which culminated with the eradication of the disease in 1980. To date, smallpox remains the only human disease that has been eradicated. Although smallpox was eradicated, and vaccination was discontinued, the biological and immunomodulatory properties of VACV, the virus used as vaccine against smallpox, generated significant interest in poxviruses among scientists worldwide. Soon after the eradication of smallpox, a few studies describing genetically engineered VACV and the use of vaccinia as a eukaryotic expression system were published [7, 8]. In addition, recombinant VACV expressing single or multiple heterologous viral antigens were developed establishing the foundation for the use of poxvirus as vaccine delivery vectors [9–11].

Some of the features that made VACV a well-received and widely used vector are (i) the large genome size (190 kb), with the presence of many non-essential genes, which could be manipulated without severely impacting virus replication; (ii) the ability of VACV to tolerate insertion of up to 25,000 bp of foreign DNA [12]; (iii) the fact that the virus is a potent inducer of both humoral and cell-mediated immunities [13]; (iv) the ease of administration and its efficacy through different immunization routes [13]; and (v) the stability of the virus at room temperature when lyophilized, which obviates the need for cold chain [14]. Given that poxviruses share many common properties, the features described for VACV above also apply to other viruses in the family. The immunomodulatory properties of poxviruses and the efficacy of VACV as a vector platform led several groups to explore other poxvirus vector alternatives. Several studies showed the potential of other members of the family *Poxviridae*, including avipoxviruses (fowlpox [FWPV] and canarypox virus [CNPV]), swinepox virus (SWPV), and Orf virus (ORFV), as vectors for human and veterinary applications. These vectors are described below with the primary focus on their use to deliver veterinary vaccines.

## 3 Strategies Used in Developing Recombinant Poxvirus Vectors

Earlier studies in the 1960s showed that genetic recombination can occur between two different strains of related poxviruses when both viruses infect a single cell [15, 16]. This process known as homologous recombination involves the exchange of nucleotide sequences between two similar or identical DNA molecules [17]. Homologous recombination is now widely used for the generation of recombinant poxviruses [18–21]. This method requires the construction of a transfer plasmid (recombination plasmid) containing the foreign gene insert (heterologous gene) and the left and right homology DNA sequences flanking the insertion site from the parent poxvirus genome (Fig. 2). Homologous recombination and recombinant poxvirus generation are achieved by infecting permissive cells with the parent poxvirus and subsequently transfecting these cells with the recombination plasmid (infection/transfection). Within cells that were infected and transfected, homologous recombination between the parental virus and the recombination plasmid takes place, resulting in a new chimeric recombinant poxvirus. The recombinant poxvirus is purified by multiple rounds of limiting dilution and/or plaque assay (Fig. 2).

There are several factors that need to be considered to achieve homologous recombination, including the homology length and the DNA structure [22]. Higher recombination frequencies were obtained, for example, when homologous flanks with at least 100–350 bp and linear plasmid DNA were used in infection/transfection experiments with VACV [22]. In addition to the insertion site and homology length, the following factors need to be considered to design and generate poxvirus-based vectors:

1. *Promoters.* Given the temporal regulation of poxvirus gene transcription (early, intermediate, and late), selection of the promoter that will drive expression of the heterologous gene is a critical aspect of the design of poxvirus vectors. In general, promoters with both early and late activities are ideal for expression of foreign genes because they drive expression of the heterologous genes throughout the vector infection cycle, promoting sustained expression of the antigen and consequent stimulation of the immune system. Early promoters would also be preferable when the poxvirus vector is replication defective or when the vector is to be used in a non-permissive animal species, both of which preclude expression driven by late promoter, which takes place after virus replication. The most commonly used promoters to drive expression of heterologous genes by poxviruses include the native VACV early/late promoters ($P_{7.5}$ or $VV_{7.5}$), the modified early promoter (mH5), or synthetic promoters such as PrS, for which expression has been optimized by mutagenesis [23].

2. *Termination signal.* The presence of poxvirus early termination signal TTTTTTNT within the sequence of heterologous genes could potentially lead to premature transcription termination and consequently low expression levels or expression of a truncated protein [24]. Therefore, termination signals should be removed through site-directed mutagenesis or synthetic biology from the heterologous gene sequence before inserting the gene into the vector [20].

3. *Codon optimization.* Codon optimization of the heterologous gene may help to achieve higher expression levels especially when the recombinant is to be used in non-target animal species, in which replication and late gene expression are impaired. Codon optimization helps in the stability of the recombinant vector by removing non-desirable sequences [23]. Additionally, it may also be used when multivalent heterosubtypic viral vectors containing two or more viral genes from closely related virus strains are designed. Codon optimization and changes in the nucleotide sequence of one of the genes increase the stability of the vector by preventing or reducing the risk of intramolecular homologous recombination.

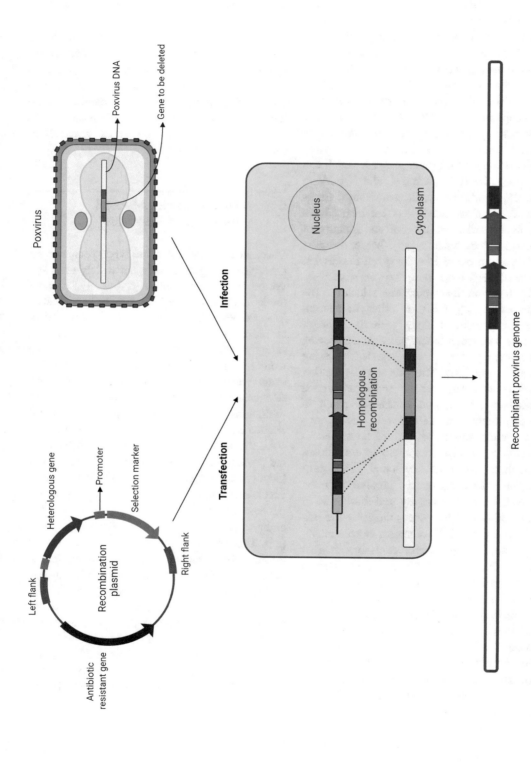

**Fig. 2** Schematic representation of homologous recombination. Cells are infected with parental poxvirus and transfected with recombination plasmid. Homologous recombination between the plasmid and parental poxvirus genome occurs within the cytoplasm resulting in a chimeric recombinant poxvirus.

4. *Selection method*. Selection of recombinant poxviruses is one of the most time-consuming steps in generating recombinant poxvirus-based vectors. Conventionally, selection of recombinant poxviruses has been based on expression of the β-galactosidase reporter gene. The 5-bromo-4-chloro-3-indolyl-β-D-galactopyranoside (X-gal) substrate is incorporated into the agarose overlay during plaque assay, and recombinants expressing β-galactosidase form blue plaques, which can be selectively picked and purified [18, 25]. Additionally, drug resistance genes like neomycin resistance gene can be used as a selectable marker for selection and isolation of poxvirus recombinants [26]. More recently, fluorescent proteins like the green fluorescent protein (GFP) have also been used successfully in recombinant poxvirus selection. The gene expressing GFP or other fluorescent proteins is inserted along with the gene of interest. The recombinant poxvirus expressing fluorescent protein can be selected by using plaque assay [20]. The presence of marker genes is not always recommended, as tandem expression of multiple genes can result in lower protein expression levels due to promoter interference. Therefore, strategies to develop markerless recombinant poxviruses have been recently developed. The most straightforward approaches involve selection of recombinant viruses by real-time PCR or immunofluorescence assays targeting the heterologous genes [27]. Whereas, more sophisticated approaches using excisable marker systems based on Cre/loxP recombination, which facilitate selection and subsequent removal of marker gene, were also developed and provide an efficient means to create markerless recombinants [28]. Recently, a marker-free system for construction of vaccinia virus vectors using CRISPR (clustered regularly interspaced short palindromic repeat)-Cas9 has also been reported [29].

## 4 Application of Poxvirus Vectors

Several poxvirus platforms have been developed and used as vaccine delivery vectors in veterinary species. Table 1 summarizes the poxvirus-vectored vaccines that are currently licensed and commercially available for use in veterinary medicine. Below we present a brief discussion of the main poxvirus vectors and their applications and/or uses in animals.

## 5 Orthopoxvirus-Based Veterinary Vaccines

Vaccinia virus (VACV), the type species of the *Orthopoxvirus* genus, has been widely used as a vector for vaccine delivery. Initially, parental moderately virulent VACV strains like Western Reserve (WR), Copenhagen, and Lister were used to develop recombinant vaccines. However, safety concerns with the use of these strains were raised especially in immunocompromised hosts which usually experienced moderate-to-severe adverse vaccine reactions [30]. These limitations led to the development of highly attenuated VACV strains. For example, the VACV strain LC16m8 was developed by sequential passage of the Lister strain in primary rabbit kidney (PRK) cells at 30 °C [31]. The modified vaccinia Ankara (MVA) has been developed by passage of VACV strain Ankara in chicken embryo fibroblasts (CEF) for 516 times [32]. The resulting virus lost ~15% of its genome during cell passaging [33], and it is replication deficient in most mammalian cells [34]. Another highly attenuated vaccinia virus strain NYVAC was derived from plaque-cloned isolate of the Copenhagen vaccine strain which contains select deletion of 15 non-essential genes [35]. The NYVAC strain is less pathogenic and has greatly reduced ability to replicate in a variety of mammalian cells (human, mice, and equine cells), but

**Table 1** Licensed and commercially available poxvirus-vectored vaccines

| Vaccine trade name[a] | Target pathogen | Target species | Insert gene | Poxvirus vector | References |
|---|---|---|---|---|---|
| RABORAL V-RG | Rabies | Wildlife canines | Glycoprotein | Vaccinia virus | [39, 166] |
| ProteqFlu | Equine influenza | Horses | HA | Canarypox | [167, 168] |
| RecombiTek Equine Influenza | Equine influenza | Horses | HA | Canarypox | [167, 168] |
| Recombitek West Nile Virus | West Nile virus | Horses | prM/E | Canarypox | [91, 169] |
| PUREVAX Feline Rabies | Rabies | Cats | Glycoprotein | Canarypox | [169] |
| PUREVAX Recombinant FeLV | Feline leukemia virus | Cats | Env, Gag/pol | Canarypox | [93, 169] |
| Recombitek Distemper | Canine distemper virus | Dogs | HA and F | Canarypox | [169, 170] |
| PUREVAX Ferret Distemper | Canine distemper virus | Ferrets | HA and F | Canarypox | [169] |
| Vectormune FP-MG | *Mycoplasma gallisepticum* | Poultry | 40k and mgc | Fowlpox | [84] |
| Vectormune FP-LT | Laryngotracheitis | Poultry | | Fowlpox | [171] |
| Vectormune FP-ND | Newcastle disease virus | Poultry | HN and F | Fowlpox | [73, 169] |
| TROVAC-AIV H5 | Avian influenza | Poultry | HA | Fowlpox | [66] |
| TROVAC-NDV | Newcastle disease virus | Poultry | HN and F | Fowlpox | [76] |

[a]Trade name might differ according to country and manufacturer

it retained the ability to induce immune response [36]. Although most of these highly attenuated vaccinia virus strains are known for their safety profile, their immunogenicity is often compromised due to high level of attenuation [37]. For example, higher doses or multiple doses of MVA-based vectored vaccines are required to achieve immune responses similar to wild-type VACV strains [38]. Nevertheless, both parental (e.g., Copenhagen, WR) and highly attenuated strains (e.g., LC16m8, MVA, NYVAC) have been used to develop recombinant vectored vaccines for veterinary use.

The first recombinant poxvirus licensed to be used as vaccine is a VACV-based vectored vaccine for rabies. This recombinant was constructed by inserting the rabies virus (RabV) glycoprotein (G) gene in the thymidine kinase (TK) locus of the Copenhagen strain of vaccinia virus [39, 40]. It has been used to control rabies in red foxes in several European countries, in coyotes and raccoons in the USA, and in raccoons in Canada [41, 42]. The vaccine is used as an oral bait which is dispersed in the wild habitat of the target species by hand or airplanes. This vaccine is safe and effective in foxes, raccoons, and coyotes [43–45]. It has been shown to be effective in vampire bats which are important reservoir for rabies virus [46]. However, it is less effective in skunks and in dogs when administered orally [41, 47, 48]. Also, as it is a live attenuated vaccine, safety concerns regarding exposure of live virus-based vaccine to non-target species have been raised. To develop safer alternatives to this vaccine, recombinant MVA, a highly attenuated VACV strain, expressing the RabV G was developed [38]. This recombinant vector was immunogenic in mice, dogs, and raccoons upon parenteral immunization. However, it was less immunogenic than the VACV Copenhagen-based recombinant vector and required a higher dose to induce immune response equivalent to the Copenhagen-based recombinant. Furthermore, the MVA recombinant failed to induce humoral immune response when immunized orally making it unsuitable to use in wild animal populations [38]. These observations highlight the fact that there is a fine balance between protective efficacy and attenuation of poxvirus vectors.

Recombinant VACV vectors expressing the hemagglutinin (H) and fusion (F) proteins of rinderpest virus have been developed. Two vaccinia recombinants were generated by inserting the H or F gene into the TK locus of VACV Wyeth strain. Immunization of cattle with either recombinant or with the mixture of the two recombinants provided 100% protection even when the immunized animals were challenged with 1000 times the lethal dose of rinderpest virus [49]. There was no transmission of the VACV vector from vaccinated animals to contact animals. Moreover, cattle vaccinated with the mixture of recombinant vectors presented solid immunity as indicated by absence of amnestic response after challenge infection with rinderpest virus [49]. The immunized animals, however, developed pock lesions at the site of immunization indicating that the vector was not completely attenuated. Additionally, the use and production of the mixed vector formulation, containing equivalent doses of two different recombinants, was cumbersome [50].

To address these drawbacks, a recombinant of VACV strain Wyeth expressing both H and F genes (vRVFH) was developed. The H gene was inserted into TK locus, and F gene was inserted into HA locus in the VACV genome. The insertional inactivation of TK and HA genes led to further attenuation of the vector. Consequently, no pock lesions were observed after intradermal immunization in cattle. The protective efficacy of the vaccine was not affected, and sterilizing immunity was observed in cattle against rinderpest virus challenge [50]. Later, another VACV-based recombinant expressing H and F genes (v2RVFH) was constructed using TK locus of the VACV Copenhagen strain. A strong synthetic VACV promoter was used instead of the natural $P_{7.5}$ VACV promoter which was used in previous constructs. This resulted in a threefold increase in the expression level of H and F genes as compared to vRVFH. Intramuscular vaccination of

v2RVFH with a dose of $10^8$ PFU provided sterilizing immunity in cattle for at least 16 months [51].

Interestingly, VACV strain Wyeth expressing H and F genes of rinderpest virus (vRVFH) provides protection to goats against peste des petits ruminants virus (PPRV) challenge [52]. Despite the inability of vRVFH to induce anti-PPRV neutralizing antibodies, complete protection was observed in goats against PPR [52]. Cell-mediated immunity or non-neutralizing antibodies might be responsible for the protection elicited by this recombinant vector in goats. Similarly, cross-protection has also been demonstrated for canine distemper virus (CDV) vectored by VACV vectors. Vaccinia virus recombinants expressing either the measles virus fusion (F) or hemagglutinin (H) glycoprotein have been shown to protect dogs against CDV [53]. The recombinants were generated by inserting measles virus F or H gene in the TK locus of VACV Copenhagen strain and using the H6 synthetic promoter. These recombinants fail to induce CDV neutralizing antibodies in dogs. However, inoculation of dogs with the recombinant VACV expressing H gene or co-immunization with the recombinant VACV expressing H and the recombinant expressing F protein was shown to protect dogs from lethal CDV challenge [53].

Vaccinia virus recombinant expressing glycoprotein (G) gene of vesicular stomatitis virus (VSV) induces protective neutralizing antibody responses in cattle. Neutralizing antibody levels increased by several fold after boosting. This recombinant provides partial protection to VSV challenge in cattle, and protection is correlated with neutralizing antibody levels [54]. The utility of VACV-based vectors in animals has also been demonstrated in chickens. A VACV expressing the Newcastle disease virus (NDV) F glycoprotein has been shown to protect chickens against live virulent NDV challenge [55]

Recombinant VACV vectors have also been used to develop vaccine candidates against protozoan parasites. A VACV expressing the LACK protein of *Leishmania* using the Western Reserve (rVV-LACK) or MVA (MVA-LACK) strains has been developed [56]. Prime-boost immunization of dogs with a plasmid DNA expressing the LACK protein (DNA vaccination) followed by booster with rVV-LACK or MVA-LACK incudes both humoral and cellular immunities in dogs. Priming with DNA and boosting with rVV-LACK provided 50% protection in dogs, whereas boosting with MVA-LACK provided 75% protection in dogs. This study showed that boosting with non-replicative MVA vector elicited higher immune response than replication-competent Western Reserve (WR) vector [56]. Later, another MVA construct expressing TRYP protein of *Leishmania* was developed [57, 58]. Dogs receiving a DNA-TRYP/MVA-TRYP prime-boost vaccination strategy produced higher levels of TRYP-specific type 1 cytokine IFN-gamma and TRYP-specific IgG antibody in comparison to MVA-LACK construct [57]. These studies provided evidence for applicability of poxvirus vectors in prime-boost vaccination regimens. In fact, VACV can be used both for priming and boosting. This has been demonstrated for a VACV recombinant expressing Gn and Gc glycoproteins of Rift Valley fever virus (RVFV) [37]. The recombinant VACV vector was generated using the Copenhagen strain, and two virulence genes (B8R and TK) of the virus were inactivated. The RVFV Gn and Gc proteins were inserted in TK locus. A single vaccination with this recombinant provided 50% protection in mice, whereas animals immunized twice with this vaccine showed 90% survival rate after challenge. This recombinant was also tested in baboons (nonhuman primate model). All animals immunized with this recombinant mounted a strong anamnestic response to booster immunization [37].

# 6    Avipoxvirus-Based Vectors

Avipoxviruses naturally infect chickens, turkeys, and many other species of pet and wild birds. Currently there are ten species of avipoxviruses recognized by ICTV (International Committee on Taxonomy of Viruses) [59]. Given that avipoxviruses have been isolated from a wide range of hosts including crows, peacock, and ostrich, among others, there are many avipoxviruses which have been tentatively proposed as new species but remain officially unclassified [60]. Avipoxviruses share many characteristics with other poxviruses. Our understanding of molecular and biological characteristics of avipoxviruses comes mainly from fowlpox virus (FWPV) and canarypox virus (CNPV), which infect domestic poultry and canaries, respectively [61].

As described in the previous section, VACV has been widely used as a vaccine vector for human and veterinary applications. However, VACV-based recombinant vaccines developed for animals pose a risk of infection to humans because of the virus broad host range. Thus, it was desirable to construct host-restricted vectors, for instance, using recombinant avipoxviruses for use in mammalian species as these viruses only cause productive infection and disease in avian species [61]. The use of host-restricted or replication-incompetent poxvirus vectors would avoid the risk of genetic recombination and disease transmission between vaccinated animal species or humans. Avipoxviruses were initially proposed as vectors for vaccine delivery in poultry [62]. Later, the findings that recombinant FWPV initiate an abortive infection in non-avian tissue culture cells and express foreign antigens capable of inducing immune response in mammals sparked interest in using avipoxviruses as vectors for humans and other animal species [63]. Additionally, pre-existing immunity to orthopoxviruses does not affect immunogenicity of FWPV and canarypox virus (CNPV), which means they could be used as vectors in humans exposed to vaccinia virus or vaccinated against smallpox [60]. As a result, a large number of avipoxvirus recombinants based on FWPV and CNPV have been developed for use in humans and animals.

## 6.1    Fowlpox Virus-Based Vectors

Several fowlpox virus recombinant constructs targeting avian influenza (AI) have been developed. A fowlpox virus recombinant expressing the influenza virus HA protein at the TK locus was able to induce hemagglutinin-inhibiting (HI) antibodies in chickens. These antibodies were detected as early as 9 days post immunization, and a boost effect was seen when chickens were re-immunized [62]. Interestingly, protection against AI has been observed in birds in presence of very low levels or even the absence of HI or neutralizing antibodies [64]. Cell-mediated immunity induced by immunization with the FWPV vector has been suggested as the effector mechanism of protection against AI in birds without significant levels of HI or neutralizing antibodies [64]. To enhance cell-mediated immunity induced by FWPV vectors, attempts were made to insert nucleoprotein gene (NP) of influenza virus along with HA gene. The co-expression of HA and NP by the FWPV vector, however, did not improve the efficacy of vaccine [65]. The FWPV-vectored avian influenza vaccine known as TROVAC-H5 has been licensed for emergency use in the USA and has full registration in Mexico, Guatemala, El Salvador, and Vietnam [66]. The vaccine consists of FWPV recombinant expressing the H5 gene from highly pathogenic AI isolate A/turkey/Ireland/1378/83 H5N8. This vaccine provides 90–100% protection against highly pathogenic Mexican avian influenza H5N8-type isolates. After a single immunization on day 1 of life, it confers protection for at least 20 weeks [67]. Good levels of protection have also been observed against some of the recent H5N1 Asian AI isolates A/chicken/South Korea/03 and A/chicken/Vietnam/04 [66].

Another recombinant FWPV vector co-expressing HA (H5 subtype) and neuraminidase (N1 subtype) can provide complete protection to chickens against AI H5N1 challenge. Protection was accompanied by high levels of HA- and N1-specific antibodies [68]. Notably, this recombinant is able to provide cross-protection against H5N1 and H7N1 highly pathogenic avian influenza (HPAI) virus challenge, presumably due to cross-reactive immunity conferred by the common N1 protein between these two HPAI types. This vaccine was licensed in China, and over 600 million doses of this

vaccines were sold by 2009 [69]. Attempts have been made to co-express cytokines along with HA protein to enhance immunogenicity of FWPV-influenza recombinants. Improvement in protective efficacy has been reported by co-expression of cytokines like chicken interleukin-2 and chicken interleukin-18 along with the HA protein [70, 71].

A fowlpox vaccine recombinant expressing the H5 hemagglutinin gene provides protection against clinical signs and mortality in chicken following challenge by nine diverse highly pathogenic avian influenza viruses [67]. This vaccine overcomes limitations of many FWPV-influenza recombinants which fail to provide cross-protection to different influenza subtypes. The protection was correlated with the amino acid sequence similarity of H5 gene of challenge virus and the H5 gene inserted in the recombinant FWPV vector [67].

A recombinant FWPV expressing either Newcastle disease virus (NDV) hemagglutinin-neuraminidase (HN) or fusion (F) proteins or recombinants expressing both proteins have been developed [72–75]. Most of the recombinants use the VV7.5 or H6 early-late promoters to drive expression of HN or F gene in FWPV. Fowlpox virus recombinant vectors expressing HN and F proteins provide 100% protection to chickens against lethal velogenic NDV challenge [72, 73, 75]. One of these FWPV-based recombinant vectors, designated as TROVAC-NDV, has been licensed for commercial use in the poultry industry. A single dose of TROVAC-NDV can induce high levels of hemagglutination-inhibiting antibodies in chickens for up to 8 weeks post immunization [73]. Recombinant FWPV expressing the NDV fusion (F) gene is also capable of inducing anti-F protein antibody responses which can provide protection to the chickens from lethal NDV challenge [72, 76]. The expression levels of NDV fusion (F) gene can be increased by inserting F protein into non-essential genes in the inverted terminal repeats (ITR) of FWPV. This occurs because a foreign gene inserted into ITR region of a poxvirus vector is usually duplicated by homologous recombination, with one copy of the gene being inserted in the 5' and 3' ITR, thus increasing the expression levels of the foreign protein [72]. This strategy, however, can also lead to loss of the gene insert due to increased instability of the resultant vector. Fowlpox-NDV recombinants administered through the intramuscular or wing-web method induce stronger immune response than that of oral or ocular inoculation [75, 76]. Several-fold increase in NDV HI titers is seen when chickens are primed with live or inactivated NDV vaccine and boosted with recombinant fowlpox virus expressing HN proteins [77].

Fowlpox vectors expressing the envelope glycoprotein of reticuloendotheliosis virus (REV) induce neutralizing antibodies and protection in chicken from viremia and runting-stunting syndrome following REV challenge [78]. Synthetic promoter $P_s$ induces higher level of expression of envelope glycoprotein than vaccinia $P_{7.5}$ promoter. Similarly, recombinant FWPV expressing multiple genes of Marek's disease virus (MDV) have been constructed. Of these FWPV-MDV constructs, vectors containing one gene of MDV provide less than 50% protection in chickens. Whereas, a synergistic effect was observed when multiple genes of MDV are expressed resulting in increased protection up to 72%. Additionally, enhanced protection (94%) was seen when FWPV-MDV recombinants were given along with the MDV closely related turkey herpesvirus (HVT) [79].

Fowlpox vector expressing the VP2 protein of infectious bursal disease virus (IBDV) can protect chickens from mortality [80, 81]. However, these recombinants cannot protect chickens against damage to the bursa of Fabricius [80], and protection levels are lower than the oil-adjuvanted inactivated whole virus vaccine [82]. Recombinant FWPV (rFWPV) expressing VP2-VP4-VP3 polyprotein of IBDV inserted within TK gene under vaccinia P.L11 late promoter fails to develop protective antibodies against IBDV, whereas rFWPV expressing only VP2 under fowlpox early/late promoter inserted immediately downstream of TK gene can express five times more VP2 protein than the former construct and also induces antibodies to IBDV [82]. Therefore,

the choice of promoter and insertion site can significantly affect the immunogenicity of poxvirus vectors. In addition to viral diseases, recombinant FWPV-vectored vaccines have been developed for non-viral diseases of poultry, including coccidiosis and mycoplasma [83, 84]. The Vectormune FP-MG consists of a FWPV recombinant vector expressing the 40k and *mgc* genes of *M. gallisepticum*, which is licensed in the USA for use in chickens and turkeys [84].

## 6.2 Canarypox Virus-Based Vectors

Another avipoxvirus that has been widely used as a vaccine vector is canarypox virus (CNPV). A plaque purified clone of CNPV designated as ALVAC is widely used as vector. This clone was obtained after serial passage of wild-type CNPV for 200 passages in CEF [85]. The safety and immunogenicity profile of CNPV ALVAC vector led to its use in human clinical trials as an HIV/AIDS vaccine candidate [86]. Additionally, there are several ALVAC-based vectored vaccines licensed for veterinary use, for example, ALVAC-AI-H5 (influenza virus), ALVAC-RV (rabies virus), and ALVAC-CDV-H/F (canine distemper virus) [87].

Canarypox virus recombinants expressing the RabV G are known to elicit high levels of neutralizing antibodies in mice, cats, and dogs. The level of protection observed after challenge infection was comparable to that induced by replication-competent VACV vector [88]. A CNPV vector expressing hemagglutinin (HA) gene of equine influenza virus (rCNPV-EIV) induces both humoral and cellular immune responses against EIV. Cellular immune response is characterized by increased levels of IFN-$\gamma$ in vaccinated ponies. Clinical signs and virus shedding were significantly reduced in rCNPV-EIV-vaccinated group after challenge infection [89]. A CNPV recombinant expressing the prM/E proteins of West Nile virus (WNV) has been licensed for use in horses [90]. A single dose of this vaccine protects horses against viremia caused by challenge with WNV-infected mosquitoes [91]. Two doses of this vaccine can provide protection for at least 1 year post vaccination [90, 92]. Similarly, rCNPV expressing *env* and *gag* genes of feline leukemia virus (FeLV) provides protection to cats against oronasal challenge with FeLV. The cats are protected from contact challenge for at least 1 year [93, 94]. The rCNPV-FeLV vaccine has been licensed for commercial use under trade name E URIFEL FeLV [93]. Both rCNPV-WNV and rCNPV-FeLV can provide protection despite absence of measurable antibody responses [91, 94]. The protection observed might be related to the activation of cell-mediated immunity which requires relatively lower antigen load/ dose. This phenomenon has been observed in rCNPV expressing glycoprotein (G) and fusion gene (F) of Hendra virus (HeV). The higher tested dose of rCNPV-HeV recombinant induced strong neutralizing antibodies in horses and hamsters, whereas at lower doses, partial protection was observed in hamsters despite the absence of detectable HeV-specific antibodies [95]. Canarypox virus vaccine vectors expressing glycoprotein (G) and fusion (F) gene of Nipah virus (NiV), when given in combination, can induce high levels of neutralizing antibodies in pigs and can provide solid protection from NiV challenge. In addition, vaccinated pigs show balanced Th1 and Th2 response with the induction of TNF-α, IL-10, and IFN-γ cytokines [96]. These rCNPV-HeV and rCNPV-NiV recombinants induce cross-neutralizing antibody against closely related Nipah virus (NiV) and Hendra virus (HeV), respectively.

The impact of maternal antibodies on the efficacy of avipoxvirus-vectored vaccines has been a subject of constant debate. This is important because layers are routinely immunized against FWPV which can impact FWPV vector-based vaccination in day-old chicks. It has been demonstrated that recombinant FWPV expressing the HN gene of NDV failed to induce immune response in chickens previously vaccinated with FWPV [77]. In contrast, there are reports that demonstrate that the FWPV or CNPV vectors remain effective in the presence of pre-existing immunity and can be used repeatedly without an adverse effect on the vaccine potency [60, 97,

98]. Additionally, pre-existing immunity against the heterologous gene inserted into avipox vector may interfere with the efficacy of the vaccine. For example, in the presence of maternally derived NDV antibodies, the humoral response provided by FWPV-NDV recombinant expressing HN protein of NDV was dampened [73]; nevertheless significant level of protection against NDV was still achieved.

# 7  Parapoxvirus-Based Vectors

The genus *Parapoxvirus* includes four species – *Bovine popular stomatitis virus* (BPSV), *Orf virus* (ORFV), *Parapoxvirus of red deer in New Zealand* (PVNZ), and *Pseudocowpox virus* (PCPV). The type species parapoxvirus ORFV has been widely used as vector. Some of the features that make ORFV an attractive candidate vector are (1) its restricted host range (sheep and goat), (2) its ability to induce humoral and cellular immune response even in non-permissive hosts [20, 27, 99], (3) its tropism that is restricted to the skin and the absence of systemic infection, (4) the fact that ORFV induces short-lived ORFV-specific immunity and does not induce neutralizing antibodies which allows repeated immunizations [20, 100, 101], and (5) the immunomodulatory properties of the virus [102]. In fact, inactivated ORFV is used as an immunomodulator in horses and has been shown to be effective in reducing clinical signs and shedding related to equine herpes virus type 1 (EHV-1) and *Streptococcus equi* (*S. equi*) infections [103]. There are several well-characterized immunomodulatory proteins (IMPs) present in Orf virus. These IMPs include an interleukin-10 homologue (vIL-10) [104], a chemokine-binding protein (CBP) [105], an inhibitor of granulocyte-monocyte colony-stimulating factor (GMC-CSF) [106], an interferon resistance gene (VIR) [107], a homologue of vascular endothelial growth factor (VEGF) [108], and at least four inhibitors of nuclear factor-kappa (NF-κB) signaling pathway [109–112]. The presence of these well-characterized IMPs provides unique opportunities for rational engineering of ORFV-based vectored vaccines.

Two strains of ORFV, D1701 and OV-IA82, have been explored as vectors for veterinary application.

The highly attenuated ORFV strain D1701 was obtained after serial cell culture passage of an ORFV isolate from sheep in African green monkey kidney cells (Vero cells). This virus is apathogenic in sheep and is well adapted to grow in cell culture [113]. Adaptation of ORFV strain D1701 in Vero cells led to further attenuation of the virus because of additional genomic deletions. This virus, designated D1701-V, is non-pathogenic even in immunosuppressed natural host sheep [114]. The utility of D1701-V strain as a vector has been explored in permissive and non-permissive animal species. In most constructs, the VEGF-E locus of D1701-V has been used to insert heterologous genes utilizing the early promoter of the VEGF-E gene. The D1701-V recombinant expressing the rabies G can stimulate high levels of rabies virus-specific neutralizing antibodies in mice, cats, and dogs [115]. Another recombinant expressing p40 protein of Borna virus provides protection to mice against Borna virus challenge and leads to the virus clearance from the infected brain eliminating persistent virus infection [116]. Similarly, ORFV recombinant (D1701-V) expressing the HA protein of influenza virus (H5N1) provides solid protection to mice against influenza A virus H5N1 and heterologous influenza A H1N1 challenge in a dose-dependent manner [117]. Orf virus recombinant (D1701-V) expressing the major capsid protein VP1 of rabbit hemorrhagic disease virus (RHDV) protected rabbits against lethal RHDV infection with a single immunization with a dose as low as $10^5$ PFU [118]. Higher or multiple doses were required to induce significant humoral response; nevertheless single dose of $10^5$ PFU was enough to provide protection. This dose is significantly lower than that of VACV or canarypox recombinants expressing RHDV VP1 which require $10^7$–$10^9$ PFU for protection when given subcutaneously, orally, or intradermally [118–120].

The D1701-V has also been used as a vector in pigs. The D1701-V recombinant expressing glycoproteins gC and gD of pseudorabies virus (PRV) induces strong cellular and humoral immune response when used to boost the pigs primed with *Sindbis virus*-derived plasmid expressing gC and gD [121]. This type of heterologous prime-boost strategy, using DNA vaccine or baculovirus-expressed protein for priming followed by boosting with ORFV-based vaccine, has been shown to be more effective than homologous prime-boost strategy [122, 123]. Orf virus recombinant expressing the E2 glycoprotein of classical swine fever virus (CSFV) confers protection against CSFV challenge in pigs [123]. A single intramuscular immunization of this recombinant induces high levels of CSFV-specific neutralizing antibodies and IFN-γ production. Interestingly, multiple-site application was shown to be superior to single-site injection with the same dose. This might be due to effective antigen processing and presentation occurring at different lymph nodes at the same time [123].

Another ORFV strain that has been used as a vector is OV-IA82. ORFV strain IA82 (OV-IA82) was obtained from the nasal secretion of a lamb at the Iowa Ram Test Station during an Orf outbreak in 1982 and was isolated in ovine fetal turbinate cells (OFTu) [124]. Four genes of ORFV (ORFV002, ORFV024, ORFV121, ORFV073) that are involved in inhibition of host nuclear factor-kappa (NF-κB) pathway have been well-characterized [109–112]. Deletion of ORFV121 from viral genome attenuates the virus as evidenced by decreased pathogenesis in sheep which makes ORFV121 deletion mutant virus an attractive vector candidate for use in livestock [109]. An ORFV (OV-IA82) recombinant expressing the full-length spike (S) glycoprotein of porcine epidemic diarrhea virus (PEDV), containing a deletion of ORFV121 gene, induces PEDV-specific neutralizing antibodies in pigs and protects pigs from clinical signs of PED [20]. This ORFV-PEDV-S recombinant was also shown to induce passive immunity and transfer of PEDV-specific IgG, IgA, and neutralizing antibody to piglets via milk and colostrum. Upon challenge with virulent PEDV, decreased clinical signs and reduced mortality was observed in piglets born from ORFV-PEDV-S-vaccinated sows. Additionally, increased protection with 100% survival was obtained when sows were primed with live

PEDV and then boosted with the ORFV-PEDV-S recombinant [101].

In addition to ORFV121, another NF-κB inhibitor, ORFV024, has also been used as a site for foreign gene insertion. The immunogenicity of two ORFV-IA82 recombinants expressing the RabV G either in the ORFV121 or ORFV024 gene loci was evaluated in pigs and cattle [27]. Both recombinants induced robust neutralizing antibody response against RabV in pigs and cattle. Additionally, both viruses induced long-lasting memory B cell responses to RabV, as evidenced by anamnestic responses following a single-dose booster on day 390 post primary immunization [27]. Notably, the neutralizing antibody titers induced by ORFV121 deletion mutant were higher than that of ORFV024 deletion mutant [27]. This type of differential regulation of innate and adaptive immune response has also been reported for NF-κB inhibitor proteins encoded by vaccinia virus (VACV), where deletion of one NF-κB inhibitor (A52R) leads to higher immune response against heterologous antigen when compared to other NF-κB inhibitors (B15, K7) [125].

Despite restricted host range, infections of humans have been reported for parapoxviruses. The virus can cause self-limiting infections in immunocompromised people or farmers who work in close contact with infected animals and present abrasions or cuts in the skin. These infections are restricted to the areas surrounding the sites of virus entry and usually involve the hands [126–128]. Since both recombinant ORFV vectors that have been used as platforms (D1701 and OV-IA82Δ121) are known to be non-pathogenic in the natural sheep and goat hosts and the strain IA82 presents marked growth impairment in human cells, it is safe to assume that the risk of human infections with these recombinants should be very low.

## 8    Swinepox Virus-Based Vector

Swinepox virus (SPV) is the only member of genus *Suipoxvirus*. SPV causes mild self-limiting infection in pigs. Because of its narrow host range

restricted to pigs, SPV-based recombinant vectored vaccine candidates have been developed mostly targeting pigs. One of the first attempts to use SPV as a vector involved a recombinant SPV targeting Aujeszky's disease (pseudorabies virus (PRV)) [129]. The gp40 and gp63 genes of PRV were inserted into TK locus of SPV under the early/late $VV_{7.5}$ promoter. At 21 days post immunization, 90% of the pigs vaccinated by scarification developed serum neutralizing antibody against pseudorabies virus, whereas 100% of the animals vaccinated by the intramuscular route developed neutralizing antibodies. Significant level of protection was observed in pigs upon challenge with virulent PRV [129]. A recombinant SPV expressing the E2 glycoprotein of classical swine fever virus (CSFV) in the TK locus expresses the E2 protein in a dimeric form in the cytoplasm of the infected cells [130]. However, this recombinant was not tested in animal model. SPV recombinant expressing HA1 gene of swine influenza virus (SIV; H1N1) elicits humoral and cellular immune responses and provides complete protection against SIV in swine and mice [131]. Although neutralizing antibody titers were low (1:8 to 1:32), potent Th1 and Th2 responses were observed as evidenced by increased levels of IL-4 and IFN-γ, which may have contributed to protection against SIV challenge. Co-expression of the HA gene of H1N1 and H3N2 subtype provides complete protection against H1N1 and H3N2 challenge in pigs [132].

Swinepox virus recombinants expressing immunodominant epitopes of certain viral pathogens have also been shown effective. A recombinant SPV expressing the A epitope of transmissible gastroenteritis virus (TGEV) spike protein induced neutralizing antibodies and strong Th1 and Th2 cytokine responses against TGEV. Notably, when neutralizing antibodies purified from vaccinated animals were fed to piglets, they were protected against severe disease and mortality after challenge with TGEV [133]. Similarly, a recombinant SPV expressing three repeats of a conserved six-amino acid epitope present in the N-terminal ectodomain of the GP3 protein of porcine reproductive and respiratory syndrome virus (PRRSV) induced cellular and humoral response in pigs [134]. Swinepox virus has also been used to develop recombinant vaccine candidates against bacterial diseases. A recombinant swinepox virus expressing M-like protein (SzP) of *Streptococcus equi* spp. *zooepidemicus* (SEZ) provided significant protection against SEZ infection [135].

Swine are the major reservoir of SPV, and the virus does not infect other mammalian or avian species. Initially, SPV was known to infect only cells of porcine origin; however, recent findings have shown that SPV exhibit a relatively broad cell culture host range in vitro [136, 137]. Recombinant SPV can infect, replicate, and express foreign genes in cells of non-porcine origin including human, monkey, hamster, and rabbit cell lines [137]. Moreover, when inoculated intradermally, SPV causes productive infection in rabbits [138]. These findings have opened opportunities for the use of SPV vector in different animal species. A SPV recombinant expressing the *gag* and *env* proteins of feline leukemia virus (FeLV) fails to replicate in feline cells, but FeLV gag virus-like particles were produced in feline cells and incorporated into SPV intracellular mature virion (IMV) [139]. The immunogenicity of this recombinant virus, however, was not tested in cats.

Deletion of the TK gene has been widely used in recombinant poxvirus vector construction. However, deletion of the TK gene may not be the best strategy to generate SPV-based recombinant vectors as it results in severe attenuation of the vector and consequently in lower vaccine efficiency. It has been shown that deletion of the TK gene results in decreased neutralizing antibody levels and in Th1- and Th2-mediated immune responses [140]. The ability of the virus to replicate in non-permissive cells also decreases significantly in TK deletion mutants [140]. Insertion of foreign antigens in intergenic or non-coding regions of the SPV genome could be used as an alternative approach. A SPV recombinant expressing the porcine IL-18, the capsid protein of porcine circovirus 2, and the SzP protein of SEZ in the intergenic region between SPV020 and SPV021 open reading frames induces immune responses against both PCV2

and SEZ, which was comparable to the immune responses elicited by commercial vaccines [141]. These findings suggest that SPV can be used as a vector platform for multivalent vaccines against diseases of swine.

## 9    Capripoxvirus-Based Vector

The genus *Capripoxvirus* includes *Goatpox virus* (GTPV), *Sheeppox virus* (SPPV), and *Lumpy skin disease virus* (LSDV). SPPV and GTPV infect sheep and goats, respectively, with some isolates being able to infect both species. LSDV causes disease in cattle and buffalo [142]. These three species of CPV share 96–97% nucleotide identity [143, 144]. Because of the high degree of sequence conservation, cross-immunity is observed among the three viruses. An attenuated LSDV deficient of an IL-10 gene homologue (ORF005) has been shown to provide protective immunity against virulent capripoxvirus (CPV) challenge in sheep and goats [145]. Thus, theoretically, an attenuated strain of any capripoxvirus should be able to protect against SPPV, GTPV, and LSDV [146]. Using these viruses as vectors, it is possible to generate multivalent vaccines against CPV and other target pathogens of ruminants. Moreover, the replication of capripoxviruses is restricted to ruminants with no evidence of human infections. These traits make CPVs good candidates for developing recombinant vectored vaccines.

Capripoxviruses have been mainly used to develop recombinant vaccines for use in ruminants. Most of the recombinants have been generated using Kenya strain-1 (KS-1), which was isolated from sheep and passaged in lamb testis and baby hamster kidney (BHK) cells [147]. Recent molecular studies have shown a close relationship between KS-1 and LSDV, suggesting KS-1 may actually be LSDV [143]. The majority of KS-1-based constructs have been generated by inserting the foreign gene into the TK locus. Two individual KS-1 recombinants generated by insertion of the fusion (F) protein or hemagglutinin (H) protein of rinderpest virus (RPV) into TK gene locus under the control of vaccinia virus late promoter P11 protect cattle against rinderpest virus after lethal challenge with a virulent RPV isolate. The recombinant with the H gene insert showed better protective efficacy than the vector expressing the F gene. Both recombinants protected cattle against LSDV challenge in addition to rinderpest [148, 149]. Interestingly, these recombinants protect goats from lethal challenge with peste des petits ruminants virus (PPRV) challenge [150]. Another recombinant capripoxvirus expressing the H and F gene of PPRV has been developed using AV41 strain of GTPV. A single dose of this recombinant elicits seroconversion in ~80% of immunized sheep and goats. Neutralizing antibodies are detected up to 6 months after vaccination. Two doses of this vaccine completely overcome the interference caused by pre-existing immunity to caprinepox virus [151]. Another recombinant CPV expressing both H and F genes of PPRV confers an earlier and stronger immune response against PPR and GTPV [152].

A KS-1 recombinant expressing the VP7 protein of bluetongue virus (BTV) provides partial protection in sheep [153]. Partial protection against BTV was achieved even when sheep were immunized with the combination of KS-1 recombinants individually expressing four proteins (NS1, NS3, VP2, VP7) of BTV [154]. Capripoxvirus has been used to develop recombinant vaccine candidates against Rift Valley fever virus (RVFV) [155, 156]. A recombinant KS-1 virus expressing Gn and Gc glycoproteins of RVFV induced neutralizing antibodies to RVFV in sheep, and two doses of the vaccine candidate provided significant protection against RVFV and SPPV challenge in sheep.

In addition to the KS-1 strain, an attenuated strain of LSDV, Neethling strain, has also been used as vector. This strain is used as vaccine against lumpy skin disease in Africa. The ribonucleotide reductase gene of Neethling strain has been identified as potential insertion site for recombinant vector generation [157]. The Neethling strain has been used to develop recombinant vaccines against RabV and bovine ephemeral fever and RVFV viruses [142]. Recombinant

LSDV (Neethling strain) generated by inserting RabV G in the ribonucleotide reductase gene locus induced strong cellular and humoral response in cattle [157]. Neutralizing antibody titers as high as 1513 IU/mL were observed in cattle. This recombinant also induced robust humoral and cellular immunity in non-permissive hosts (mice and rabbits). Immunization of mice with this recombinant vector protected the animals from an aggressive intracranial RabV challenge [158].

Because of the cross-protection offered by CPVs, theoretically it should be possible to develop a universal recombinant CPV vector that would provide protection against all capripoxviruses in addition to target pathogen. However, the geographical distribution of the different CPV limits that possibility. Sheeppox and goatpox viruses are endemic to Asia, Middle East, and Africa south of the equator, whereas LSDV is mainly present in sub-Saharan Africa [142]. A country would refuse to use capripoxvirus-vectored vaccine if the vector is not endemic. Future research should aim at identifying immunomodulatory genes and virulence factors encoded by CPV, which would allow the development of safer recombinant CPV-based vectors that could potentially be used in both endemic and non-endemic countries.

## 10    Leporipoxvirus as Vector

Myxoma virus (MYXV), the type species of genus *Leporipoxvirus*, specifically infects rabbits and hare (leporides). MYXV was used as biological agent to control European rabbit population in Australia. This method, however, was not sustainable because of the coevolution of the virus and the rabbit which led to the adaptation of the virus to the novel host species [159].

MYXV has been used as vector for both leporide and non-leporide species. A recombinant myxoma virus expressing capsid protein (VP60) of rabbit hemorrhagic disease virus (RHDV) has been shown to protect rabbit from myxomatosis and RHDV [120]. MYXV expressing hemagglutinin (HA) of influenza virus can induce high levels of anti-HA antibodies in rabbits as efficiently as VACV vector [160]. MYXV recombinant has been shown to be effective in protecting cats from feline calicivirus [161]. The possibility of using MYXV as a non-replicative vector in small ruminants (sheep) has been demonstrated [162]. Moreover, the use of MYXV-vectored immunocontraception as a means to control wildlife species has been assessed [163]. Myxoma virus expressing immunocontraceptive antigen (zona pellucida 3 [rZp3] glycoprotein of rabbit) was able to induce autoimmune infertility in 70% of rabbits at the first breeding [164].

MYXV encodes several proteins that are known to cause immunosuppression in rabbits [165]. Deletion of those immunosuppressive genes is essential to enhance the safety and immunogenicity of the MYXV vector.

## 11 Future Directions and Potential for Other Applications

Poxvirus-based vectors have long been used as vaccine delivery platforms in many animal species. The overall immunogenicity, safety, and broad disease and species applicability of these viral vectors make them especially attractive for vaccine delivery in veterinary medicine. After decades of research, there is a general consensus that replication-competent poxvirus vectors can induce better immune responses in target species, but they have potential for infecting non-target hosts (e.g., vaccinia virus). In contrast, replication-deficient poxviruses are safer but induce comparatively lower immune responses. Hence, future studies focusing on rationally designed poxvirus vector platforms could lead to more balanced vectors, with an improved safety profile and immunogenicity. Poxviruses are known for encoding several immunomodulatory proteins (IMPs) that target host immune responses to allow efficient virus infection and replication. Most importantly, several of these genes encode for virulence determinants that contribute to poxvirus disease pathogenesis. By targeting those genes, one could expect to attenuate a given poxvirus vector and, perhaps,

simultaneously enhance its immunogenicity in target animal species. As additional poxviral IMPs are identified and characterized, we are likely to see the development and refinement of poxviral vectors.

In addition to modulation of poxvirus vector safety and immunogenicity, the field would greatly benefit from additional studies focused on identifying better promoters for expression of heterologous genes. Promoters that would allow sustained gene expression following immunization would likely result in more robust and long-lasting immune responses. There is also a need for better recombinant selection methods. The use of CRISPR/Cas9 in combination with fluorescent activated cell sorting is a promising approach that may facilitate and speed up the selection process of poxvirus vectors. Additional research assessing the effect of dose and route of immunization, the stability of the recombinants, the sustained heterologous gene expression, and perhaps even the use of host cytokines to enhance T-cell responses will be pivotal in developing safe and immunogenic poxvirus-based vectored vaccines in the future.

## 12 Summary

Since the first demonstration of recombinant vaccinia virus as an expression vector in 1982, we have seen a huge improvement in different aspects of poxvirus recombinant technology from generating safer poxvirus strains to improved methods of recombinant selection. We now have better understanding of poxvirus infection biology which has enabled us to generate safer and more immunogenic vectored vaccines. There are many poxvirus-based recombinant vaccines that have been licensed for commercial use, and clinical trials are in progress for many other infectious diseases of humans and animals. Finally, given their utility, broad species applicability, and immunogenicity, one can expect that poxviruses will continue to evolve and remain as one of the main vaccine delivery platforms both in human and veterinary medicine.

# References

1. ICTV. Virus taxonomy: Online (10th) Report of the International Committee on Taxonomy of Viruses. 2017
2. Moss B. Poxviridae. In: Knipe D, Howley P, editors. Fields Virology. 6th ed. Philadelphia: Lippincott Williams and Wilkins; 2013. p. 2129–59.
3. Bhanuprakash V, Hosamani M, Venkatesan G, Balamurugan V, Yogisharadhya R, Singh RK. Animal poxvirus vaccines: a comprehensive review. Expert Rev Vaccine. 2012;11:1355–74.
4. Murphy F, Gibbs E, Horzinek M, Studdert M. Veterinary virology. 3rd ed. San Diego: Academic; 1999.
5. Baldick CJ, Moss B. Characterization and temporal regulation of mRNAs encoded by vaccinia virus intermediate-stage genes. J Virol. 1993;67:3515–27.
6. Lakhani S. Early clinical pathologists: Edward Jenner (1749–1823). J Clin Pathol. 1992;45:756.
7. Mackett M, Smith GL, Moss B. Vaccinia virus: a selectable eukaryotic cloning and expression vector. Proc Natl Acad Sci. 1982;79:7415–9.
8. Panicali D, Paoletti E. Construction of poxviruses as cloning vectors: insertion of the thymidine kinase gene from herpes simplex virus into the DNA of infectious vaccinia virus. Proc Natl Acad Sci. 1982;79:4927–31.
9. Smith GL, Mackett M, Moss B. Infectious vaccinia virus recombinants that express hepatitis B virus surface antigen. Nature. 1983;302:490–5.
10. Smith GL, Murphy BR, Moss B. Construction and characterization of an infectious vaccinia virus recombinant that expresses the influenza hemagglutinin gene and induces resistance to influenza virus infection in hamsters. Proc Natl Acad Sci U S A. 1983;80:7155–9.
11. Perkus ME, Piccini A, Lipinskas BR, Paoletti E. Recombinant vaccinia virus: immunization against multiple pathogens. Science. 1985;229:981–4.
12. Smith GL, Moss B. Infectious poxvirus vectors have capacity for at least 25 000 base pairs of foreign DNA. Gene. 1983;25:21–8.
13. Gherardi MM, Esteban M. Mucosal and systemic immune responses induced after oral delivery of vaccinia virus recombinants. Vaccine. 1999;17:1074–83.
14. Collier LH. The development of a stable smallpox vaccine. Epidemiol Infect. 1955;53:76–101.
15. Woodroofe GM, Fenner F. Genetic studies with mammalian poxviruses: IV. Hybridization between several different poxviruses. Virology. 1960;12:272–82.
16. Fenner F. Genetic studies with mammalian poxviruses: II. Recombination between two strains of vaccinia virus in single HeLa cells. Virology. 1959;8:499–507.
17. Cooper GM. Recombination between homologous DNA sequences. 2000.
18. Mackett M, Smith GL. Vaccinia virus expression vectors. 1986.
19. Wyatt LS, Earl PL, Moss B. Generation of recombinant vaccinia viruses. Curr Protoc Protein Sci. 2017;89:5.13.1–5.13.18.
20. Hain KS, Joshi LR, Okda F, et al. Immunogenicity of a recombinant parapoxvirus expressing the spike protein of Porcine epidemic diarrhea virus. J Gen Virol. 2016;97:2719–31. https://doi.org/10.1099/jgv.0.000586.
21. Panicali D, Davis SW, Weinberg RL, Paoletti E. Construction of live vaccines by using genetically engineered poxviruses: biological activity of recombinant vaccinia virus expressing influenza virus hemagglutinin. Proc Natl Acad Sci U S A. 1983;80:5364–8.
22. Yao X-D, Evans DH. Effects of DNA structure and homology length on vaccinia virus recombination. J Virol. 2001;75:6923–32.
23. García-Arriaza J, Esteban M. Enhancing poxvirus vectors vaccine immunogenicity. Hum Vaccin Immunother. 2014;10:2235–44.
24. Earl PL, Hügin AW, Moss B. Removal of cryptic poxvirus transcription termination signals from the human immunodeficiency virus type 1 envelope gene enhances expression and immunogenicity of a recombinant vaccinia virus. J Virol. 1990;64:2448–51.
25. Chakrabarti S, Brechling K, Moss B. Vaccinia virus expression vector: coexpression of beta-galactosidase provides visual screening of recombinant virus plaques. Mol Cell Biol. 1985;5:3403–9.
26. Franke CA, Rice CM, Strauss JH, Hruby DE. Neomycin resistance as a dominant selectable marker for selection and isolation of vaccinia virus recombinants. Mol Cell Biol. 1985;5:1918–24.
27. Martins M, Joshi LR, Rodrigues FS, Anziliero D, Frandoloso R, Kutish GF, Rock DL, Weiblen R, Flores EF, Diel DG. Immunogenicity of ORFV-based vectors expressing the rabies virus glycoprotein in livestock species. Virology. 2017;511:229–39. https://doi.org/10.1016/j.virol.2017.08.027.
28. Rintoul JL, Wang J, Gammon DB, Van Buuren NJ, Garson K, Jardine K, Barry M, Evans DH, Bell JC, Halford WP. A selectable and excisable marker system for the rapid creation of recombinant poxviruses. 2011; https://doi.org/10.1371/journal.pone.0024643.
29. Yuan M, Gao X, Chard LS, et al. A marker-free system for highly efficient construction of vaccinia virus vectors using CRISPR Cas9. Mol Ther Methods Clin Dev. 2015;2:15035.
30. Parrino J, Graham BS. Smallpox vaccines: past, present, and future. J Allergy Clin Immunol. 2006;118:1320–6.
31. Kenner J, Cameron F, Empig C, Jobes DV, Gurwith M. LC16m8: an attenuated smallpox vaccine. Vaccine. 2006;24:7009–22.
32. McCurdy LH, Larkin BD, Martin JE, Graham BS. Modified vaccinia Ankara: potential as an

alternative smallpox vaccine. Clin Infect Dis. 2004;38:1749–53.

33. Meyer H, Sutter G, Mayr A. Mapping of deletions in the genome of the highly attenuated vaccinia virus MVA and their influence on virulence. J Gen Virol. 1991;72:1031–8.

34. Blanchard TJ, Alcami A, Andrea P, Smith GL. Modified vaccinia virus Ankara undergoes limited replication in human cells and lacks several immunomodulatory proteins: implications for use as a human vaccine. J Gen Virol. 1998;79:1159–67.

35. Tartaglia J, Perkus ME, Taylor J, Norton EK, Audonnet J-C, Cox WI, Davis SW, Van Der Hoeven J, Meignier B, Riviere M. NYVAC: a highly attenuated strain of vaccinia virus. Virology. 1992;188:217–32.

36. Tartaglia J, Cox WI, Pincus S, Paoletti E. Safety and immunogenicity of recombinants based on the genetically-engineered vaccinia strain, NYVAC. Dev Biol Stand. 1994;82:125–9.

37. Papin JF, Verardi PH, Jones LA, Monge-Navarro F, Brault AC, Holbrook MR, Worthy MN, Freiberg AN, Yilma TD. Recombinant Rift Valley fever vaccines induce protective levels of antibody in baboons and resistance to lethal challenge in mice. Proc Natl Acad Sci U S A. 2011;108:14926–31.

38. Weyer J, Rupprecht CE, Mans J, Viljoen GJ, Nel LH. Generation and evaluation of a recombinant modified vaccinia virus Ankara vaccine for rabies. Vaccine. 2007;25:4213–22.

39. Desmettre P, Languet B, Chappuis G, et al. Use of vaccinia rabies recombinant for oral vaccination of wildlife. Vet Microbiol. 1990;23:227–36.

40. Wiktor TJ, Macfarlan RI, Reagan KJ, Dietzschold B, Curtis PJ, Wunner WH, Kieny M-P, Lathe R, Lecocq J-P, Mackett M. Protection from rabies by a vaccinia virus recombinant containing the rabies virus glycoprotein gene. Proc Natl Acad Sci. 1984;81:7194–8.

41. Weyer J, Rupprecht CE, Nel LH. Poxvirus-vectored vaccines for rabies – a review. Vaccine. 2009;27:7198–201.

42. Cliquet F, Aubert M. Elimination of terrestrial rabies in Western European countries. Dev Biol (Basel). 2004;119:185–204.

43. Blancou J, Kieny MP, Lathe R, Lecocq JP, Pastore PP, Soulebot JP, Desmettre P. Oral vaccination of the fox against rabies using a live recombinant vaccinia virus. Nature. 1986;322:373.

44. Rupprecht CE, Wiktor TJ, Johnston DH, Hamir AN, Dietzschold B, Wunner WH, Glickman LT, Koprowski H. Oral immunization and protection of raccoons (Procyon lotor) with a vaccinia-rabies glycoprotein recombinant virus vaccine. Proc Natl Acad Sci. 1986;83:7947–50.

45. Maki J, Guiot A-L, Aubert M, et al. Oral vaccination of wildlife using a vaccinia–rabies-glycoprotein recombinant virus vaccine (RABORAL V-RG®): a global review. Vet Res. 2017;48:57.

46. Aguilar-Setién A, Campos YL, Cruz ET, Kretschmer R, Brochier B, Pastoret P-P. Vaccination of vampire bats using recombinant vaccinia-rabies virus. J Wildl Dis. 2002;38:539–44.

47. Grosenbaugh DA, Maki JL, Rupprecht CE, Wall DK. Rabies challenge of captive striped skunks (Mephitis mephitis) following oral administration of a live vaccinia-vectored rabies vaccine. J Wildl Dis. 2007;43:124–8.

48. Rupprecht CE, Hanlon CA, Blanton J, Manangan J, Morrill P, Murphy S, Niezgoda M, Orciari LA, Schumacher CL, Dietzschold B. Oral vaccination of dogs with recombinant rabies virus vaccines. Virus Res. 2005;111:101–5.

49. Yilma T, Hsu D, Jones L, Owens S, Grubman M, Mebus C, Yamanaka M, Dale B. Protection of cattle against rinderpest with vaccinia virus recombinants expressing the HA or F gene. Science. 1988;242:1058–61.

50. Giavedoni L, Jones L, Mebus C, Yilma T. A vaccinia virus double recombinant expressing the F and H genes of rinderpest virus protects cattle against rinderpest and causes no pock lesions. Proc Natl Acad Sci U S A. 1991;88:8011–5.

51. Verardi PH, Aziz FH, Ahmad S, Jones LA, Beyene B, Ngotho RN, Wamwayi HM, Yesus MG, Egziabher BG, Yilma TD. Long-term sterilizing immunity to rinderpest in cattle vaccinated with a recombinant vaccinia virus expressing high levels of the fusion and hemagglutinin glycoproteins. J Virol. 2002;76:484–91.

52. Jones L, Giavedoni L, Saliki JT, Brown C, Mebus C, Yilma T. Protection of goats against peste des petits ruminants with a vaccinia virus double recombinant expressing the F and H genes of rinderpest virus. Vaccine. 1993;11:961–4.

53. Taylor J, Pincus S, Tartaglia J, Richardson C, Alkhatib G, Briedis D, Appel M, Norton E, Paoletti E. Vaccinia virus recombinants expressing either the measles virus fusion or hemagglutinin glycoprotein protect dogs against canine distemper virus challenge. J Virol. 1991;65:4263–74.

54. Mackett M, Yilma T, Rose JK, Moss B. Vaccinia virus recombinants: expression of VSV genes and protective immunization of mice and cattle. Science. 1985;227:433–5.

55. Meulemans G, Letellier C, Gonze M, Carlier MC, Burny A. Newcastle disease virus f glycoprotein expressed from a recombinant vaccinia virus vector protects chickens against live-virus challenge. Avian Pathol. 1988;17:821–7.

56. Ramos I, Alonso A, Marcen JM, Peris A, Castillo JA, Colmenares M, Larraga V. Heterologous prime-boost vaccination with a non-replicative vaccinia recombinant vector expressing LACK confers protection against canine visceral leishmaniasis with a predominant Th1-specific immune response. Vaccine. 2008;26:333–44.

57. Carson C, Antoniou M, Ruiz-Argüello MB, Alcami A, Christodoulou V, Messaritakis I, Blackwell JM, Courtenay O. A prime/boost DNA/modified vaccinia virus Ankara vaccine expressing recombinant Leishmania DNA encoding TRYP is safe and immunogenic in outbred dogs, the reservoir of zoonotic visceral leishmaniasis. Vaccine. 2009;27:1080–6.

58. Stober CB, Lange UG, Roberts MTM, Alcami A, Blackwell JM. Heterologous priming-boosting with DNA and modified vaccinia virus Ankara expressing tryparedoxin peroxidase promotes long-term memory against Leishmania major in susceptible BALB/c Mice. Infect Immun. 2007;75:852–60.

59. ICTV. International Committee on Taxonomy of Viruses ICTV. 2017;5:1–32.

60. Weli SC, Tryland M. Avipoxviruses: infection biology and their use as vaccine vectors. Virol J. 2011;8:49.

61. Boyle DB. Genus Avipoxviruses. In: Mercer A, Schmidt A, Weber O, editors. Poxviruses. Basel: Birkhauser Verlag; 2007. p. 217–39.

62. Boyle DB, Coupar BEH. Construction of recombinant fowlpox viruses as vectors for poultry vaccines. Virus Res. 1988;10:343–56.

63. Taylor J, Weinberg R, Languet B, Desmettre P, Paoletti E. Recombinant fowlpox virus inducing protective immunity in non-avian species. Vaccine. 1988;6:497–503.

64. Taylor J, Weinberg R, Kawaoka Y, Webster RG, Paoletti E. Protective immunity against avian influenza induced by a fowlpox virus recombinant. Vaccine. 1988;6:504–8.

65. Webster RG, Kawaoka Y, Taylor J, Weinberg R, Paoletti E. Efficacy of nucleoprotein and haemagglutinin antigens expressed in fowlpox virus as vaccine for influenza in chickens. Vaccine. 1991;9:303–8.

66. Bublot M, Pritchard N, Swayne DE, Selleck P, Karaca K, Suarez DL, J-C A, Mickle TR. Development and use of fowlpox vectored vaccines for avian influenza. Ann N Y Acad Sci. 2006;1081:193–201.

67. Swayne DE, Garcia M, Beck JR, Kinney N, Suarez DL. Protection against diverse highly pathogenic H5 avian influenza viruses in chickens immunized with a recombinant fowlpox vaccine containing an H5 avian influenza hemagglutinin gene insert. Vaccine. 2000;18:1088–95.

68. Qiao C, Yu K, Jiang Y, Jia Y, Tian G, Liu M, Deng G, Wang X, Meng Q, Tang X. Protection of chickens against highly lethal H5N1 and H7N1 avian influenza viruses with a recombinant fowlpox virus co-expressing H5 haemagglutinin and N1 neuraminidase genes. Avian Pathol. 2003;32:25–31.

69. Suarez DL, Pantin-Jackwood MJ. Recombinant viral-vectored vaccines for the control of avian influenza in poultry. Vet Microbiol. 2017;206:144–51.

70. Yun S-L, Zhang W, Liu W-J, Zhang X-R, Chen S-J, Wu Y-T, Peng D-X, Liu X-F. Construction of recombinant fowlpox virus coexpressing HA gene from H5N1 avian influenza virus and chicken interleukin-2 gene and assessment of its protective efficacy. Chinese. J Virol. 2009;25:430–6.

71. Chen H-Y, Shang Y-H, Yao H-X, Cui B-A, Zhang H-Y, Wang Z-X, Wang Y-D, Chao A-J, Duan T-Y. Immune responses of chickens inoculated with a recombinant fowlpox vaccine coexpressing HA of H9N2 avian influenza virus and chicken IL-18. Antiviral Res. 2011;91:50–6.

72. Boursnell ME, Green PF, Campbell JI, Deuter A, Peters RW, Tomley FM, Samson AC, Emmerson PT, Binns MM. A fowlpox virus vaccine vector with insertion sites in the terminal repeats: demonstration of its efficacy using the fusion gene of Newcastle disease virus. Vet Microbiol. 1990;23:305–16.

73. Taylor J, Christensen L, Gettig R, Goebel J, Bouquet J-F, Mickle TR, Paoletti E. Efficacy of a recombinant fowl pox-based newcastle disease virus vaccine candidate against velogenic and respiratory challenge. Avian Dis. 1996;40:173.

74. Ogawa R, Yanagida N, Saeki S, Saito S, Ohkawa S, Gotoh H, Kodama K, Kamogawa K, Sawaguchi K, Iritani Y. Recombinant fowlpox viruses inducing protective immunity against Newcastle disease and fowlpox viruses. Vaccine. 1990;8:486–90.

75. Edbauer C, Weinberg R, Taylor J, Rey-Senelonge A, Bouquet JF, Desmettre P, Paoletti E. Protection of chickens with a recombinant fowlpox virus expressing the Newcastle disease virus hemagglutinin-neuraminidase gene. Virology. 1990;179:901–4.

76. Taylor J, Edbauer C, Rey-Senelonge A, Bouquet JF, Norton E, Goebel S, Desmettre P, Paoletti E. Newcastle disease virus fusion protein expressed in a fowlpox virus recombinant confers protection in chickens. J Virol. 1990;64:1441–50.

77. Iritani Y, Aoyama S, Takigami S, Hayashi Y, Ogawa R, Yanagida N, Saeki S, Kamogawa K. Antibody response to Newcastle disease virus (NDV) of recombinant fowlpox virus (FPV) expressing a hemagglutinin-neuraminidase of NDV into chickens in the presence of antibody to NDV or FPV. Avian Dis. 1991;35:659–61.

78. Calvert JG, Nazerian K, Wi'ter RL, Yanagidat N. Fowlpox virus recombinants expressing the envelope glycoprotein of an avian reticuloendotheliosis retrovirus induce neutralizing antibodies and reduce viremia in chickens. 1993.

79. Lee LF, Witter RL, Reddy SM, Wu P, Yanagida N, Yoshida S. Protection and synergism by recombinant fowl pox vaccines expressing multiple genes from Marek's disease virus. Avian Dis. 2003;47:549–58.

80. Bayliss CD, Peters RW, Cook JK, Reece RL, Howes K, Binns MM, Boursnell ME. A recombinant fowlpox virus that expresses the VP2 antigen of

infectious bursal disease virus induces protection against mortality caused by the virus. Arch Virol. 1991;120:193–205.

81. Heine H-G, Boyle DB. Infectious bursal disease virus structural protein VP 2 expressed by a fowlpox virus recombinant confers protection against disease in chickens. Arch Virol. 1993;131:277–92.

82. Heine HG, Boyle DB. Infectious bursal disease virus structural protein VP2 expressed by a fowlpox virus recombinant confers protection against disease in chickens. Arch Virol. 1993;131:277–92.

83. Yang G, Li J, Zhang X, Zhao Q, Liu Q, Gong P. Eimeria tenella: construction of a recombinant fowlpox virus expressing rhomboid gene and its protective efficacy against homologous infection. Exp Parasitol. 2008;119:30–6.

84. Zhang GZ, Zhang R, Zhao HL, Wang XT, Zhang SP, Li XJ, Qin CZ, Lv CM, Zhao JX, Zhou JF. A safety assessment of a fowlpox-vectored Mycoplasma gallisepticum vaccine in chickens. Poult Sci. 2010;89:1301–6.

85. Paoletti E, Perkus M, Taylor J, Tartaglia J, Norton E, Riviere M, de Taisne C, Limbach K, Johnson G, Pincus S. ALVAC canarypox virus recombinants comprising heterologous inserts. 1998.

86. Sanchez-Sampedro L, Perdiguero B, Mejias-Perez E, Garcia-Arriaza J, Di Pilato M, Esteban M. The evolution of poxvirus vaccines. Viruses. 2015;7:1726–803.

87. Poulet H, Minke J, Pardo MC, Juillard V, Nordgren B, Audonnet JC. Development and registration of recombinant veterinary vaccines. The example of the canarypox vector platform. Vaccine. 2007;25:5606–12.

88. Taylor J, Trimarchi C, Weinberg R, Languet B, Guillemin F, Desmettre P, Paoletti E. Efficacy studies on a canarypox-rabies recombinant virus. Vaccine. 1991;9:190–3.

89. Paillot R, Kydd JH, Sindle T, Hannant D, Edlund Toulemonde C, Audonnet JC, Minke JM, Daly JM. Antibody and IFN-γ responses induced by a recombinant canarypox vaccine and challenge infection with equine influenza virus. Vet Immunol Immunopathol. 2006;112:225–33.

90. Minke JM, Siger L, Karaca K, Austgen L, Gordy P, Bowen R, Renshaw RW, Loosmore S, Audonnet JC, Nordgren B. Recombinant canarypoxvirus vaccine carrying the prM/E genes of West Nile virus protects horses against a West Nile-mosquito challenge. Arch Virol Suppl. 2004:221–30.

91. Siger L, Bowen RA, Karaca K, Murray MJ, Gordy PW, Loosmore SM, Audonnet J-CF, Nordgren RM, Minke JM. Assessment of the efficacy of a single dose of a recombinant vaccine against West Nile virus in response to natural challenge with West Nile virus-infected mosquitoes in horses. Am J Vet Res. 2004;65:1459–62.

92. Seino KK, Long MT, Gibbs EPJ, Bowen RA, Beachboard SE, Humphrey PP, Dixon MA, Bourgeois MA. Comparative efficacies of three commercially available vaccines against West Nile Virus (WNV) in a short-duration challenge trial involving an equine WNV encephalitis model. Clin Vaccine Immunol. 2007;14:1465–71.

93. Poulet H, Brunet S, Boularand C, Guiot AL, Leroy V, Tartaglia J, Minke J, Audonnet JC, Desmettre P. Efficacy of a canarypox virus-vectored vaccine against feline leukaemia. Vet Rec. 2003;153:141–5.

94. Tartaglia J, Jarrett O, Neil JC, Desmettre P, Paoletti E. Protection of cats against feline leukemia virus by vaccination with a canarypox virus recombinant, ALVAC-FL. J Virol. 1993;67:2370–5.

95. Guillaume-Vasselin V, Lemaitre L, Dhondt KP, Tedeschi L, Poulard A, Charreyre C, Horvat B. Protection from Hendra virus infection with Canarypox recombinant vaccine. Nat Publ Gr. 2016;1:1.

96. Weingartl HM, Berhane Y, Caswell JL, Loosmore S, Audonnet J-C, Roth JA, Czub M. Recombinant Nipah virus vaccines protect pigs against challenge. J Virol. 2006;80:7929–38.

97. Paoletti E. Applications of pox virus vectors to vaccination: an update. Proc Natl Acad Sci U S A. 1996;93:11349–53.

98. Slack Tidwell RS, Marshall JL, Gulley JL, et al. Phase I study of sequential vaccinations with Fowlpox-CEA(6D)-TRICOM alone and sequentially with Vaccinia-CEA(6D)-TRICOM, with and without granulocyte-macrophage colony-stimulating Phase I study of sequential vaccinations with Fowlpox-CEA (6D)- TRICOM alone. Artic J Clin Oncol. 2005; https://doi.org/10.1200/JCO.2005.10.206.

99. Fischer T, Planz O, Stitz L, Rziha H-J. Novel recombinant parapoxvirus vectors induce protective humoral and cellular immunity against lethal herpesvirus challenge infection in mice. J Virol. 2003;77:9312–23.

100. Haig D, Mcinnes C, Deane D, Reid H, Mercer A. The immune and inflammatory response to orf virus. Comp Immunol Microbiol Infect Dis. 1997;20:197–204.

101. Joshi LR, Okda FA, Singrey A, et al. Passive immunity to porcine epidemic diarrhea virus following immunization of pregnant gilts with a recombinant orf virus vector expressing the spike protein. Arch Virol. 2018; https://doi.org/10.1007/s00705-018-3855-1.

102. Friebe A, Siegling A, Friederichs S, Volk H-D, Weber O. Immunomodulatory effects of inactivated parapoxvirus ovis (ORF virus) on human peripheral immune cells: induction of cytokine secretion in monocytes and Th1-like cells. J Virol. 2004;78:9400–11.

103. Ons E, Van Brussel L, Lane S, King V, Cullinane A, Kenna R, Lyons P, Hammond T-A, Salt J, Raue R. Efficacy of a Parapoxvirus ovis-based immunomodulator against equine herpesvirus type 1 and

Streptococcus equi equi infections in horses. Vet Microbiol. 2014;173:232–40.

104. Fleming SB, Anderson IE, Thomson J, Deane DL, McInnes CJ, McCaughan CA, Mercer AA, Haig DM. Infection with recombinant orf viruses demonstrates that the viral interleukin-10 is a virulence factor. J Gen Virol. 2007;88:1922–7.

105. Seet BT, McCaughan CA, Handel TM, Mercer A, Brunetti C, McFadden G, Fleming SB. Analysis of an orf virus chemokine-binding protein: shifting ligand specificities among a family of poxvirus viroceptors. Proc Natl Acad Sci U S A. 2003;100:15137–42.

106. Deane D, McInnes CJ, Percival A, Wood A, Thomson J, Lear A, Gilray J, Fleming S, Mercer A, Haig D. Orf virus encodes a novel secreted protein inhibitor of granulocyte-macrophage colony-stimulating factor and interleukin-2. J Virol. 2000;74:1313–20.

107. McInnes CJ, Wood AR, Mercer AA. Orf virus encodes a homolog of the vaccinia virus interferon-resistance gene E3L. Virus Genes. 1998;17:107–15.

108. Westphal D, Ledgerwood EC, Hibma MH, Fleming SB, Whelan EM, Mercer AA. A novel Bcl-2-like inhibitor of apoptosis is encoded by the parapoxvirus ORF virus. J Virol. 2007;81:7178–88.

109. Diel DG, Luo S, Delhon G, Peng Y, Flores EF, Rock DL. Orf virus ORFV121 encodes a novel inhibitor of NF-kappaB that contributes to virus virulence. J Virol. 2011;85:2037–49.

110. Diel DG, Luo S, Delhon G, Peng Y, Flores EF, Rock DL. A nuclear inhibitor of NF-kappaB encoded by a poxvirus. J Virol. 2011;85:264–75.

111. Diel DG, Delhon G, Luo S, Flores EF, Rock DL. A novel inhibitor of the NF-κB signaling pathway encoded by the parapoxvirus orf virus. J Virol. 2010;84:3962–73.

112. Khatiwada S, Delhon G, Nagendraprabhu P, Chaulagain S, Luo S, Diel DG, Flores EF, Rock DL. A parapoxviral virion protein inhibits NF-κB signaling early in infection. PLoS Pathog. 2017;13: e1006561.

113. Mayr A, Herlyn M, Mahnel H, Danco A, Zach A, Bostedt H. Control of ecthyma contagiosum (pustular dermatitis) of sheep with a new parenteral cell culture live vaccine. Zentralbl Veterinarmed B. 1981;28:535–52.

114. Rziha H-J, Henkel M, Cottone R, Bauer B, Auge U, Götz F, Pfaff E, Röttgen M, Dehio C, Büttner M. Generation of recombinant parapoxviruses: non-essential genes suitable for insertion and expression of foreign genes. J Biotechnol. 2000;83:137–45.

115. Amann R, Rohde J, Wulle U, Conlee D, Raue R, Martinon O, Rziha H-J. A new rabies vaccine based on a recombinant ORF virus (parapoxvirus) expressing the rabies virus glycoprotein. J Virol. 2013;87:1618–30.

116. Henkel M, Planz O, Fischer T, Stitz L, Rziha H-J. Prevention of virus persistence and protection against immunopathology after Borna disease virus infection of the brain by a novel Orf virus recombinant. J Virol. 2005;79:314–25.

117. Rohde J, Amann R, Rziha H-J. New Orf virus (Parapoxvirus) recombinant expressing H5 hemagglutinin protects mice against H5N1 and H1N1 influenza A virus. PLoS One. 2013;8:e83802.

118. Rohde J, Schirrmeier H, Granzow H, Rziha H-J. A new recombinant Orf virus (ORFV, Parapoxvirus) protects rabbits against lethal infection with rabbit hemorrhagic disease virus (RHDV). Vaccine. 2011;29:9256–64.

119. Fischer L, Le Gros FX, Mason PW, Paoletti E. A recombinant canarypox virus protects rabbits against a lethal rabbit hemorrhagic disease virus (RHDV) challenge. Vaccine. 1997;15:90–6.

120. Bertagnoli S, Gelfi J, Petit F, Vautherot JF, Rasschaert D, Laurent S, Le Gall G, Boilletot E, Chantal J, Boucraut-Baralon C. Protection of rabbits against rabbit viral haemorrhagic disease with a vaccinia-RHDV recombinant virus. Vaccine. 1996;14:506–10.

121. Dory D, Fischer T, Béven V, Cariolet R, Rziha H-J, Jestin A. Prime-boost immunization using DNA vaccine and recombinant Orf virus protects pigs against Pseudorabies virus (Herpes suid 1). Vaccine. 2006;24:6256–63.

122. van Rooij EMA, Rijsewijk FAM, Moonen-Leusen HW, Bianchi ATJ, Rziha H-J. Comparison of different prime-boost regimes with DNA and recombinant Orf virus based vaccines expressing glycoprotein D of pseudorabies virus in pigs. Vaccine. 2010;28:1808–13.

123. Voigt H, Merant C, Wienhold D, Braun A, Hutet E, Le Potier M-F, Saalmüller A, Pfaff E, Büttner M. Efficient priming against classical swine fever with a safe glycoprotein E2 expressing Orf virus recombinant (ORFV VrV-E2). Vaccine. 2007;25:5915–26.

124. Delhon G, Tulman ER, Afonso CL, Lu Z, de la Concha-Bermejillo A, Lehmkuhl HD, Piccone ME, Kutish GF, Rock DL. Genomes of the parapoxviruses ORF virus and bovine papular stomatitis virus. J Virol. 2004;78:168–77.

125. Di Pilato M, Mejías-Pérez E, Sorzano COS, Esteban M. Distinct roles of vaccinia virus NF-κB inhibitor proteins A52, B15, and K7 in the immune response. J Virol. 2017;91:e00575–17.

126. Uzel M, Sasmaz S, Bakaris S, Cetinus E, Bilgic E, Karaoguz A, Ozkul A, Arican O. A viral infection of the hand commonly seen after the feast of sacrifice: human orf (orf of the hand). Epidemiol Infect. 2005;133:653–7.

127. Lederman ER, Austin C, Trevino I, et al. Orf virus infection in children: clinical characteristics, transmission, diagnostic methods, and future therapeutics. Pediatr Infect Dis J. 2007;26:740–4.

128. Geerinck K, Lukito G, Snoeck R, De Vos R, De Clercq E, Vanrenterghem Y, Degreef H, Maes B. A case of human orf in an immunocompromised patient

treated successfully with cidofovir cream. J Med Virol. 2001;64:543–9.

129. van der Leek ML, Feller JA, Sorensen G, Isaacson W, Adams CL, Borde DJ, Pfeiffer N, Tran T, Moyer RW, Gibbs EP. Evaluation of swinepox virus as a vaccine vector in pigs using an Aujeszky's disease (pseudorabies) virus gene insert coding for glycoproteins gp50 and gp63. Vet Rec. 1994;134:13–8.

130. Hahn J, Park SH, Song JY, An SH, Ahn BY. Construction of recombinant swinepox viruses and expression of the classical swine fever virus E2 protein. J Virol Methods. 2001;93:49–56.

131. Xu J, Huang D, Liu S, Lin H, Zhu H, Liu B, Lu C. Immune responses and protective efficacy of a recombinant swinepox virus expressing HA1 against swine H1N1 influenza virus in mice and pigs. Vaccine. 2012;30:3119–25.

132. Xu J, Huang D, Xu J, Liu S, Lin H, Zhu H, Liu B, Chen W, Lu C. Immune responses and protective efficacy of a recombinant swinepox virus co-expressing HA1 genes of H3N2 and H1N1 swine influenza virus in mice and pigs. Vet Microbiol. 2013;162:259–64.

133. Yuan X, Lin H, Fan H. Efficacy and immunogenicity of recombinant swinepox virus expressing the A epitope of the TGEV S protein. Vaccine. 2015;33:3900–6.

134. Lin H, Ma Z, Hou X, Chen L, Fan H. Construction and immunogenicity of a recombinant swinepox virus expressing a multi-epitope peptide for porcine reproductive and respiratory syndrome virus. Sci Rep. 2017;7:43990.

135. Lin H, Huang D, Wang Y, Lu C, Fan H. A novel vaccine against Streptococcus equi ssp. zooepidemicus infections: the recombinant swinepox virus expressing M-like protein. Vaccine. 2011;29:7027–34.

136. Kasza L, Bohl EH, Jones DO. Isolation and cultivation of swine pox virus in primary cell cultures of swine origin. Am J Vet Res. 1960;21:269–73.

137. Bárcena J, Blasco R. Recombinant swinepox virus expressing β-galactosidase: investigation of viral host range and gene expression levels in cell culture. Virology. 1998;243:396–405.

138. Delhon G, Tulman ER, Afonso CL, Rock DL. Genus swinepox virus. In: Schmidt A, Wolff M, Kaufmann S, editors. Poxviruses; 2006. p. 203–15.

139. Winslow BJ, Cochran MD, Holzenburg A, Sun J, Junker DE, Collisson EW. Replication and expression of a swinepox virus vector delivering feline leukemia virus Gag and Env to cell lines of swine and feline origin. Virus Res. 2003;98:1–15.

140. Yuan X, Lin H, Li B, He K, Fan H. Swinepox virus vector-based vaccines: attenuation and biosafety assessments following subcutaneous prick inoculation. Vet Res. 2018;49:14.

141. Lin H, Ma Z, Yang X, Fan H, Lu C. A novel vaccine against Porcine circovirus type 2 (PCV2) and Streptococcus equi ssp. zooepidemicus (SEZ) co-infection. Vet Microbiol. 2014;171:198–205.

142. Diallo A, Viljoen G. Genus capripoxvirus. In: Mercer A, Schmidt A, Weber O, editors. Poxviruses; 2006. p. 167–81.

143. Tulman E, Lu Z, Zsak L, Sur J, Sandybaev NT, Kerembekova U, Zaitsev V, Kutish G, Rock D. The genomes of sheeppox and goatpox viruses. J Virol. 2002;76:6054–61.

144. Tulman E, Afonso C, Lu Z, Zsak L, KUtish G, Rock D. Genome of lumpy skin disease virus. J Virol. 2001.

145. Boshra H, Truong T, Nfon C, et al. A lumpy skin disease virus deficient of an IL-10 gene homologue provides protective immunity against virulent capripoxvirus challenge in sheep and goats. Antiviral Res. 2015;123:39–49.

146. Boshra H, Truong T, Nfon C, Gerdts V, Tikoo S, Babiuk LA, Kara P, Mather A, Wallace D, Babiuk S. Capripoxvirus-vectored vaccines against livestock diseases in Africa. Antiviral Res. 2013;98:217–27.

147. Kitching R, Hammond J, Taylor W. A single vaccine for the control of capripox infection in sheep and goats. Res Vet Sci. 1987;42:53–60.

148. Romero CH, Barrett T, Evans SA, Kitching RP, Gershon PD, Bostock C, Black DN. Single capripoxvirus recombinant vaccine for the protection of cattle against rinderpest and lumpy skin disease. Vaccine. 1993;11:737–42.

149. Romero CH, Barrett T, Chamberlain RW, Kitching RP, Fleming M, Black DN. Recombinant capripoxvirus expressing the hemagglutinin protein gene of rinderpest virus: protection of cattle against rinderpest and lumpy skin disease viruses. Virology. 1994;204:425–9.

150. Romero CH, Barrett T, Kitching RP, Bostock C, Black DN. Protection of goats against peste des petits ruminants with recombinant capripoxviruses expressing the fusion and haemagglutinin protein genes of rinderpest virus. Vaccine. 1995;13:36–40.

151. Chen W, Hu S, Qu L, Hu Q, Zhang Q, Zhi H, Huang K, Bu Z. A goat poxvirus-vectored peste-des-petits-ruminants vaccine induces long-lasting neutralization antibody to high levels in goats and sheep. Vaccine. 2010;28:4742–50.

152. Fakri F, Bamouh Z, Ghzal F, Baha W, Tadlaoui K, Fihri OF, Chen W, Bu Z, Elharrak M. Comparative evaluation of three capripoxvirus-vectored peste des petits ruminants vaccines. Virology. 2018;514:211–5.

153. Wade-Evans AM, Romero CH, Mellor P, Takamatsu H, Anderson J, Thevasagayam J, Fleming MJ, Mertens PPC, Black DN. Expression of the major core structural protein (VP7) of bluetongue virus, by a Recombinant capripox virus, provides partial protection of sheep against a virulent heterotypic bluetongue virus challenge. Virology. 1996;220:227–31.

154. Perrin A, Albina E, Bréard E, et al. Recombinant capripoxviruses expressing proteins of bluetongue virus: evaluation of immune responses and protection in small ruminants. Vaccine. 2007;25:6774–83.

155. Soi RK, Rurangirwa FR, McGuire TC, Rwambo PM, DeMartini JC, Crawford TB. Protection of sheep against Rift Valley fever virus and sheep poxvirus with a recombinant capripoxvirus vaccine. Clin Vaccine Immunol. 2010;17:1842–9.

156. Wallace DB, Ellis CE, Espach A, Smith SJ, Greyling RR, Viljoen GJ. Protective immune responses induced by different recombinant vaccine regimes to Rift Valley fever. Vaccine. 2006;24:7181–9.

157. Aspden K, van Dijk AA, Bingham J, Cox D, Passmore J-A, Williamson A-L. Immunogenicity of a recombinant lumpy skin disease virus (neethling vaccine strain) expressing the rabies virus glycoprotein in cattle. Vaccine. 2002;20:2693–701.

158. Aspden K, Passmore J-A, Tiedt F, Williamson A-L. Evaluation of lumpy skin disease virus, a capripoxvirus, as a replication-deficient vaccine vector. J Gen Virol. 2003;84:1985–96.

159. Kerr P, Liu J, Cattadori I, Ghedin E, Read A, Holmes E. Myxoma virus and the leporipoxviruses: an evolutionary paradigm. Viruses. 2015;7:1020–61.

160. Kerr PJ, Jackson RJ. Myxoma virus as a vaccine vector for rabbits: antibody levels to influenza virus haemagglutinin presented by a recombinant myxoma virus. Vaccine. 1995;13:1722–6.

161. McCabe VJ, Spibey N. Potential for broad-spectrum protection against feline calicivirus using an attenuated myxoma virus expressing a chimeric FCV capsid protein. Vaccine. 2005;23:5380–8.

162. Pignolet B, Boullier S, Gelfi J, Bozzetti M, Russo P, Foulon E, Meyer G, Delverdier M, Foucras G, Bertagnoli S. Safety and immunogenicity of myxoma virus as a new viral vector for small ruminants. J Gen Virol. 2008;89:1371–9.

163. Tyndale-Biscoe C. Virus-vectored immunocontraception of feral mammals. Reprod Fertil Dev. 1994;6:281.

164. Mackenzie SM, McLaughlin EA, Perkins HD, et al. Immunocontraceptive effects on female rabbits infected with recombinant myxoma virus expressing rabbit ZP2 or ZP3. Biol Reprod. 2006;74:511–21.

165. Kerr P, McFadden G. Immune responses to myxoma virus. Viral Immunol. 2002;15:229–46.

166. Kieny MP, Lathe R, Drillien R, Spehner D, Skory S, Schmitt D, Wiktor T, Koprowski H, Lecocq JP. Expression of rabies virus glycoprotein from a recombinant vaccinia virus. Nature. 1984;312:163.

167. Minke JM, Audonnet J-C, Fischer L. Equine viral vaccines: the past, present and future. Vet Res. 2004;35:425–43.

168. Soboll G, Hussey SB, Minke JM, Landolt GA, Hunter JS, Jagannatha S, Lunn DP. Onset and duration of immunity to equine influenza virus resulting from canarypox-vectored (ALVAC®) vaccination. Vet Immunol Immunopathol. 2010;135:100–7.

169. Meeusen ENT, Walker J, Peters A, Pastoret P-P, Jungersen G. Current status of veterinary vaccines. Clin Microbiol Rev. 2007;20:489–510, table of contents

170. Stephensen CB, Welter J, Thaker SR, Taylor J, Tartaglia J, Paoletti E. Canine distemper virus (CDV) infection of ferrets as a model for testing Morbillivirus vaccine strategies: NYVAC- and ALVAC-based CDV recombinants protect against symptomatic infection. J Virol. 1997;71:1506–13.

171. Davison S, Gingerich EN, Casavant S, Eckroade RJ. Evaluation of the efficacy of a live fowlpox-vectored infectious laryngotracheitis/avian encephalomyelitis vaccine against ILT viral challenge. Avian Dis. 2006;50:50–4.

# The Construction and Evaluation of Herpesvirus Vectors

K. E. Robinson and T. J. Mahony

**Abstract**

Viruses belonging to the *Herpesviridae* family are widely distributed in nature and are associated with many diseases. Herpesvirus infections and associated diseases of production animals are of particular importance due to economic losses incurred by industry and significant animal health and welfare issues. To address herpesvirus-associated diseases in production animals, the use of vaccination has been widely applied. To further enhance vaccine efficiency, concerted efforts to develop vaccination systems that provide optimal safety profiles while providing protective immunological responses is an area of intense study. This includes research to exploit herpesviruses as viral vector systems that can deliver heterologous antigens from other viral and bacterial pathogens as a multivalent viral vector system. This chapter describes the construction and development of several species-specific *Alphaherpesvirinae*-derived vector platforms which have been exploited as multivalent live viral vaccines against pathogens affecting horse, swine, poultry and cattle industries.

**Keywords**

Herpesvirus · Pseudorabies virus · Turkey herpesvirus-1 · Marek's disease virus-1 · Bovine herpesvirus 1 · Recombinant vaccine · Viral vector

**Learning Objectives** After reading this chapter you should be able to:

- Describe the key features of the herpesviruses
- Describe the properties of the herpesviruses which are suited to vector applications
- Explain the commonly used strategies used to generate herpesvirus vectors
- Describe the advantages of using a herpesvirus vector
- List important diseases for which herpesvirus vectors have been developed
- Describe some limitations of herpesvirus vectors

## 1 Introduction

The family *Herpesviridae* is a large and diverse family encompassing over 200 of the most successful and evolutionarily ancient viruses. Species-specific herpesviruses are pervasive in nature and have been isolated from diverse hosts including reptilian, mammalian, amphibian, fish and molluscum species [1]. *Herpesviridae*

K. E. Robinson (✉) · T. J. Mahony
The University of Queensland, Queensland Alliance for Agriculture and Food Innovation, St Lucia, Australia
e-mail: k.robinson2@uq.edu.au

© Springer Nature Switzerland AG 2021
T. Vanniasinkam et al. (eds.), *Viral Vectors in Veterinary Vaccine Development*,
https://doi.org/10.1007/978-3-030-51927-8_7

members are further segregated into three major subfamilies, namely, *Alphaherpesvirinae*, *Betaherpesvirinae* and *Gammaherpesvirinae*, based on their genomic structural organisation. Morphologically, herpesvirus particles are composed of four distinct structures designated: core, capsid, tegument and envelope [1]. The core contains the viral double-stranded DNA and is encased within an icosahedral nucleocapsid structure, consisting of 162 capsomers (150 hexameric and 12 pentameric) with a diameter of 95–110 nm. Residing adjacent to the nucleocapsid of the herpesvirus virion is the tegument. The tegument is an amorphous proteinaceous layer of variable thickness that is anchored to the vertices of the nucleocapsid. The envelope enclosing the tegument is a lipid bilayer mostly derived from of the nucleus and plasma membranes and contains several integral viral glycoproteins that appear as spikes in electron micrographs [1, 8, 13]. Due to the highly pleomorphic nature of the envelope and variable thickness of the tegument, the dimensions of herpesvirus particles can range from 150 to 200 nm in diameter [1, 13].

The replication pathway of *Herpesviridae* can be segregated into three distinct phases. The primary phase includes contact, attachment and entry into susceptible cells resulting in the localisation of the viral genome to the cell's nucleus. The second phase incorporates gene transcription, translation and replication of the genome via a cascade of sequentially transcribed genes ($\alpha$-, $\beta$-, $\gamma$-stage induction), with the final phase incorporating capsid assembly, DNA packaging and release [13]. A defining characteristic of *Herpesviridae* family members is the capacity of these viruses to persist in the host in a non-replicating latent form. The periodic reactivation of the virus from the latent state allows for the transmission, potentially in the absence of clinical signs, of the virus to naïve hosts ensuring its persistence within the host population. Infection or reactivation of herpesviruses can be associated with several diseases in important production animals, and significant economic losses can be incurred due to lost production, additional animal therapeutics and management. To address animal welfare and reduce economic losses

resulting from herpesvirus-associated disease, vaccination is widely applied.

Vaccination has proved to be one of the most important health and welfare advances for human and animals. In veterinary applications, the use of vaccines (concurrent with other disease management practices) is a cost-effective method of improving the health and welfare of animals [2]. Numerous vaccine preparations encompassing traditionally derived inactivated/killed whole-pathogen, live attenuated and subunit vaccines are available to producers with each displaying advantages and disadvantages depending on the species to be vaccinated, vaccination route, required immunological response, disease incidence and, importantly, cost per dose.

Live attenuated viral vaccines (LAVV), consisting of a pathogen that has reduced virulence/replicative capacity in its host, and thus not able to cause disease, can induce a protective immune response. The effectiveness of LAVV lies in the capacity of the vaccine to display the complete complement of viral antigens to the immune system, be administered by the natural route of infection and, depending on the nature of the attenuation, mimic a natural infection of the host [3]. These characteristics are likely to induce the innate, humoral and cell-mediated immune responses in the vaccinated host to protect against the wild-type virus.

Development of LAVV aims to derive a virus that is non-virulent, retains antigenicity and can elicit a robust protective immune response. Traditionally, the development of a LAVV candidate was achieved by using either chemical, mechanical, or temperature stresses or serial passage of virus in semi-permissive, non-host species or cells. Subsequent selection of appropriate mutants and evaluation of the vaccinal capacity was then assessed by measuring immune responses and protective efficacy via vaccination/challenge models. Many successful vaccines have been developed using this approach. One disadvantage in using this approach is that the attenuation genotype of the virus of interest is undefined and the possibility of reversion to wild-type virulence is possible. However, advances in next-generation sequencing technologies and the associated costs should address this issue. Therefore, LAVV displaying reduced virulence offers advantages for further exploitation as through targeted genetic modification(s). Current molecular technologies, including homologous recombination (Fig.1) or CRISPR [4], provide the capacity to develop LAVV with defined alterations, mutations and insertions, thus generating LAVV that are defined in their mode of attenuation while being optimised to deliver heterologous antigens.

With respect to herpesviruses, considerable resources have been directed towards their development as heterologous antigen delivery systems, as species-specific herpesviruses are exceptional delivery vehicles able to accommodate, efficiently present antigens and induce protective immune responses that can limit the severity of disease in host animals. Herpesvirus-derived LAVV-polyvalent vaccines may be engineered for the specific targeting of pathogens, immunological responses, sites and hosts.

Many of the herpesviruses are uniquely suited to development as vaccine vectors due to their inherent biological and physical characteristics. These characteristics include (i) the capacity of herpesviruses to be propagated in a broad range of cell types, resulting in high titre stocks. This property can reduce vaccine dose costs that can be restrictive in intensively farmed animals such as chicken meat/egg production systems. (ii) Genetically, herpesviruses contain a double-stranded DNA genome that is stable and can be easily manipulated by routine molecular biological techniques. A significant advance was the capacity to clone the complete genomes of herpesviruses, which allows the viral genome to be maintained as a bacterial artificial chromosome (BAC) and has facilitated the application of bacterial gene mutagenesis techniques to viral genomes [5, 6]. (iii) Many genes of herpesviruses are nonessential for in vitro growth. Although which genes are nonessential has not been established for the majority of herpesviruses, the gene requirement of several important human and animal herpesviruses has been assessed [7, 8]. The strong conservation of herpesvirus gene function allows for the extrapolation of

gene requirements from one virus to another, though exceptions to this rule do exist [7, 8]. The consensus from these studies suggests that approximately half of the genes encoded by herpesviruses are not required for in vitro growth. An important aspect of utilising herpesviruses as vaccine vectors is that nonessential genes can be utilised as insertion sites for genes encoding antigens from other pathogens. Further, if the deleted herpesvirus gene generates a detectable immunological response in the host, the absence of this response may be useful for differentiating infected versus vaccinated animals (DIVA) approaches. (iv) Finally, herpesviruses can stably accommodate large amounts of exogenous DNA well in excess of other viral vector systems, thus improving the likelihood of successfully developing multivalent vaccines. These are just some of the biological characteristics which make the herpesviruses ideal vaccine vectors.

This chapter describes the construction and development of herpesvirus vectors, followed by a generic overview in the recombination technology dominantly used in constructing multivalent herpesvirus vectors, with a final summary of selected alphaherpesvirus LAVV derived from *Bovine herpesvirus* 1 (BoHV-1), *Equine herpesvirus* 1 (EHV-1), *Turkey herpesvirus* (HVT) and *Pseudorabies virus* (PRV) that have been exploited as vaccines for the protection of cattle, horse, poultry and swine, respectively, targeting a number of important diseases.

## 2 Considerations for the Construction of Herpesvirus Vectors

The generation of efficacious herpesvirus vaccine vector candidates encompasses several important considerations. Primarily, the vector must be appropriate for the target host. Growth properties that are either restrictive or flexible in a host range can endow specific- or cross-protective immune responses. For example, the use of host-restricted BoHV-1 vector for *Bovine viral diarrhoea virus* (BVDV) vaccination [9, 10] versus the use of flexible host range EHV-1 vector also utilised to deliver BVDV antigens to cattle [11]. The advantage of using a non-host vector system for vaccination is that an immune response can be induced to a pathogen in the presence of maternal or pre-existing immunity.

Secondly, the importance of amenable genomic sites for insertion of exogenous antigen sequences is a significant determinant of the success of a vector. Although approximately half of the genes of herpesviruses can be considered nonessential in vitro, this is most likely not the case in vivo. The complete gene requirement of BoHV-1 has been assessed [7] and showed that approximately half of the genes encoded are nonessential for in vitro growth. However, growth kinetic data suggested that deletion of certain genes can cause moderate replicative deficiency that would be detrimental to the development of a vaccine vector [7]. Therefore, it is imperative that any insertion site within a vector be assessed for growth efficiency both in vitro and in vivo.

Thirdly, once an appropriate site has been identified, the expression of genetic information inserted into the vector requires that the transcripts be efficiently transcribed, translated and presented to the immune system. RNA virus antigen sequences are generally not appropriate for expression in DNA-based vector systems, as observed for the use of *Bovine respiratory syncytial virus* (BRSV) antigens [12] and BVDV antigens [9]. Therefore, the expression of cDNA transcripts may require the use of synthetic and/or codon-optimised open reading frames (ORFs) under the control of efficient transcriptional promoter elements. And depending on these requirements, these antigens can be engineered to transcribe inserts continuously upon infection or be induced at specific stages of productive herpesvirus infection (i.e. α-, β-, γ-stage induction) [13].

A fourth consideration in the development of viral vectors is the ease of genomic manipulation. Technological advances allowing the stable maintenance of a herpesvirus genome in bacteria as a BAC have facilitated the rapid advancement herpesvirus biology and vector development. Investigation and manipulation of the herpesvirus genome can now be conducted independent of cell culture systems and coupled with bacterial recombination technologies; it is now possible to incorporate several foreign exogenous gene sequences into the genome with ease. These powerful technologies open the possibility of developing herpesviruses vaccine vectors that may induce protective immunity to several pathogens simultaneously.

Lastly, determining the upper limit to the amount of genetic material that can be added to a herpesvirus genome without any deleterious effects is an important consideration. Studies using human adenovirus have demonstrated that foreign DNA equating to approximately 5% of the viral genome can be added without detriment to virus replication. With regard to alphaherpesviruses, it has been suggested the genome of *Human herpesvirus 1* (HHV-1) can accommodate up to 30 Kbp of exogenous sequence representing a 20% increase in size of viral genome [14]. Mutagenesis studies on HHV-1 suggest the virus can readily accommodate insertion of at least 10 Kbp without affecting replication [15]. Expression studies evaluating conditional expression switches in a HHV-1 BAC vector have reported the successful accommodation and reconstitution of virus with an 8.1% increase in genomic size [15].

Herpesvirus-derived vectors can provide efficacious multivalent vaccine strategies. Numerous herpesviruses have been investigated as vaccine vectors to delivery immunogenic, apoptotic, immune-modulating or therapeutic genes in veterinary applications [16]. Despite the potential of herpesvirus vectors, some obstacles are still to be resolved. Most significant is the prevention of virus transmission via shedding and overcoming the capacity of herpesviruses to enter the latent state and persist within animals [17]. Advances in the understanding of basic herpesvirus biology, molecular biology, immunology and vaccine adjuvant formulations herald a plethora of vaccines that may provide avenues to overcome these obstacles: in this respect, technologies such as the advances in RNA interference (RNAi) and its potential in combating herpesvirus-associated diseases [18, 19]. While at the same time, immune stimulation by the viral vector antigens and/or heterologous antigens, herpesviruses and their associated diseases can be better controlled and provide new avenues for the treatment of animal disease and has the potential to transform animal disease control.

## 3 Construction Strategy for Herpesvirus-Based Vaccine Vectors

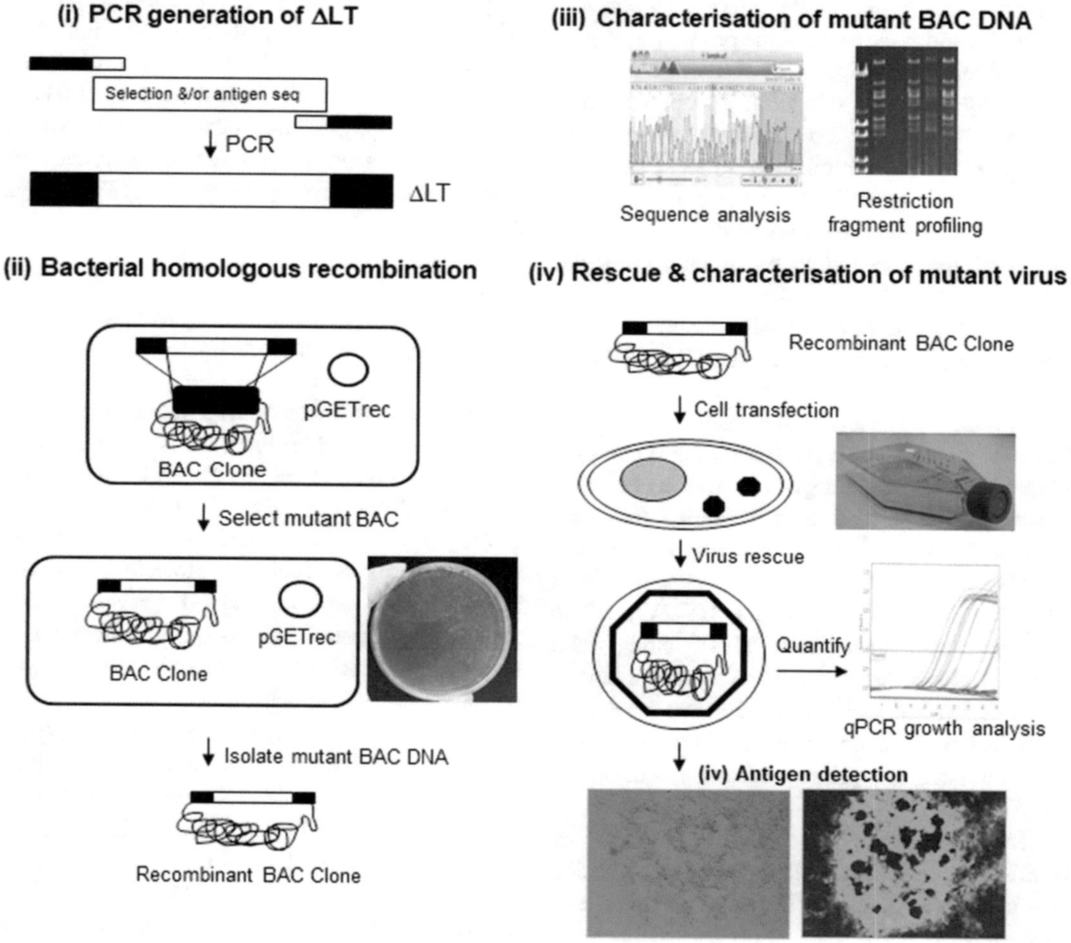

### (i) PCR generation of ΔLT

### (iii) Characterisation of mutant BAC DNA

Sequence analysis

Restriction fragment profiling

### (ii) Bacterial homologous recombination

### (iv) Rescue & characterisation of mutant virus

**Fig. 1** Schematic representation illustrating the generic steps required to delete/replace an element from a herpesvirus genome maintained as an infectious bacterial artificial chromosome (BAC) by homologous recombination and characterisation of the mutant. (i) Generation of the linear transgene (LT) molecule using PCR and chimeric primers. At the end of the PCR process, the LT encodes a bacterial resistance cassette (Kan$^R$) and sequence with homology (black segments) to the 5′ and 3′ of the sequence to be deleted/replaced. (ii) The LT is electroporated into bacterial cells transiently enabled for homologous recombination by the induction of recombination proteins encoded by extrachromosomal plasmid. (iii) Recombinant BAC clones are selected for on LB media containing antibiotics. (iv) DNA from putatively recombinant BAC clones were further characterised by DNA sequence analyses to confirm the presence of the LT. The integrity of the modified BAC is assessed using restriction endonuclease digestion in comparison with the parental BAC clone. (v) The effect of the introduced mutation on the rescue of infectious virus is then determined by transfection of the recombinant BAC into virus susceptible cells. The replication capacity and characterisation of insert expression in the rescued virus can be determined using quantitative PCR (qPCR), protein/immunological detection and/or visualisation (e.g. IFA, TEM) relative to the unmodified parental clone (adapted from [20])

## 4    Bovine Herpesvirus 1

BoHV-1 is a globally distributed virus of cattle and associated with respiratory and reproductive diseases in feedlots, dairy and other intensively farmed bovids. BoHV-1 is classified in the *Varicellovirus* genus of the *Alphaherpesvirinae* subfamily and is considered the prototype herpesvirus of ruminants. The virus contains a genome 135.3 kbp in length encoding 73 open reading frames (ORF). Common clinical syndromes are associated with uncomplicated BoHV-1 infection of cattle, including infectious bovine rhinotracheitis and infectious pustular vulvovaginitis, with the latter known as infectious pustular balanoposthitis (IPB) in bulls.

In addition, BoHV-1 is also associated with bovine respiratory disease (BRD). BRD is characterised by severe respiratory distress, mucopurulent nasal and ocular discharge, resulting from a complex interplay of factors that encompass viral and bacterial pathogens as well as environmental conditions and animal stresses that may cause and exacerbate the severity of BRD. BRD is considered the most important cause of illness and mortality in feedlot herds. BVDV, *Bovine parainfluenza-3* virus and BRSV along with BoHV-1 are considered the main viruses linked to BRD. Typically, BRD manifests as a result of secondary bacterial colonisation of the respiratory epithelium due to a primary infection with one or more viruses.

Here we summarise the development of BoHV-1 as a multivalent recombinant vaccine vector for use against BRD and other significant pathogens.

### (a)   Bovine Herpesvirus 1 Vectors

BoHV-1 has long been identified as a virus that displays immense potential as a vaccine vector for the delivery of antigenic determinants to the bovine immune system [21, 22]. Traditionally, the use of BoHV-1 as a vaccine vector was reliant on the attenuation of the virus by mutation, deletion or disruption of genes individually, such as

the thymidine kinase [21, 22] and, glycoprotein E (gE) or in multiples such as the dual gG-TK deletion mutant, gE-TK or gE-UL49.5-US9 triple deletion mutants. Several BoHV-1-derived vaccine vectors have been constructed that express the major immunogens using expression cassettes of either genomic or synthetic coding sequences of several significant pathogens of ruminants including BVDV, BRSV and *Foot and mouth disease virus* (FMDV). Furthermore, BoHV-1 has been investigated as a novel oncolytic vector for human cancers [23]. The following sections summarise the development of BoHV-1 as a polyvalent vaccine vector.

### (b)  BoHV-1 Vector Expressing Bovine Respiratory Syncytial Virus Antigens

BRSV is a member of the *Pneumovirus* genus within the family *Paramyxoviridae* and is a significant pathogen of cattle causing lower respiratory tract disease in calves [13]. In an effort to develop a safe, efficient and efficacious vaccine that may provide challenge protection to cattle against BRSV, several BoHV-1-derived vaccine vectors expressing a modified, chemically synthesised attachment protein G ORF (G-syn) of BRSV have been constructed [24].

The use of G-syn was devised from experiments conducted by Kühnle et al. [12] which showed genomic RNA-derived cDNA of the G ORF of BRSV was incompatible with the codon usage patterns of BoHV-1, resulting in unstable expression of the cDNA transcript in the nucleus. To achieve stable expression of the BRSV G ORF, G-syn was constructed in which the splice-donor and polyadenylation motifs were removed to further reflect the codon usage pattern of BoHV-1 Recombinant BoHV-1 constructs harbouring G-syn stably express the transcript and are post-transcriptionally modified with the addition of N- and O-linked glycosides as well as their incorporation into the BoHV-1 envelope [12]. The advance was an important aspect of designing LAVV incorporating RNA virus antigens.

Subsequent experiments assessing the immunogenicity of two recombinant BoHV-1 vaccine vectors expressing G-syn were assessed [24, 25]. A BoHV-1 gE negative vaccine vector harbouring the G-syn of BRSV (BoHV1/BRSV-G) induced a high degree of protection against both BRSV and BoHV-1 when compared to commercial vaccine preparation containing live attenuated BoHV-1, BRSV and BVDV [25]. BRSV challenge studies of BoHV1/BRSV-G-vaccinated calves reduced clinical signs with temperatures not exceeding 39.5 °C. A significant reduction in viral shedding was observed as well as the induction of antibodies directed against the BRSV-G protein. Following BoHV-1 challenge of BoHV1/BRSV-G-vaccinated calves, clinical signs were reduced with temperatures also not exceeding 39.5 °C Nasal lesions were mild with a significant reduction in the amount (~100-fold lower) and duration of virus shedding when compared to control calves. High titres of neutralizing antibodies were induced to BoHV-1, and it was concluded that the BoHV-1 vector expressing G-syn induced a significant protective response that may effectively protect cattle from both BoHV-1- and BRSV-induced disease [25].

A second BoHV-1-derived vector harbouring BRSV G-syn antigen inserted adjacent to the gE promoter (BHV-1/G-syn) [24] showed that in comparison with a frameshifted gE negative clone mutant of BoHV-1 (BoHV-1/gEfs), an actual increase in virulence was produced [24]. BRSV challenge of gnotobiotic calves vaccinated with either BoHV-1/G or BoHV-1/gEfs showed that calves initially vaccinated with BoHV-1/gEfs exhibited mild clinical signs. In contrast, calves vaccinated with BoHV-1/G-syn developed more severe clinical disease with a higher recovery of BoHV-1/G virus from the lungs of calves. However, vaccination with BoHV-1/G-syn was shown to induce BRSV-specific serum and respiratory secretion antibodies. Furthermore, BoHV-1/G-syn vaccinated calves showed a significant reduction in the shedding of challenge virus, with BRSV only recoverable from the lungs of one calf. Overall, pneumonic lesions were significantly reduced in challenged calves immunised with BoHV-1/G-syn compared to control calves

vaccinated with BoHV-1/gEfs [24]. In comparison with the previous construct, BoHV-1/G-syn was considered to induce protection against BRSV infection of the lower and to some extent the upper respiratory tract of calves. However, the perceived increase in virulence displayed by BoHV-1/G-syn was speculated to result from an enhanced ability to infect respiratory cells or induce suppression of the pulmonary defence mechanisms by protein G [24].

### (c) BoHV-1 Vector Expressing Bovine Viral Diarrhoea Virus Antigens

BVDV is a member of the *Pestivirus* genus of the *Flaviviridae* family and is associated with respiratory, reproductive and enteric diseases of ruminants. BVDV infection of pregnant heifers during the first trimester of pregnancy can result in foetal infection, potentially resulting in calves that are born persistently infected with BVDV [13]. Persistently infected calves fail-to-thrive continuously shed virus and may eventually develop fatal mucosal disease [13]. To develop vaccines for the immunisation of cattle against BVDV, BoHV-1-derived vectors expressing either synthetic [9] or genomic [10] forms of the major BVDV epitope, E2 gene, have been constructed.

To develop a bivalent LAVV to protect against BVDV, Kühnle et al. [12] utilised a synthetic form of the BVDV E2 ORF. However, no expression of genomic RNA-derived cDNA E2 ORF was detected, and while the exact reasons for this were not determined, the authors suggest that cryptic RNA signalling motifs (such as splice-donor sites and/or polyadenylation signals) and codon usage destabilised the E2 transcripts in the nucleus. Assuming this premise for BVDV, a BoHV-1 vector harbouring a chemically synthesised BVDV E2 ORF (E2-syn) preceded by the pestivirus $E^{rns}$ signal peptide was constructed [9]. The E2-syn was expressed and stably incorporated into the viral envelope of the BoHV-1-derived vector [9].

Studies conducted by Kweon et al. [26] showed that the genomic form of the BVDV E2, under the control of the BoHV-1 TK promoter (pBHVE2), could be expressed. However, growth kinetic studies comparing unmodified BoHV-1 vector virus to the pBHVE2 construct showed that the construct growth, while inducing a cytopathic effect, was delayed by 4 days. Immunogenicity studies using a Guinea pig model showed that immunisation with the pBHVE2 construct induced neutralising antibodies to BVDV E2. Subsequent studies in cattle immunised with pBHVE2 were inconclusive. Vaccination with pBHVE2 demonstrated that three out of four immunised cows recorded temperature increases above 39 °C for 2–7 days postinfection with associated clinical signs of BoHV-1 disease which resolved 10 days postinfection. Increases in antibody titres to E2 of BVDV due to an immunostimulatory effect of pBHVE2 were observed for three cows; however, one showed no detectable antibody response. Subsequent studies where cattle were vaccinated with pBHVE2 and subsequently challenged by intravenous and intramuscular inoculation BoHV-1 and BVDV, respectively, showed no signs of infection with either BoHV-1 or BVDV [26].

### (d) BoHV-1 Vector Expressing Foot-and-Mouth Disease Virus Antigens

BoHV-1 has been used as a vector to present FMDV immunogenic epitopes in an effort to protect cattle from foot-and-mouth disease (FMD) [27]. These studies show that BoHV-1 may be used to incorporate and express antigenic sequences of FMDV and induce a protective response to this important pathogen. The BoHV-1-derived vector was constructed to incorporate either the monomeric or dimeric form of the FMDV VP1 protein with the insert sequence preceded by the bovine growth hormone signal sequence with the complete cassette under the control of the native BoHV-1 gC promoter. Recovery and visualisation of the recombinant BoHV-1 virus by immunoelectron microscopy studies showed that the FMDV-VP1 (dimer) fusion protein was expressed as a repeated

structure on the surface of virions. The inoculation of calves with the recombinant virus resulted in no adverse clinical signs. Subsequent challenge with virulent BoHV-1 demonstrated that vaccinated calves were protected from BoHV-1 clinical disease. The vaccine also induced anti-VP1 antibodies to levels that would be protected from FMDV infection [27].

In summary, it is evident that BoHV-1 is amenable for use as a vaccine vector for the delivery of immunogenic epitopes from those pathogens that are implicated in the BRDC and other significant pathogens including parasites. These studies show that BoHV-1 expressing exogenous DNA sequences can induce immunological responses to both vector and expressed antigens that may protect from associated virus-induced disease in cattle.

## 5 Equine Herpesvirus 1 Vectors

EHV-1 is a major cause of respiratory disease, abortion and neurological disease in horses worldwide [28]. EHV-1 is a member of the *Alphaherpesvirinae* subfamily residing within the *Varicellovirus* genus. EHV-1 is an excellent candidate for development as a viral vaccine vector due to its capacity to infect a broad range of cell types and induce strong humoral and cellular immune responses in a wide range of species [29]. Several EHV-1 vectors have been developed for use in both equine and other veterinary species. The potential use of EHV-1-derived vectors to human cancer and gene therapies has also been explored [29]. For veterinary applications, EHV-1 has been used as a bivalent vector to express antigens from *West Nile virus* (WNV), *Equine influenza virus* (EIV), *Swine influenza virus* (SIV), BVDV, *Rift Valley fever virus* (RVFV) and *Canine influenza virus* (CIV) for use in various animals.

### (a) EHV-1 Vectors Expressing West Nile Virus Antigens

WNV is an arbovirus of the *Flaviviridae* family and is a common cause of epidemic meningoencephalitis in North America [30]. Horses and humans are incidental WNV hosts, as the natural life cycle of the virus is between birds and mosquitoes. However, in horses, WNV replicates with enough efficiency to titres that facilitate zoonotic transmission [30]. An EHV-1 vector was demonstrated to stably express and incorporate the prM and E antigens of WNV into the viral envelope with no impact on the in vitro growth properties of the virus [31]. Immunisation of horses with the recombinant viral vector resulted in the production of virus neutralising (VN) antibodies and a substantial increase in WNV-specific IgG antibodies [31]. The immune responses generated in response to the EHV-1 vector suggested that a degree of protection from WNV challenge would be expected, therefore reducing the risk of horses as a zoonotic WNV reservoir [31].

### (b) EHV-1 Vectors Expressing Rift Valley Fever Virus Antigens

RVFV is an arthropod-borne bunyavirus of the *Phlebovirus* genus that can cause fatal disease in humans and animals. Infection of ruminants is characterised by acute hepatitis, abortion in pregnant animals and high mortality rates, particularly in neonates. RVFV can be transmitted from infected animals to humans [30]. The utilisation of EHV-1 (RacH) vaccine strain, maintained as a BAC clone, as a vector to deliver synthetic and codon-optimised immunodominant RVFV glycoproteins Gn and Gc, showed strong expression and no negative effects on the replication of resulting vaccine vector [32]. Inoculation of sheep produced neutralising antibody responses that were able to reduce viral plaque size by 50%, a known correlate with protection from RVFV infection [32].

### (c) EHV-1 Vectors Expressing BVDV Antigens

For the immunisation of cattle against BVDV, an EHV-1 vector incorporating synthetically open reading frame encoding the BVDV structural proteins C, $E^{rns}$, E1 and E2 was constructed [11]. The EHV-1 viral vector stably expressing the inserted ORFs with the translated BVDV protein products has no detectable effects on the replication capacity of the vector [11]. Intramuscular (i.m.) immunisation of cattle induced a strong BVDV neutralising antibody responses, as determined using in vitro virus neutralisation assays. Challenge of vaccinated cattle with BVDV-1 was shown to induce an anamnestic neutralizing antibody response resulting in significantly reduced viremia and virus shedding in nasal secretions [11].

As outline previously, the ideal choice of vector to deliver BVDV antigens to cattle would be BoHV-1, as it could serve as protection from two important pathogens. However, the use of BoHV-1 as the vaccine vector may not be possible in cattle that have previously been vaccinated or naturally exposed to BoHV-1. A BoHV-1 vector might also affect eradication campaigns. This highlights the flexibility of using EHV-1 as a vaccine vector due to its capacity to induce protective immunological responses to heterologous antigens in non-host species in the presence pre-existing immunological responses to the homologous vector for the species of interest. The other advantage of using an EHV-1 vector instead of a BoHV-1 viral vector for cattle is it would not interfere with BoHV-1 control and/or eradication programmes.

### (d) EHV-1 Vectors Expressing Influenza Virus Antigens

EHV-1 has been developed as a polyvalent vaccine vector expressing the haemagglutinin (HA) gene of EIV subtype H3N8 for immunization of horses [33] and dogs [34] and the H1N1 influenza A virus of swine origin for the immunisation of pigs.

EIV is a leading cause of respiratory, neurological and late-term abortion in mares [13]. To develop a polyvalent vaccine candidate representative of the circulating strains of EHV-1, Van de Walle et al. [33] used an abortion-associated strain to express a codon-optimised H3 gene from EIV (H3N8). The resulting viral vector was evaluated by immunisation of horses. Anti-EHV-1 antibodies were detected 2 weeks after the initial vaccination. Vaccinated animals were protected from intranasal challenge with neurovirulent EHV-1 [33]. While the vaccinated horses were not challenged with EIV, EIV-specific antibodies were detected at levels likely to confer protection against EIV [33].

CIV is an emerging infection of canines resulting from the horizontal transmission of equine influenza virus (EIV) (H3N8) from horses to dogs [35]. A bivalent vaccine utilising EHV-1 (RacH) vector expressing a codon-optimised H3 gene from EIV (H3N8) (EHV-H3) was tested in both mice and dogs [34]. Mice inoculated i.n. three times at seven-day intervals with EHV-H3 showed a dose-dependent antibody response the EIV H3 antigen. In canines, subcutaneous (s.c.) immunisation with EHV-H3 induced strong H3 antibody responses. Challenge of vaccinated canines with CIV (A/Canine/PA/10915–07) showed a significant reduction and early resolution of CIV associated clinical signs [34]. The was also a significant reduction in CIV shedding, likely to reduce transmission to susceptible hosts.

As discussed above, EHV-1-derived vectors expressing exogenous antigens are promising prophylactic agents that can confer a significant level of protection in several species of veterinary importance. The application and general utility of EHV-1 as a vaccine vector for the delivery of immunogenic epitopes coupled with the capacity of the virus to infect a range of cell types present opportunities that are promising for the further development of EHV-1 as a vaccine vector.

## 6    Herpesvirus of Turkey Vectors

Herpesvirus of Turkeys (HVT), with the correct species name *Meleagrid herpesvirus*-1 (MeHV-1), is a member of the *Madivirus* genus within the *Alphaherpesvirinae* subfamily. Marek's disease (MD) is a highly contagious, lymphoproliferative and tumorigenic disease caused by Gallid herpesvirus 2 or Marek's disease virus serotype

1 (MDV-1). HVT is non-pathogenic and non-contagious in chickens and can elicit protective immune responses against MDV-1. These characteristics coupled with those inherent to herpesviruses have led to its development as a multivalent recombinant vaccine vector to many avian pathogens.

### (a) HVT Vectors Expressing Marek's Disease Virus Antigens

Several HVT-1 vectors derived from the vaccinal strain FC-126 have been constructed to express either single or multiple exogenous sequences such as the immunodominant glycoprotein B (gB) [36] and/or gC [37] genes of MDV-1 with gB of MDV-1 shown to induce robust immune responses that can protect chickens from tumour formation [38]. Exploiting this phenomenon, a recombinant vaccine vector was constructed by inserting the gB of MDV-1 alongside the native gB of HVT with the former shown to stably process and efficiently express with the vector displaying no replicative detriment to growth [36]. Subsequent testing of the HVT vector by intramuscular vaccination of 1-day-old chicks showed the potential vaccine to be non-pathogenic and potentially protective as a strong antibody response was induced to both the HVT antigens and the gB of MDV-1 as assessed by serology. In this study, vaccinated chickens were not challenged with MDV-1, but the potential use of an HVT vector to deliver exogenous antigens of a significant viral pathogen to the avian immune system and induce a robust immune response to that may provide enhanced protection against MD was a significant step [36].

### (b) HVT Vectors Expressing Infectious Bursal Disease Virus Antigen

IBDV, a member of the *Avibirnavirus* genus within the *Birnaviridae* family, is associated with disease in chicken flocks by inducing immunosuppression, bursal destruction, impaired growth and high mortality rates. HVT vectors expressing the IBDV virion protein 2 (VP2) gene have been constructed and have been shown to elicit either partial or complete protect chickens from lethal IBDV. The disparity observed between the protective effects highlights the dependency on choosing the correct promoter to drive exogenous antigen expression in some viral vectors. Darteil et al. [39] constructed two rHVT vectors expressing the VP2 gene of the IBDV inserted into either the ribonucleotide reductase (RR) ORF of HVT under the transcriptional control of the native RR promoter (ß-gene) (rHVT001) or the glycoprotein I (gI) locus under the control of the human cytomegalovirus (CMV) immediate early promoter (pCMV) (rHVT002). rHVT002 was shown to induce 100% protection in challenged birds [39]. However, partial protection was observed upon IBVD challenge with chickens inoculated with rHVT001. The reduced protection was attributed to lower RR promoter activity of the rHVT001 vector. Furthermore, the partial protection from the development of MD following MDV-1 challenge of chickens vaccinated with either rHVT001 or rHVT002 was suggested to be a consequence of limited or no replication of the rHVT001 or rHVT002 vectors in vivo, perhaps due to deletion of the RR or gI loci [39].

### (c) HVT Polyvalent Vaccine Vectors

HVT has also been evaluated for use as a trivalent vaccine for protection against MDV and Newcastle disease virus (NDV), a vector expressing the gB and gC homologues of MDV-1 as well as the fusion (F) and the hemagglutinin-neuraminidase (HN) of NDV (rHVT-F or rHVT-HN) for *in ovo* [37] and post-hatch [40]. Heckert et al. [40] demonstrated that complete protection against NDV-1 challenge could be achieved within 2 weeks when vaccinated with recombinant HVT expressing F and HN of NDV at 1-day post-hatch. For *in ovo* vaccination, Reddy et al. [37] showed that a single inoculation of 18-day-old embryonated eggs could induce immune responses that protected chickens from challenge with very virulent MDV-1 or NDV [37]. Furthermore, the rHVT vaccine was shown to persist in inoculated birds as compared to an inactivated NDV vaccine,

which could not be detected 1 week p.i. suggesting that active persistence of the rHVT vaccine may elicit a stronger cell-mediated immune response and therefore greater protective effect [37].

A polyvalent HVT vector expressing a fusion (F) gene of NDV and gD and gI genes of ILTV has been developed to protect chickens from ILTV and NDV infection. ILTV and NDV infections can occur early post-hatch and cause considerable mortality and morbidity. However, traditional vaccines for these diseases when given in concert have significant reduction in efficacy. Therefore, the application of a safe, efficacious single polyvalent HVT-based recombinant vaccine to protect chicks against these diseases at the same time would be ideal. A single dose of rHVT-NDV-ILTV via *in ovo* vaccination at embryonic development day 18 followed by subsequent challenge with NDV, ILTV or MDV at day-1 post-hatch showed protection levels of 97%, 94% and 97%, respectively. Comparatively, a single subcutaneous vaccination with rHVT-NDV-ILTV at day-1 post-hatch followed by respective virus challenge induced high levels of protection 100%, 100% and 88%, respectively.

The studies summarised above regarding vaccination of poultry have further highlighted the importance of continuous and early induction of gene expression to elicit protective responses in vaccinated animals. The widely utilised HVT-based vaccine is an efficacious vector that is amenable for use as a recombinant multivalent vaccine vector for the protection of chickens against significant avian pathogens. Of interest is the *in ovo* application of a polyvalent HVT vector that can robust immune responses conferring excellent protection that may potentially provide complete protection from significant flock pathogens from the moment of hatch.

## 7    Pseudorabies Virus Vectors

Suid herpesvirus-1 (SuHV-1), commonly referred to as PRV or Aujeszky's disease virus, is a member of the *Varicellovirus* genus of the *Alphaherpesvirinae* subfamily and is a significant pathogen of swine. PRV has been studied extensively as a model system to delineate the biology of herpesviruses, as expertly reviewed in [41]. Due to its status as a model for herpesviruses, PRV has long been identified as a vaccine vector and has been extensively utilised for the delivery of one or more immunogenic epitopes of significant viral pathogens of swine.

(a)  PRV Vector Expressing Porcine Reproductive and Respiratory Syndrome Virus Antigens

Porcine reproductive and respiratory syndrome virus (PRRSV) is a member of the *Arterivirus* genus within the order *Nidovirales*, associated with respiratory distress and reproductive failure in swine. To limit PRRSV-induced diseases, PRV-derived vectors have been constructed that express the major membrane-associated protein GP5 (rPRV-GP5) [42] or co-express both GP5 (modified) (GP5m) and M proteins (rPRV-GP5m-M) of PRRSV [43]. Immunisation of piglets with rPRV-GP5 and subsequent challenge with PRRSV demonstrated significant reductions in clinical signs and reduced viremia after the challenge virulent PRRSV [42].

Further, studies conducted by Jiang et al. [43] utilising PRV as a bivalent vector construct (rPRV-GP5m-M) demonstrated the vector was able to induce a PRV-specific humoral response in mice that completely protected them against lethal PRV challenge. Although mice are refractory to PRRSV infection, rPRV-GP5m-M was able to induce PRRSV-specific neutralising antibodies and cell-mediated responses in the mouse model. Subsequent studies utilising the rPRV-GP5m-M construct in piglets demonstrated that the viral vector was able to induce PRRSV-specific neutralising antibodies and enhanced cell-mediated immune responses. Subsequent challenge of piglets with PRRSV suggested reductions in clinical signs, viremia and virus excretion in piglets vaccinated with the viral vector [43].

(b)  PRV Vector Expressing Porcine Circovirus Type 2 Antigens

Porcine circovirus type 2 (PCV-2) is a recently emergent (circa 1997) single-stranded DNA virus of pigs belonging to the *Circoviridae* family, strongly associated with the development of post-weaning multi-systemic wasting syndrome (PMWS) in piglets. A PRV viral vector was engineered to express the major immunogenic epitopes, ORF1 and/or ORF2 of PCV-2 [44, 45]. The rPRV vector constructed by Ju et al. [45] incorporated ORF1 and ORF2 of PCV-2 (rPRV-ORF1/2) and induced strong anti-PRV and anti-PCV2 antibodies that protected mice from lethal PRV challenge. Subsequent studies in pigs utilising rPRV-ORF1/2demonstrated a significant immune response could be induced; however, the level of protection from challenge afforded by rPRV-ORF1/2 in pigs was not assessed in this study [45]. A second PRV-derived vaccine vector expressing the PCV-2 ORF2 (rPRV-ORF2), constructed by Song et al. [44], induced significant humoral responses to PRV and PCV-2 with PCV-2-specific lymphoproliferative responses detected at day 49 [44]. While challenge studies were not conducted, the immunological responses to the vaccine were considered to have the capacity to protect vaccinated animals.

(c)  PRV Vector Expressing Classical Swine Fever Virus Antigens

CSFV is the causative agent for hog cholera, one of the most important diseases of swine worldwide [13]. CSFV is a member of the *Pestivirus* genus within the *Flaviviridae* family, and due to the economic losses associated with CSFV infection, the utility of PRV-derived vaccine vectors expressing the immunodominant epitopes of CSFV has been constructed and evaluated. PRV-derived vaccine vectors developed against CSFV incorporating the E1 gene had the capacity to completely protect swine from both PRV and CSFV challenge [46]. Peeters et al. [47], utilising a gD/gE double-deletion PRV-derived viral vector, incorporated the E2 antigen of CSFV (PRV-E2); when used to inoculate pigs, it elicited strong neutralising antibodies to CSFV E2 that completely protected pigs from challenge with CSFV. Furthermore, the challenge of vaccinated pigs virulent PRV was shown to produce no clinical signs of PRV infection and concurrent to a reduction in viral shedding.

### (d) PRV Vector Expressing Foot-and-Mouth Disease Virus Antigens

To protect animal herds from disease caused by FMDV, PRV-derived vectors have been constructed to express the VP1 capsid gene of FMDV (rPRV-VP1) or to co-express the FMDV precursor P1-2A capsid protein and the VP2 capsid protein of *porcine parvovirus* (PPV) (rPRV-P1-2A-VP2) [48].

Vaccination of pigs with the rPRV-VP1 construct showed only limited success in protecting pigs from subsequent challenge with 20 minimal infecting doses of type O FMDV [49]. It was shown that rPRV-VP1 was able to induce an antibody response to both PRV and VP1, however, with respect to the latter antigen, at a much lower level than those control animals vaccinated with a commercial vaccine preparation. This low level of antibody induction was purported to be a result of the insertion position of the VP1 gene, which was fused to the N-terminus of the gG gene of PRV, a γ-transcribed gene induced late in the replication cycle. The deficiency of the gG promoter to provide the required amount of antigen to elicit a protective immune response is analogous to that observed by the HVT constructs of Darteil et al. [39] and Tsukamoto et al. [50]. It is possible that a protective immune response could be achieved for rPRV-VP1 if the VP1 insert was under the transcriptional control of a strong α-phase, β-phase or chimeric transcription promoter.

From the summary above, it is evident that PRV is an extensively utilised virus amenable for use as a vaccine vector for the delivery of immunogenic epitopes from several significant pathogens that affect swine. These studies show that PRV-derived vaccine vectors expressing exogenous DNA sequences can induce strong neutralizing antibody responses to both vector and exogenous antigens with efficacies able to protect animals from subsequent pathogen challenge.

## 8 Concluding Remarks

Here, we have focused on those alphaherpesviruses that naturally infect, are strongly associated with economically damaging diseases in important production animals and have been investigated as potential multivalent vaccine vector candidates. From the summaries above, it is evident that alphaherpesviruses are amenable for use as vaccine vectors for the delivery of antigens from a broad range of pathogens. Increasing the efficiency of vaccination with multivalent vaccine vectors now depends on the protective immune responses to the antigens included in the vector, determining the best vaccination regimen and understanding the impact of vaccination on the animal's stage of development. For example, are calves better protected against respective pathogens when the dam is vaccinated prior to implant, pre- or postbirth or weaning stages? As improved levels of protection can be attained with the correct timing of vaccinations. In most instances, however, the respective herpesvirus vector used can induce robust immune responses that provide challenge protection in the target animal and consequentially reduce the adverse health, welfare and economic burdens in intensive animal production systems.

While there is a considerable body of literature pertaining to the successful development and application of viral vaccines using herpesviruses vectors, it is likely that we are on the cusp of a new era based on several recently emerged or emerging technologies. As next-generation sequencing technologies continue to improve and become more cost-effective, we will begin to better understand the genetics of the viruses associated with livestock diseases. This will also extend to the viral vaccine vectors we used to combat these diseases. When combined with host sequencing data (both genomic and transcriptomic), we will begin to fully understand the complex interplay between the host and the pathogen to identify new and more effective intervention strategies. At the same time, genome editing tools will provide the capacity to manipulate the genomes of herpesvirus vectors with greater precision. This capacity will not only improve our understanding of herpesvirus biology but also facilitate the development of new viral vectors with enhanced properties, such as robust transgene expression or targeting the vector to specific cell types in the host. Finally, synthetic biology and associated technologies may facilitate the construction of herpesvirus vectors which combine the most desirable attributes of well-characterised vectors into one ideal viral vaccine vector with highly malleable attributes to facilitate rapid adaptation to the disease or host interest.

# 9 Summary

Herpesvirus-based vectors are uniquely suited to development as vaccine vectors due to their characteristics such as the capacity to be propagated in a broad range of cell types and possessing a stable DNA genome that is relatively easy to manipulate. Based on studies to date, herpesvirus vectors are a promising platform for the development of veterinary vaccines. Recently emerged technologies such as next-generation sequencing will play an important role in the development of efficacious veterinary vaccines based upon the herpesvirus vector in the future.

# References

1. Pellett PE, Roizman B. The family *herpesviridae*: a brief introduction. In: Knipe DM, Howley PM, Griffin DE, Lamb RA, Martin MA, Roizman B, et al., editors. Fields virology. 2. 5th ed. Philadelphia: Lippincott Willims & Wilkins; 2007. p. 2479–99.

2. van Drunen Littel-van den Hurk S. Rationale and perspectives on the success of vaccination against bovine herpesvirus-1. Vet Microbiol. 2006;113 (3–4):275–82.

3. Potter A, Gerdts V, Littel-van den Hurk SvD. Veterinary vaccines: alternatives to antibiotics? Anim Health Res Rev. 2008;9(Special Issue 02):187–99.

4. Tang N, Zhang Y, Pedrera M, Chang P, Baigent S, Moffat K, et al. A simple and rapid approach to develop recombinant avian herpesvirus vectored vaccines using CRISPR/Cas9 system. Vaccine. 2018;36(5):716–22.

5. Brune W, Ménard C, Hobom U, Odenbreit S, Messerle M, Koszinowski UH. Rapid identification of essential and nonessential herpesvirus genes by direct transposon mutagenesis. Nat Biotechnol. 1999;17:360–4.

6. Mahony TJ, McCarthy FM, Gravel JL, West L, Young PL. Construction and manipulation of an infectious clone of the Bovine herpesvirus 1 genome maintained as a bacterial artificial chromosome. J Virol. 2002;76 (13):6660–8.

7. Robinson KE, Meers J, Gravel JL, McCarthy FM, Mahony TJ. The essential and non-essential genes of Bovine herpesvirus 1. J Gen Virol. 2008;89 (11):2851–63.

8. Roizman B. The function of herpes simplex virus genes: a primer for genetic engineering of novel vectors. Proc Natl Acad Sci U S A. 1996;93 (21):11307–12.

9. Schmitt J, Becher P, Thiel H, Keil GM. Expression of Bovine viral diarrhoea virus glycoprotein E2 by bovine herpesvirus-1 from a synthetic ORF and incorporation of E2 into recombinant virions. J Gen Virol. 1999;80(11):2839–48.

10. Lingshu W, Whitbeck JC, William CL, Denys VV, Leonard JB. Expression of the genomic form of the Bovine viral diarrhea virus E2 ORF in a Bovine herpesvirus-1 vector. Virus Genes. 2003;27(1):83–91.

11. Rosas CT, Konig P, Beer M, Dubovi EJ, Tischer BK, Osterrieder N. Evaluation of the vaccine potential of an Equine herpesvirus type 1 vector expressing Bovine viral diarrhea virus structural proteins. J Gen Virol. 2007;88(3):748–57.

12. Kühnle G, Heinze A, Schmitt J, Giesow K, Taylor G, Morrison I, et al. The class II membrane glycoprotein G of Bovine respiratory syncytial virus, expressed from a synthetic open reading frame, is incorporated into virions of recombinant Bovine herpesvirus 1. J Virol. 1998;72(5):3804–11.

13. Fenner FJ, Gibbs EPJ, Murphy FA, Rott R, Studdert MJ, White DO. Veterinary virology. 2nd ed. Sydney: Academic Press; 1997.

14. McGeoch DJ, Dalrymple MA, Davison AJ, Dolan A, Frame MC, McNab D, et al. The complete DNA sequence of the long unique region in the genome of Herpes simplex virus type 1. J Gen Virol. 1988;69 (7):1531–74.

15. Knopf C, Zavidij O, Rezuchova I, Rajčáni J. Evaluation of the T-REx™ transcription switch for

conditional expression and regulation of HSV-1 vectors. Virus Genes. 2008;36(1):55–66.

16. Souza APD, Haut L, Reyes-Sandoval A, Pinto AR. Recombinant viruses as vaccines against viral diseases. Braz J Med Biol Res. 2005;38(4):509–22.

17. Mahony TJ. Understanding the molecular basis of disease is crucial to improving the design and construction of herpesviral vectors for veterinary vaccines. Vaccine. 2015;33(44):5897–904.

18. Lambeth LS, Zhao Y, Smith LP, Kgosana L, Nair V. Targeting Marek's disease virus by RNA interference delivered from a herpesvirus vaccine. Vaccine. 2009;27(2):298–306.

19. Baigent SJ, Smith LP, Nair VK, Currie RJW. Vaccinal control of Marek's disease: current challenges, and future strategies to maximize protection. Vet Immunol Immunopathol. 2006;112(1–2):78–86.

20. Robinson KE, Mahony TJ. Herpesvirus mutagenesis facilitated by infectious bacterial artificial chromosomes (iBACs). Methods Mol Biol. 2015;1227:181–97.

21. Bello LJ, Whitbeck JC, Lawrence WC. Bovine herpesvirus 1 as a live virus vector for expression of foreign genes. Virology. 1992;190(2):666–73.

22. Kit S, Qavi H. Thymidine kinase (TK) induction after infection of TK-deficient rabbit cell mutants with Bovine herpesvirus type 1 (BHV-1): isolation of TK-BHV-1 mutants. Virology. 1983;130(2):381–9.

23. Rodrigues R, Cuddington B, Mossman K. Bovine herpesvirus type 1 as a novel oncolytic virus. Cancer Gene Ther. 2009;17:1–12.

24. Taylor G, Rijsewijk FA, Thomas LH, Wyld SG, Gaddum RM, Cook RS, et al. Resistance to Bovine respiratory syncytial virus (BRSV) induced in calves by a recombinant Bovine herpesvirus-1 expressing the attachment glycoprotein of BRSV. J Gen Virol. 1998;79(7):1759–67.

25. Schrijver RS, Langedijk JPM, Keil GM, Middel WGJ, Maris-Veldhuis M, Van Oirschot JT, et al. Immunization of cattle with a BHV1 vector vaccine or a DNA vaccine both coding for the G protein of BRSV. Vaccine. 1997;15(17–18):1908–16.

26. Kweon C-H, Kang S-W, Choi E-J, Kang Y-B. Bovine herpes virus expressing envelope protein (E2) of Bovine viral diarrhea virus as a vaccine candidate. J Vet Med Sci. 1999;61(4):395–401.

27. Kit S, Kit M, DiMarchi RD, Little SP, Gale C. Modified-live infectious bovine rhinotracheitis virus vaccine expressing monomer and dimer forms of foot-and-mouth disease capsid protein epitopes on surface of hybrid virus particles. Arch Virol. 1991;120(1–2):1–17.

28. Telford EAR, Watson MS, McBride K, Davison AJ. The DNA sequence of Equine herpesvirus-1. Virology. 1992;189(1):304–16.

29. Trapp S, von Einem J, Hofmann H, Kostler J, Wild J, Wagner R, et al. Potential of Equine herpesvirus 1 as a vector for immunization. J Virol. 2005;79(9):5445–54.

30. Gyure KAMD. West Nile virus infections. [review]. J Neuropathol Exp Neurol. 2009;68(10):1053–60.

31. Rosas CT, Tischer BK, Perkins GA, Wagner B, Goodman LB, Osterrieder N. Live-attenuated recombinant equine herpesvirus type 1 (EHV-1) induces a neutralizing antibody response against West Nile virus (WNV). Virus Res. 2007;125(1):69–78.

32. Said A, Elmanzalawy M, Ma G, Damiani AM, Osterrieder N. An equine herpesvirus type 1 (EHV-1) vector expressing Rift Valley fever virus (RVFV) Gn and Gc induces neutralizing antibodies in sheep. Virol J. 2017;14(1):154.

33. Van de Walle GR, May MA, Peters ST, Metzger SM, Rosas CT, Osterrieder N. A vectored equine herpesvirus type 1 (EHV-1) vaccine elicits protective immune responses against EHV-1 and H3N8 Equine influenza virus. Vaccine. 2010;28(4):1048–55.

34. Rosas C, Van de Walle GR, Metzger SM, Hoelzer K, Dubovi EJ, Kim SG, et al. Evaluation of a vectored equine herpesvirus type 1 (EHV-1) vaccine expressing H3 haemagglutinin in the protection of dogs against canine influenza. Vaccine. 2008;26(19):2335–43.

35. Crawford PC, Dubovi EJ, Castleman WL, Stephenson I, Gibbs EPJ, Chen L, et al. Transmission of Equine influenza virus to dogs. Science. 2005;310(5747):482–5.

36. Ross LJN, Binns MM, Tyers P, Pastorek J, Zelnik V, Scott S. Construction and properties of a Turkey herpesvirus recombinant expressing the Marek's disease virus homologue of glycoprotein B of herpes simplex virus. J Gen Virol. 1993;74(3):371–7.

37. Reddy SK, Sharma JM, Ahmad J, Reddy DN, McMillen JK, Cook SM, et al. Protective efficacy of a recombinant herpesvirus of turkeys as an in ovo vaccine against Newcastle and Marek's diseases in specific-pathogen-free chickens. Vaccine. 1996;14(6):469–77.

38. Nazerian K, Lee LF, Yanagida N, Ogawa R. Protection against Marek's disease by a fowlpox virus recombinant expressing the glycoprotein B of Marek's disease virus. J Virol. 1992;66(3):1409–13.

39. Darteil R, Bublot M, Laplace E, Bouquet J-F, Audonnet J-C, Riviere M. Herpesvirus of Turkey recombinant viruses expressing infectious bursal disease virus (IBDV) VP2 immunogen induce protection against an IBDV virulent challenge in chickens. Virology. 1995;211(2):481–90.

40. Heckert RA, Riva J, Cook S, McMillen J, Schwartz RD. Onset of protective immunity in chicks after vaccination with a recombinant Herpesvirus of Turkeys vaccine expressing Newcastle Disease Virus fusion and hemagglutinin-neuraminidase antigens. Avian Dis. 1996;40(4):770–7.

41. Nauwynck H, Glorieux S, Favoreel H, Pensaert M. Cell biological and molecular characteristics of pseudorabies virus infections in cell cultures and in pigs with emphasis on the respiratory tract. Vet Res. 2007;38:229–41.

42. Qiu H-J, Tian Z-J, Tong G-Z, Zhou Y-J, Ni J-Q, Luo Y-Z, et al. Protective immunity induced by a recombinant pseudorabies virus expressing the GP5 of porcine reproductive and respiratory syndrome virus in piglets. Vet Immunol Immunopathol. 2005;106(3–4):309–19.

43. Jiang Y, Fang L, Xiao S, Zhang H, Pan Y, Luo R, et al. Immunogenicity and protective efficacy of recombinant pseudorabies virus expressing the two major membrane-associated proteins of porcine reproductive and respiratory syndrome virus. Vaccine. 2007;25 (3):547–60.

44. Song Y, Jin M, Zhang S, Xu X, Xiao S, Cao S, et al. Generation and immunogenicity of a recombinant Pseudorabies virus expressing cap protein of Porcine circovirus type 2. Vet Microbiol. 2007;119 (2–4):97–107.

45. Ju C, Fan H, Tan Y, Liu Z, Xi X, Cao S, et al. Immunogenicity of a recombinant Pseudorabies virus expressing ORF1-ORF2 fusion protein of Porcine circovirus type 2. Vet Microbiol. 2005;109 (3–4):179–90.

46. Hofft van Iddekinge BJL, de Wind N, Wensvoort G, Kimman TG, Gielkens ALJ, Moormann RJM. Comparison of the protective efficacy of recombinant Pseudorabies viruses against pseudorabies and classical swine fever in pigs; influence of different promoters on gene expression and on protection. Vaccine. 1996;14(1):6–12.

47. Peeters B, Bienkowska-Szewczyk K, Hulst M, Gielkens A, Kimman T. Biologically safe, non-transmissible pseudorabies virus vector vaccine protects pigs against both Aujeszky's disease and classical swine fever. J Gen Virol. 1997;78(12):3311–5.

48. Hong Q, Qian P, Li X-M, Yu X-L, Chen H-C. A recombinant pseudorabies virus co-expressing capsid proteins precursor P1-2A of FMDV and VP2 protein of porcine parvovirus: a trivalent vaccine candidate. Biotechnol Lett. 2007;29(11):1677–83.

49. Qian P, Li X-M, Jin M-L, Peng G-Q, Chen H-C. An approach to a FMD vaccine based on genetic engineered attenuated pseudorabies virus: one experiment using VP1 gene alone generates an antibody responds on FMD and pseudorabies in swine. Vaccine. 2004;22(17–18):2129–36.

50. Tsukamoto K, Saito S, Saeki S, Sato T, Tanimura N, Isobe T, et al. Complete, long-lasting protection against lethal infectious bursal disease virus challenge by a single vaccination with an Avian herpesvirus vector expressing VP2 antigens. J Virol. 2002;76(11):5637–45.

# Part III

# RNA Virus Vectors

# Paramyxoviruses as Vaccine Vectors

Siba K. Samal

## Abstract

Paramyxoviruses are a group of enveloped, non-segmented, negative-sense RNA viruses that include some of the major human, animal, and avian pathogens. These viruses cause a variety of diseases, ranging in severity from mild respiratory infections to encephalitis and death. The development of methods to recover recombinant paramyxoviruses from cDNA, known as reverse genetics systems, has made it possible not only to study the lifecycle of these viruses but also to use these viruses as vaccine vectors. Paramyxoviruses stably express a wide variety of heterologous antigens at relatively high levels in many species of animals including nonhuman primates, thus making them potential vaccine vectors. Paramyxovirus vectors have several advantages over other viral vectors. It is most likely that paramyxovirus vectors will play an important role in future human and veterinary vaccine developments. This chapter describes our current knowledge of paramyxovirus biology, genome replication, reverse genetics, construction of vaccine vectors, advantages and limitations of paramyxovirus vectors, and specific recombinant paramyxoviruses used for veterinary vaccine development.

## Keywords

Paramyxovirus · Reverse genetics · Vaccine · Vector vaccine · Paramyxovirus vector · Vaccine development · Poultry vaccine · Veterinary vaccine · Recombinant vaccine

**Learning Objectives**

After reading, this chapter you should be able to:

- Describe the basic mechanisms of paramyxovirus replication
- Explain the reverse genetics method used to generate recombinant paramyxoviruses from cloned cDNAs
- Recognize the characteristics of paramyxovirus vectors which are suitable for veterinary vaccine development
- Explain the strategy used to construct a paramyxovirus vaccine vector
- List important animal diseases for which paramyxovirus vaccine vectors have already been evaluated
- Describe some of the limitations of using paramyxoviruses as vaccine vectors

S. K. Samal (✉)
Virginia Maryland College of Veterinary Medicine, University of Maryland, College Park, MD, USA
e-mail: ssamal@umd.edu

# 1 Introduction

Paramyxoviruses are a family of enveloped viruses with non-segmented, single-stranded, and negative-sense RNA genomes. Members of

T. Vanniasinkam et al. (eds.), *Viral Vectors in Veterinary Vaccine Development*,
https://doi.org/10.1007/978-3-030-51927-8_8

this family include apathogenic to highly pathogenic viruses affecting the human population, as well as animals and birds. These viruses are predominantly responsible for respiratory diseases and are usually transmitted via airborne droplets. Some of the paramyxoviruses that cause human diseases include measles, mumps, and human parainfluenza viruses. There are also a number of paramyxoviruses that infect animals and cause notable diseases, such as rinderpest, Newcastle disease, canine distemper, peste des petits ruminants, and shipping fever in cattle. In addition, Nipah and Hendra viruses are two highly lethal zoonotic paramyxoviruses isolated in modern times.

Availability of reverse genetics systems for paramyxoviruses has allowed us to not only understand basic molecular biology and pathogenesis of these viruses but also use paramyxoviruses as vectors for vaccine development. Paramyxovirus vector vaccines represent an attractive strategy to overcome some of the limitations of conventional vaccines. These vaccines are already available in the veterinary vaccine market, and research to develop more improved paramyxovirus vector vaccines against a number of animal diseases is in progress around the world. Undoubtedly, paramyxovirus vector vaccines will play a major role in future veterinary vaccine regimens.

## 2    History

The first paramyxovirus was discovered in 1926 from outbreaks of severe respiratory and neuronal disease of chickens in Java, Indonesia [66], and in Newcastle-upon-Tyne, England [25]. The virus was named Newcastle disease virus (NDV). In 1953, another paramyxovirus was isolated in Sendai, Japan, from a mouse inoculated with an autopsy specimen from an infant with fatal pneumonia [69]. The virus was named Sendai virus (SeV) and was later identified as a murine virus, which is not pathogenic to humans. Between 1956 and 1960, four serotypes of human parainfluenza virus (HPIV-1 to HPIV-4) were recovered from children with lower respiratory

tract illnesses [13, 14]. In 1959, the name parainfluenza was first used to describe HPIV1–3 and SeV because some of the disease signs are similar to influenza. Bovine parainfluenza virus type 3 (BPIV-3), a close relative of HPIV-3, was isolated in 1959 from cattle with shipping fever [100]. PIV-5, previously known as Simian virus 5 (SV5), was first isolated in 1954 as a contaminant from primary monkey kidney cells [46]. PIV-5 was found to be the cause of Kennel cough in dogs. The measles virus was first isolated in 1954. Hendra and Nipah viruses were the first highly lethal zoonotic paramyxoviruses discovered in 1994 and 1998 from horses and pigs, respectively [115].

## 3    Host Range and Virus Propagation

Paramyxoviruses have a broad host range. They have been isolated from many different vertebrate hosts including mice, rats, bats, dogs, cattle, pigs, horses, reptiles, dolphins, seals, birds, and humans. Most paramyxoviruses are capable of infecting a wide variety of vertebrate hosts, but in very rare cases, they may cross species barrier and infect other species including humans. For example, NDV can infect and cause mild, transient conjunctivitis or flu-like symptoms in humans [89]. All morbilliviruses, except canine distemper virus (CDV), are highly host-specific. CDV has a broad host range, with infections being observed in seals, lions, and monkeys. Among members of the family *Paramyxoviridae*, henipaviruses exhibit a remarkable broad host range, with natural infections in several species of animals and humans. Typically, paramyxoviruses are attenuated in nonnatural hosts. Most paramyxoviruses grow well in many different primary and established cell lines. Some paramyxoviruses do not readily grow in cell culture and require adaptation by several passages in cell culture. Cell cultures derived from homologous species are generally used for the cultivation of paramyxoviruses. However, several paramyxoviruses grow well in cells of different host origin. For example, avian paramyxoviruses grow well in embryonated chicken eggs or cells derived from avian and non-avian species. Avirulent NDV strains require the addition of a protease, such as trypsin, α-chymotrypsin, or allantoic fluid (as a source of secreted protease) to the medium for growth in cell culture. This is necessary for cleavage activation of the viral fusion (F) protein. The characteristic cytopathic effect of paramyxovirus infection is the formation of syncytia (multinucleated giant cells).

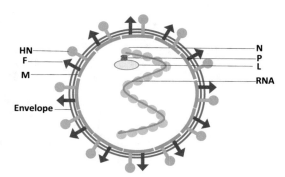

**Fig. 1** Schematic diagram of a paramyxovirus particle. The lipid bilayer is shown as an envelope, and the matrix protein is located beneath the envelope. The haemagglutinin-neuraminidase (HN) and the fusion (F) glycoproteins are inserted through the viral envelope. Inside the virus particle lies the nucleocapsid, which is made up of negative-sense RNA encapsidated with nucleocapsid protein (N) and associated with the phosphoprotein (P) and the large polymerase protein (L)

# 4    Virus Classification

All paramyxoviruses belong to the family *Paramyxoviridae*. The family *Paramyxoviridae* is one of the 11 virus families in the order *Mononegavirales*. This order comprises viruses with a non-segmented, linear, single-stranded negative-sense RNA genome [47]. The similarities in structure and in transcription and replication mechanisms of RNA genomes suggest that all members of this order may have evolved from an ancestral virus. The family *Paramyxoviridae* has four subfamilies: *Avulavirinae, Metaparamyxovirinae, Orthoparamyxovirinae*, and *Rubulavirinae* [47]. The subfamily *Avulavirinae* has three genera: *Metaavulavirus, Orthoavulavirus*, and *Paraavulavirus*. The subfamily *Metaparamyxovirinae* has one genus: *Synodonvirus*. The subfamily *Orthoparamyxovirinae* has eight genera: *Aquaparamyxovirus, Ferlavirus, Henipavirus, Jeilongvirus, Morbillivirus, Narmovirus, Respirovirus*, and *Salemvirus*. All avian paramyxoviruses (APMVs) are classified in the subfamily *Avulavirinae*.

# 5    Virus Structure

Virions are enveloped, 150–500 nm in diameter, spherical, or pleomorphic (Fig. 1). Occasionally, filamentous forms of variable length are present. The envelope contains usually two transmembrane glycoproteins, an attachment protein (HN, H, or G), and a fusion (F) protein [71]. These proteins are present as homo-oligomers and form densely packed spike-like projections on the outer surface of the virion. Some paramyxoviruses (PIV-5, MuV, and APMV-6) have a third, small transmembrane protein, known as small hydrophobic (SH) protein. Beneath the envelope lies a non-glycosylated matrix (M) protein. Inside the viral envelope lies the nucleocapsid. In the nucleocapsid, the viral genome is tightly bound along the entire length with the nucleocapsid (N) protein. The viral nucleocapsid is associated with the phosphoprotein (P) and the large polymerase protein (L). The nucleocapsid with its associated P and L proteins has RNA-dependent RNA polymerase activity and is the minimum unit necessary to accomplish an infectious paramyxovirus life cycle. Although multiploidy virions are found, most virions contain a single functional genome.

# 6    Viral RNAs

The paramyxovirus genome is a non-segmented negative-sense RNA, 14–19 kilobases in length and contains 6–10 genes encoding up to 12 different proteins (Fig. 2). At the 3′ and 5′ ends of the genome are short extragenic (noncoding) regions

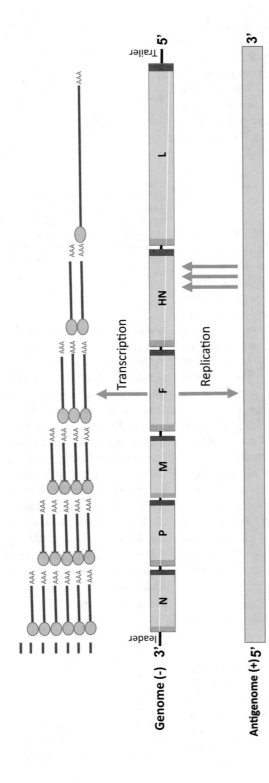

**Fig. 2** Schematic diagram showing a typical paramyxovirus genome organization. The genes are arranged in the genome as a light orange rectangle with gene start and gene end signals indicated in green and red rectangles, respectively. The leader and trailer regions at the 3′ and 5′ ends of the genome (−), respectively, are indicated. The transcription products are shown above the genome (−). The mRNAs are capped and polyadenylated, but not the leader RNA. The antigenome (+) synthesized as a result of RNA replication is shown below the genome (−)

known as the "leader" and "trailer," respectively. The leader is approximately 55 nucleotides (nt) in length, whereas the length of the trailer is variable. The leader region contains a single genomic promoter that is involved in the synthesis of the mRNAs as well as a complete positive-sense replicative intermediate called the antigenome. The trailer region contains the promoter involved in the synthesis of a complete negative-sense genomic RNA. The genes are arranged in the order N, P, M, F, HN, and L between the leader and trailer. At the beginning and at the end of each gene are 10–13 nucleotide conserved transcriptional control signals involved in initiation and termination/polyadenylation of the mRNAs. These conserved sequences are known as gene-start (GS) and gene-end (GE) sequences. The genes are separated by short intergenic sequences (IGS) that are not copied into mRNAs. Each gene encodes one protein except the P gene which encodes one or more accessory proteins using RNA editing [71]. PIV-5, MuV, and APMV-6 each contain a seventh small gene called SH. The genomic and antigenomic RNAs are always tightly bound along their entire length with the N protein at a ratio of one protein molecule per six nucleotides. The N-RNA protects the RNA from degradation by nucleases. Therefore, a genome whose length is not an even multiple of six nucleotides will not be precisely encapsidated by the N protein and will not be protected from nucleases. Consequently, all paramyxovirus genome lengths are even multiples of six, thus meeting the requirement for efficient RNA replication, which is termed as the "rule of six" [11].

# 7    Transcription and RNA Replication

Transcription and RNA replication take place in the cytoplasm of the host cell and are performed by the viral RNA-dependent RNA polymerase (RdRp), which is composed of L and P proteins. The genomic and antigenomic RNAs encapsidated by the N protein form the ribonucleoprotein (RNP) core. Only such RNP, but not free RNA, is suitable as a template for viral RdRp.

Transcription begins at the 3'-leader region. The genes are transcribed sequentially in their 3'–5' order by responding to their GS and GE signals on the genome. At a GS signal, the viral polymerase begins mRNA synthesis, adds a methylated cap to the mRNA, and when it reaches the GE signal polyadenylates and releases the mRNA. The polymerase remains attached to the viral RNA and scans to locate the next GS signal, where it reinitiates RNA synthesis (Fig. 2). Occasionally, the viral polymerase dissociates from the template leading to a gradient of transcription. Therefore, the genes at the 3' end of the genome are synthesized in large amounts than the genes present at the 5' end of the genome. This is a mechanism that paramyxoviruses and all other members of *Mononegavirales* use to control the level and type of viral RNA synthesis. After primary transcription of viral mRNAs and accumulation of sufficient viral proteins, the (−) sense genome is replicated to generate full length (+) sense antigenomes. In this case, the polymerase is highly processive and disregards all the transcription signals. The antigenomes in turn act as templates for the polymerase to synthesize (−) sense genomic RNAs, which are used for secondary transcription and for the formation of progeny virions. It is believed that the promoter in the antigenome is stronger than the promoter in the genome, which leads to the synthesis of more genomic RNAs. The genomic and antigenomic RNAs differ from mRNAs by being encapsidated with N protein as they are synthesized.

# 8    Viral Proteins

Paramyxoviruses encode eight to twelve viral proteins depending on the virus. However, there are six proteins common to all paramyxoviruses that are essential for virus replication. These include two surface glycoproteins: F and HN, H, or G. The concerted effort of both the glycoproteins is required for entry of paramyxoviruses into the host cell. The F protein mediates penetration of the host cell by inducing fusion between the viral envelope and the host cell plasma membrane. It also causes the fusion of infected and uninfected cells resulting in syncytia formation. The F protein is synthesized as an inactive precursor ($F_0$), which is activated following cleavage by a cellular protease to yield two disulfide-linked F1 and F2 subunits. Cleavage of $F_0$ is a prerequisite for paramyxovirus infectivity and pathogenesis. The attachment protein (HN, H, or G) is responsible for adsorption of the virus to the host cell by binding to a cell surface receptor. Fusion of the viral envelope with the host cell membrane then occurs by a process that is driven by large conformational changes in the F protein [71].

The M protein is the most abundant protein in the virion and plays a coordinating role in virus assembly and budding. The N protein binds tightly to the genomic and antigenomic RNAs to form an RNase-resistant helical nucleocapsid. The P protein is heavily phosphorylated and acts as a non-catalytic subunit of the viral RNA polymerase. The P protein plays an important role in tethering the L protein to the genomic template during viral RNA synthesis. The L protein is a large multifunctional protein responsible for RNA synthesis, mRNA capping, and methylation. The P gene, depending on the virus, encodes one or more accessory proteins, namely, C, V, W, I, and D through RNA editing [71]. The functions of the SH protein are not well understood.

## 9    Immune Response During Natural Infection

Paramyxovirus infections in humans and animals induce potent local and systemic humoral and cellular immune responses. The two surface glycoproteins are involved in the generation of immunity to infection. Antibodies specific to F and HN proteins have virus neutralizing activities [50]. However, the F protein induces higher levels of neutralizing antibodies compared to the HN protein [111]. Anti-HN antibodies, which block virus attachment, are effective only when there is little cell fusion and the virus spreads extracellularly, whereas, anti-F antibodies are capable of completely preventing the spread of virus in cells undergoing cell fusion, as well as

nonfusing cells [83]. Studies have demonstrated the ability of serum neutralizing antibodies to provide resistance to virus replication [83]. The local nasal secretory IgA antibodies play an important role in neutralizing virus infectivity in the respiratory tract. Although the levels of antibodies correlate well with protection, specific T cell-induced immunity also plays a critical role in the clearance of paramyxovirus infection [51].

## 10    Reverse Genetics

Reverse genetics is a method used to determine the function of a viral gene or a specific gene sequence by mutating the sequence in vitro and finding the changes at the phenotypic level. Among animal viruses, genome manipulations were first performed in DNA viruses followed next by positive-sense RNA viruses. Manipulation of the genome of negative-sense RNA viruses was challenging due to several reasons. The most notable challenge was that the naked genomic RNA of non-segmented negative-sense RNA viruses (*Mononegavirales*) is not infectious, unless it is encapsidated by the N proteins and associated with viral RdRp (P-L). Therefore, the minimum viral unit to initiate an infectious cycle is not only the genomic RNA but also the genomic RNA plus the viral N, P, and L proteins, which must be available inside the cell to start the first round of virus-specific mRNA synthesis. The first reverse genetics system for a non-segmented negative-sense RNA was achieved in 1994 for rabies virus [103]. In this study, a plasmid was constructed that expressed a full-length rabies virus RNA transcript from a T7 RNA polymerase promoter. The plasmid DNA containing the full-length copy of the viral genome was transfected into cells infected with a recombinant vaccinia virus expressing the T7 RNA polymerase. Three other plasmids expressing the rabies virus N, P, and L proteins were also co-transfected into these cells. This system was based on the use of anti-genomic (positive-sense) rather than the genomic (negative-sense) RNA because of concern that simultaneous          expression          of          naked negative-sense genomic RNA and positive-sense mRNAs would result in hybridization and generation of double-stranded RNA, which will induce interferon and may not serve as a template for transcription and replication. This approach has been applied to the development of reverse genetics systems for all families of *Mononegavirales*.

The development of reverse genetics systems for non-segmented negative-sense RNA viruses has improved our understanding of the structure-function relationships of viral components and their contributions to the pathogenicity of these viruses. It has also made it possible to engineer rational live attenuated vaccines against some of these viral agents. In addition, it has enabled us to engineer negative-sense RNA viruses as vectors to express foreign proteins. These novel vectors have potential applications in vaccine development, gene therapy, and cancer treatment.

The first paramyxovirus reverse genetics systems were developed in parallel in 1995 for measles virus [98] and for Sendai virus [30]. Since then reverse genetics systems have been developed for many paramyxoviruses representing all major genera (Table 1).

## 11 Construction Strategies of Paramyxovirus Reverse Genetics Systems

Development of a reverse genetics system for a paramyxovirus basically consists of artificially creating a viral replication cycle inside a cell by transfecting four plasmids, one plasmid containing a cDNA encoding the full-length antigenomic RNA of the virus and three plasmids containing cDNAs encoding the minimal viral proteins (N, P, and L) necessary for viral RNA transcription and replication (Fig. 3a). Construction of a reverse genetics system first requires determination of the complete genome sequence of the virus strain to be rescued. Once the complete sequence of the genome, including the exact 3′ and 5′ ends, is known, the next step is to assemble a full-length cDNA copy of the antigenome.

## 11.1 Assembly of a Full-Length cDNA Clone Encoding the Antigenome of the Virus

Several strategies are used to assemble a full-length cDNA clone of the antigenome. Among these strategies, the sequential cloning based on naturally occurring or artificially created restriction enzyme (RE) sites in the genome is the most widely used method for the rescue of paramyxoviruses. This strategy requires the use of a sequence analysis software (e.g., DNASTAR, Clone Manager) to analyze the complete genome sequence of the virus strain and identify the naturally occurring unique RE sites. In cases where the convenient unique RE sites are not available, the nucleotide sequences (preferably in the 3′ noncoding region of a gene) are modified to create unique RE sites. The total length of the cDNA is adjusted to maintain the "rule of six." The cDNA fragments are amplified from the genomic RNA by RT-PCR using specific primers and a high-fidelity polymerase. The cDNA fragments are then sequentially cloned into the multiple cloning site of a transcription vector. The transcription vector is modified to include a linker containing the RE sites necessary for assembly of the full-length antigenomic cDNA. Although this sequential cloning method is widely used, it is extremely time-consuming and requires multiple cloning steps.

Since the sequential cloning method is time-consuming, ligation-independent cloning (LIC) systems that do not require multiple cloning steps have been developed. The LIC systems have been used to assemble the antigenome of NDV using an In-Fusion® PCR (Clonetech, USA) system [40, 75]. This system is rapid and reliable compared to the traditional assembly systems using the RE sites.

New advances in chemical synthesis of long nucleic acid fragments have also made it possible to synthesize a full-length cDNA clone of the antigenome with additional RE sites to facilitate cloning into the transcription vector.

Regardless of which method is used to assemble a full-length cDNA, the sequence of the entire

**Table 1** Reverse genetics systems currently available for *Paramyxoviridae*. References for the first published reverse genetics system of the respective species

| Genus | Species | Virus (abbreviation) | Reverse genetics established | References |
|---|---|---|---|---|
| *Orthoavulavirus* | Avian avulavirus 1 | Avian paramyxovirus 1 (APMV-1) | Romer-Oberdorfer et al. (1999) | [101] |
| | Avian avulavirus 1 | Newcastle disease virus (NDV) | Krishnamurthy et al. (2000) | [67] |
| *Metaavulavirus* | Avian avulavirus 2 | Avian paramyxovirus 2 (APMV-2) | Subbiah et al. (2011) | [113] |
| | Avian avulavirus 6 | Avian paramyxovirus 6 (APMV-6) | Tsunekuni et al. (2017) | [118] |
| | Avian avulavirus 7 | Avian paramyxovirus 7 (APMV-7) | Xiao et al. (2012) | [128] |
| | Avian avulavirus 10 | Avian paramyxovirus 10 (APMV-10) | Tsunekuni et al. (2017) | [118] |
| *Paraavulavirus* | Avian avulavirus 3 | Avian paramyxovirus 3 (APMV-3) | Kumar et al. (2011) | [68] |
| | Avian avulavirus 4 | Avian paramyxovirus 4 (APMV-4) | Kim et al. (2013) | [56] |
| *Henipavirus* | Cedar henipavirus | Cedar virus (CedV) | Laing et al. (2018) | [70] |
| | Hendra henipavirus | Hendra virus (HeV) | Marsh et al. (2013) | [80] |
| | Nipah henipavirus | Nipah virus (NiV) | Yoneda et al. (2006) | [131] |
| *Morbillivirus* | Canine morbillivirus | Canine distemper virus (CDV) | Gassen et al. (2000) | [31] |
| | Measles morbillivirus | Measles virus (MeV) | Radecke et al. (1995) | [98] |
| | Small ruminant morbillivirus | Peste des petits ruminants virus (PPRV) | Hu et al. (2012) | [42] |
| | Rinderpest morbillivirus | Rinderpest virus (RPV) | Baron and Barrett (1997) | [3] |
| *Respirovirus* | Bovine respirovirus 3 | Bovine parainfluenza virus 3 (BPIV-3) | Schmidt et al. (2000) | [102] |
| | Human respirovirus 1 | Human parainfluenza virus 1 (HPIV-1) | Newman et al. (2002) | [90] |
| | Human respirovirus 3 | Human parainfluenza virus 3 (HPIV-3) | Hoffman and Banerjee (1997) | [38] |
| | Murine respirovirus | Sendai virus (SeV) | Garcin et al. (1995) | [30] |
| | Canine rubulavirus | Parainfluenza virus 5 (PIV-5) | He et al. (1997) | [37] |
| | Human rubulavirus 2 | Human parainfluenza virus 2 (HPIV-2) | Kawano et al. (2001) | [52] |
| | Mumps rubulavirus | Mumps virus (MuV) | Clarke et al. (2000) | [20] |
| | Sosuga rubulavirus | Sosuga virus | Welch et al. (2018) | [125] |
| *Narmovirus* | Tupaia narmovirus | Tupaia paramyxovirus (TPMV) | Engelond et al. (2017) | [27] |
| *Jeilongvirus* | J paramyxovirus | J paramyxovirus (JPV) | Abraham et al. (2018) | [1] |

full-length cDNA clone must be determined and compared with the sequence of the parent virus to make sure that there is no error in the sequence of the full-length cDNA. The full-length cDNA is then cloned into a plasmid vector. In theory, most cloning plasmids can carry a DNA insert of at least 15 kb in size. Although low, medium, and high copy numbers of plasmids can be used as vectors to clone a full-length cDNA insert, low-copy-number plasmids are preferred because of their ability to prevent spurious transcription of foreign DNA sequences from upstream plasmid promoters. The *Escherichia coli* cells are most commonly used for propagation of plasmids. An

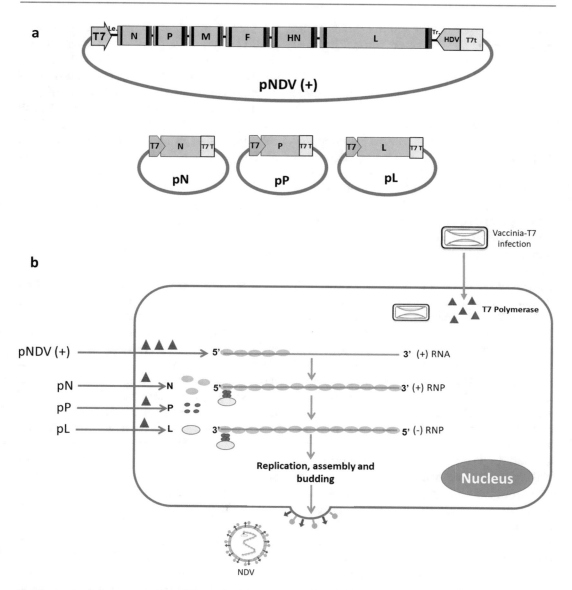

**Fig. 3** A schematic representation of the method used to rescue Newcastle disease virus (NDV) from cDNAs. (**a**) The plasmid containing the full-length cDNA of NDV antigenome and three plasmids containing the ORFs of nucleocapsid protein (N), phosphoprotein (P), and large polymerase protein (L) are constructed. (**b**) To generate recombinant NDV from cDNAs, antigenome sense (+) full-length RNAs are synthesized by T7 RNA polymerase in cDNA transfected cells. Coexpression of viral N, P, and L proteins in the same cell results in the production of (+) sense ribonucleoproteins (RNPs). Subsequent transcription to (−) sense RNPs, followed by the expression of the viral proteins from the genome leads to the production of infectious NDVs

*E. coli* strain that can propagate large fragments of DNA, does not possess specified restriction-modification systems, or grows very fast (e.g., DH10β, TOP10) is used.

## 11.2   Transcription of Infectious Virus

In order to transcribe viral RNA from cDNA inside the cell, the viral sequence needs to be cloned next to a suitable promoter, and the specific RNA polymerase enzyme needs to be present for binding to the promoter. The most commonly used promoters for paramyxovirus infectious clones are the bacteriophage promoter T7 and, to a lesser extent, the eukaryotic promoter of the human cytomegalovirus (referred to as CMV promoter). Each promoter has its own advantages and disadvantages. The bacteriophage T7 promoter is highly efficient in transcription but requires the presence of T7 polymerase inside the cell. On the other hand, the CMV promoter can produce viral RNA after direct transfection of full-length cDNA into eukaryotic cells. Although CMV promoter confers some advantages over T7 promoter, it is transcriptionally active in *E. coli* leading to the problems of the stability and the toxicity of the plasmid in the bacteria, while T7 promoter is not transcriptionally active in *E. coli*.

The T7 polymerase expression system using poxvirus has been the most frequently used method in virus rescue systems. The recombinant vaccinia virus vTF7-3 was used to rescue rabies virus in 1994 [103]. Although vTF-7-3 expresses a high level of T7 polymerase, the virus produces a high degree of cytopathic effect, and elimination of the helper virus from the recovered virus stock is time-consuming. Therefore, a highly attenuated, host-range-restricted modified vaccinia virus strain Ankara (MVA/T7) has been used to express T7 polymerase [126]. MVA/T7 can replicate in avian cells but cannot multiply in most mammalian cells. Although MVA/T7 is widely used to rescue paramyxoviruses, it is not ideal for vaccine development for several reasons. First, this method requires a stringent step to separate the recovered paramyxovirus from the T7 expressing helper virus. Secondly, poxvirus can increase the chance of recombination between transfected plasmids. Thirdly, poxvirus encodes a large number of modifying enzymes, which can affect the genes of the recovered virus. Lastly, MVA shuts down host cell protein synthesis, which can affect the rescue of the virus. Therefore, several alternative transcription systems that do not use poxvirus have been developed. In one system, RNA polymerase II under

the control of CMV promoter has been used [84]. This system has been shown to be effective in the recovery of NDV [15, 74, 122, 137]. In another approach, cell lines stably expressing T7 polymerase have been developed. The most commonly used is a baby hamster kidney (BHK) cell line expressing T7 RNA polymerase (BSRT7/5 cells) [5]. Although this system circumvents some of the issues associated with poxvirus, it is not widely used because of low efficiency in virus recovery compared to T7 expressing poxvirus systems. This is due to the lower level of T7 expression by T7 cell lines [26].

One factor that is critical for the recovery of paramyxovirus is the synthesis of antigenomic RNA with the exact 5$'$ and 3$'$ ends. T7-dependent transcription of full-length cDNA adds three nonviral G residues at the 5$'$ end of the viral antigenomic RNA. Although these nonviral residues are removed during replication, their presence can affect the efficiency of virus rescue. To address this problem, some rescue systems have introduced a hammerhead ribozyme (HamRz) sequence after the T7 promoter and the start of the antigenome sequence [4, 72], which creates a correct 5'end and improves rescue efficiency. However, this strategy is not widely used for paramyxovirus rescue. Another factor that also plays a major role in rescue efficiency is the amount of T7 polymerase inside the cell. To increase the level of expression of T7 polymerase, the sequence of the T7 polymerase gene has been codon-optimized for mammalian cells. It has been reported that addition of a HamRz sequence immediately preceding the start of the antigenome sequence and the use of codon-optimized T7 polymerase gene has substantially increased rescue efficiency for paramyxoviruses [4]. To generate the exact 3$'$ end of the antigenome after transfection from the plasmid, an 84-nt hepatitis delta virus (HDV) autocleaving ribozyme sequence is placed directly after the 3$'$ end of the cDNA [95]. The HDV ribozyme sequence is then followed by a T7 terminator sequence. Addition of the T7 terminator sequence also increases the rescue efficiency.

## 11.3 Construction of Support Plasmids

In addition to the plasmid containing the full-length antigenomic cDNA, three support plasmids expressing viral N, P, and L proteins are needed for the rescue of paramyxoviruses. Generally, viral N, P, and L ORFs flanked by the T7 promoter and T7 terminator sequences are cloned into three separate plasmid vectors (Fig. 3a). For successful virus rescue, the viral N, P, and L proteins must be expressed in a precise ratio. Therefore, the amount of each support plasmid varies considerably in different paramyxovirus rescue systems. Generally, the N plasmid is used in large amounts, followed by P and L plasmids. The optimal ratio of different support plasmids necessary for successful recovery of a paramyxovirus is generally determined through trial and error.

## 11.4 Transfection and Rescue of Infectious Virus

Once the full-length antigenome plasmid and the three support plasmids have been constructed, the next step is the rescue of infectious virus (Fig. 3b). Several mammalian cell lines which produce T7 polymerase upon infection with MVA/T7 and support the growth of the paramyxovirus are used (e.g., HEp-2, BHK-21, HEK293, Vero, and A549). For efficient rescue, the quality of plasmid, the titer of MVA/T7 virus stock, and the condition of the cell monolayer are critical. Generally, standard Midiprep or Maxiprep purification of the plasmid DNA is sufficient for transfection, but the transfection efficiency can be improved by purifying the plasmid through CsCl density gradient centrifugation. Many commercially available transfection reagents (e.g., Xfect, lipofectamine) have been successfully used to rescue paramyxoviruses. The quality of cells used for transfection is important. The cells for transfection should be regularly maintained. Subconfluent monolayers (60%–80% confluence) of overnight cells are highly supportive of transfection. For rescue of avirulent (lentogenic) NDV strains (e.g., LaSota and B1), 5% fresh allantoic fluid is supplemented in the cell culture medium after transfection, because the F protein of these strains is synthesized as an inactive precursor F0, which requires an extracellular trypsin-like protease for cleavage site activation [86]. Following transfection, the cells are maintained at 37 °C for 3–4 days. The supernatant from the transfected cells is then passaged two or three times in the same mammalian cells to remove MVA/T7 and amplify the rescued virus. In the case of NDV, the supernatant from transfected cells is amplified in embryonated chicken eggs. The presence of a rescued virus in the cell culture or allantoic fluid is determined by hemagglutination (HA) test and followed by sequencing of the viral genome. The rescued virus is further separated from any residual helper MVA/T7 virus by plaque purification.

It should be noted that reverse genetics systems for paramyxoviruses are in general inefficient and not always produce consistent results. The recovery efficiency also varies among paramyxoviruses. Rescue of some paramyxoviruses is notoriously inefficient and requires repeated rescue attempts. In general, those viruses that grow poorly in cell culture are also difficult to rescue. In order to rescue a paramyxovirus, all four plasmids must be transfected into the same cell in an optimal ratio, which happens rarely. Consequently, the efficiency of recovery of *Mononegavirales* is very low. For example, the recovery frequency of the first rabies virus rescue was ~1 focus forming unit for $10^7$ cells in the vaccinia-based system [103]. Although the recovery efficiency has since then improved significantly due to the addition of HDV ribozyme sequences to generate exact 3′ end, the recovery efficiency is still very low. Therefore, there is a need to improve the rescue efficiency, so that it will be possible to rescue some of the difficult and highly attenuated paramyxoviruses.

## 12    Paramyxoviruses as Vaccine Vectors

Paramyxoviruses have several characteristics that make them attractive as vaccine vectors for human or veterinary use [10, 58]. Most importantly, several naturally occurring or recombinant paramyxovirus strains exist which are highly safe in humans and animals. Paramyxoviruses replicate well in vivo and induce robust immune responses. They have a broad host range, which makes them suitable as a vaccine vector for multiple animal species and humans. Paramyxoviruses replicate in the cytoplasm of infected cells and, therefore, do not integrate into the host cell DNA. In contrast to adeno-, herpes-, and poxvirus vectors whose genome encodes many proteins, paramyxoviruses encode only eight to twelve proteins, and thus, there is less competition for immune responses between vector proteins and the expressed foreign antigen. Paramyxoviruses readily incorporate foreign membrane proteins into their envelope, to enhance the immune response of the foreign protein. A foreign gene is relatively stable in paramyxoviruses because there is no genome recombination or reassortment and loss of the foreign gene, as frequently happens in positive-sense RNA viruses [28, 104]. Most paramyxoviruses infect via the intranasal route and, therefore, induce both mucosal and systemic immune responses. Paramyxoviruses are pleomorphic, with sizes ranging from 150 to 500 nm in diameter. This structural flexibility allows the virus genome to accommodate foreign genes of various sizes, which is a constraint for icosahedral viruses. Paramyxoviruses grow in a wide range of cell types and in embryonated eggs with high titers, which makes vaccine production cost-effective.

## 13    Rescue of a Recombinant Paramyxovirus Containing a Foreign Gene

The modular organization of the paramyxovirus genome (Fig. 2) facilitates insertion of a foreign gene. The short GS and GE signals further facilitate engineering of the foreign gene. The general method for recovery of a paramyxovirus containing a foreign gene is described in the following sections, and the detailed step-by-step laboratory method is described in several publications [2, 60].

### 13.1    Construction of a Full-Length cDNA Clone Containing a Foreign Gene

The complete nucleotide sequence of the foreign gene is determined and analyzed using a sequence analysis software to ensure that the viral GS and GE like sequences and polymerase slippage sequences are not present, which would affect the expression of the foreign gene. If present, these sequences must be changed by silent mutagenesis. It has been found that codon optimization increases the expression of the foreign gene [78, 105]. If chosen to codon-optimize for the animal species in which it will be used as a vaccine, gene-synthesis companies can codon-optimize and synthesize the foreign gene. Then, a transcription cassette is constructed by engineering a cDNA of the foreign gene ORF so that it is flanked by virus-specific GS and GE sequences (Fig. 4). Generally, a eukaryotic Kozak consensus sequence is added before the start codon of the foreign gene for better protein expression [65]. The transcription cassette must meet the "rule of six" requirement for efficient replication [11]. Therefore, if necessary, additional nucleotides can be added downstream of the foreign gene ORF after the stop codon to make the final genome length a multiple of six.

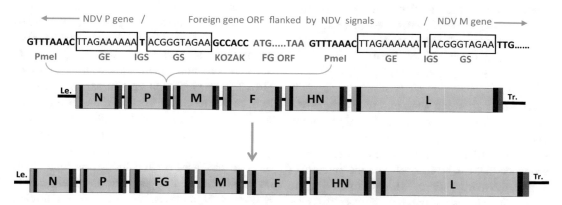

**Fig. 4** Constructions of a Newcastle disease virus (NDV) full-length cDNA containing a foreign gene (FG). The ORF of the FG was cloned by using *Pmel* sites into NDV antigenome cDNA under the control of NDV gene-start (GS) and gene-end (GE) transcription signals that direct its expression as an additional mRNA. The Kozak sequence was added in front of the ORF for efficient expression. The intergenic sequence (IGS) between every two genes is also shown

## 13.2 Insertion of the Foreign Gene into the Viral Genome

The transcription cassette is then inserted into the full-length cDNA clone at a 3′ noncoding region of a viral gene as an additional transcription unit that is transcribed as an additional mRNA. Due to the polar gradient nature of transcription, foreign genes are expressed at a higher level when placed closer to 3′ end of the genome. Although the foreign gene can be inserted between any two genes, some may affect the replication of the recombinant virus. A systematic analysis of the recombinant virus will identify the optimal insertion site for high-level expression of the foreign gene. For example, analysis of recombinant NDV revealed that the addition of a foreign gene between the N and P genes attenuates virus replication. The insertion site between the P and M genes was found optimal for efficient expression of the foreign gene with least effect on the replication of NDV [12, 87, 137, 141]. Interestingly, it was found that the insertion site between the N and P genes of APMV-3 is optimal for efficient expression of the Ebola virus glycoprotein gene [133], but the insertion site between P and M genes was found optimal for the expression of green fluorescent protein [132]. These results suggest that a new foreign gene needs to be inserted into several sites to determine the optimal site for that specific gene in the virus.

## 13.3 Rescue of a Recombinant Paramyxovirus Containing the Foreign Gene

The rescue of a paramyxovirus containing the foreign gene is carried out as described previously. The recovered virus is amplified by serial passage in cell culture or in embryonated chicken eggs. After confirmation of the rescued virus in cell culture supernatant or in the allantoic fluid by HA and by sequencing, it is triple plaque purified to remove the presence of any helper virus. Further in vitro studies are then performed to fully characterize the rescued virus. These include sequencing of the complete viral genome containing the foreign gene, growth characteristics of the rescued virus, IFA and Western blot analysis to determine the expression of the foreign gene, and any other special assay for the expressed foreign protein.

## 14    The Size of the Foreign Gene and the Number of Foreign Genes That Can Be Accommodated by a Paramyxovirus Vector

Insertion of a foreign gene into the genome of a paramyxovirus increases its length and gene number, which can inversely affect the replication of the virus [10]. The size of the foreign gene sequence that can be inserted into the genome of a paramyxovirus varies widely from virus to virus. There are some paramyxoviruses that cannot even tolerate the insertion of a small-size foreign gene, while other paramyxoviruses can tolerate the insertion of a large-size foreign gene. For example, APMV types 2, 4, and 7 could not be rescued when a small size foreign gene was inserted (our unpublished results), whereas HPIV3 and NDV can be rescued with foreign gene inserts of at least 5 kb (109, our unpublished results). It was found that, in these viruses, inserts of up to ~1.4 kb have little or no effect on virus replication in vitro and in vivo, but attenuation becomes evident and increases with the size of the insert (108, our unpublished results). Addition of a large size insert may not affect the replication of the virus in vitro but can reduce replication of the virus in vivo.

Previous research shows that up to three foreign genes can be inserted into HPIV-3 [109]. Although the number of foreign genes that can be inserted into the NDV genome is unknown, insertion of two foreign genes has little effect on virus replication in vitro and in vivo [41, 54].

## 15    Stability of the Foreign Gene in Paramyxovirus Vector

For paramyxoviruses to be useful as vaccine vectors, the foreign gene must be maintained stably both in vitro and in vivo. This is a major concern because the foreign gene is not required for the growth of the vector, and therefore, there is no selective pressure to maintain the foreign gene.

Surprisingly, the foreign genes are quite stable in paramyxovirus vectors compared to other RNA virus vectors. It was found that foreign genes in paramyxovirus vectors accumulate point and deletion mutations more slowly than expected [10]. The low mutation rates in paramyxoviruses may be due to higher fidelity of the paramyxovirus polymerase compared to other RNA virus polymerases. The long genetic stability of foreign genes in paramyxovirus vectors may be explained by the absence of recombination and tight encapsidation of the genomic and antigenomic RNAs by N proteins. We have found foreign genes to be stable in NDV vector at least 10 passages in eggs or cell culture and three passages in chickens (our unpublished results).

## 16    Incorporation of the Foreign Glycoprotein into the Envelope of Paramyxovirus Vector

Most foreign viral glycoproteins expressed by paramyxovirus vector are incorporated readily into the envelope of the vector without any modification to their sequences [10, 58]. Although the requirements for incorporation of foreign glycoproteins into paramyxovirus envelopes are not completely known, it seems that the requirements vary from glycoprotein to glycoprotein and also from vector to vector. In some cases, it was found that changing the transmembrane (TM) and cytoplasmic tail (CT) of the foreign glycoprotein to those of NDV F protein enhanced the incorporation of the foreign glycoprotein into NDV particle [10, 88, 93], but in other cases, it was observed that this change decreased incorporation compared to the incorporation of unmodified foreign glycoprotein [53, 134]. However, change of CT or TMCT might affect the conformation of the foreign glycoprotein. Therefore, the effect of changing the CT or TMCT on the immunogenicity of the foreign protein should be evaluated. It seems that the incorporation of a foreign glycoprotein into the vector virus envelope may depend on specific structural requirements of the foreign glycoprotein. Incorporation of foreign glycoprotein into the envelope of the vector might increase the immunogenicity of the foreign antigen. For example, replacement of the CT and TM domains of the HA protein of H7 avian influenza virus with those of the NDV F protein resulted in an enhanced incorporation into the NDV envelope, an increased immune response, and an increased protection against the highly pathogenic avian influenza virus compared to the recombinant NDV expressing the unmodified HA protein [93]. However, in rare cases, incorporation of the foreign glycoprotein into the envelope of the vector can alter the cell-to-cell spread of the vector virus in vitro. For example, incorporation of rabies virus glycoprotein (RVG) into the envelope of NDV vector enabled the vector virus to spread from cell to cell and altered its replication in cell culture in the absence of trypsin, but the NDV vector containing RVG was not neutralized by rabies virus neutralizing antibody, and its virulence was also not increased in poultry or mice [32]. Furthermore, there has not been a single case where the incorporation of a foreign glycoprotein into a paramyxovirus envelope has increased its tissue tropism or virulence.

## 17 Immunogenicity of Paramyxovirus-Vectored Vaccines

The design of an efficacious paramyxovirus-vectored vaccine requires a detailed knowledge of the antigen and how that antigen contributes to the induction of immune responses best suited to prevent the pathogen. It should be noted that the immune response is influenced not only by the antigen but also by the vaccine vector from which it is expressed. The immunogenicity and protective efficacy of a paramyxovirus-vectored vaccine can be affected by several factors, most importantly, the level of replication of the vector in the host. A higher level of replication will lead to increased immunogenicity, but it should not compromise the safety of the vector. Ideally, the vector vaccine should be sufficiently attenuated and still retains a level of replication enough to be satisfactorily immunogenic. The immunogenicity

and protective efficacy of the vaccine will also depend on the site of vector replication. For example, avirulent NDV strain LaSota replicates only in the respiratory tract and induces both local and systemic immune responses. Hence, it is an effective vector for respiratory pathogens, such as avian influenza virus, infectious bronchitis virus, and infectious laryngotracheitis virus [48, 88, 105], whereas APMV-3 replicates not only in the respiratory tract but also systemically, which consequently induces a robust immune response and a suitable vector for a non-respiratory pathogen, such as infectious bursal disease virus (our unpublished results). It was also recently shown that a recombinant APMV-3 vector expressing the hemagglutinin (HA) protein of HPAIV H5N1 induced a higher level of antibodies against the HA protein than recombinant NDV strain LaSota expressing the HA protein of HPAIV H5N1 [106].

An important factor that influences the immunogenicity of a paramyxovirus-vectored vaccine is the level of expression of the foreign gene. The quantity of foreign antigen expressed by the vector will determine the extent of immune response. In paramyxoviruses, the foreign gene is expressed at a high level when it is placed closer to the $3'$ end of the genome due to the gradient nature of transcription. Therefore, attempts should be made to place the foreign gene at the $3'$ most position possible. It has been found that codon optimization of the foreign gene increases its level of expression and immunogenicity [78, 105]. Another modification that can enhance the expression of the foreign gene is the addition of flanking 3' and 5' UTRs of a vector gene into the foreign gene [55]. UTRs of several vector genes need to be evaluated to identify the UTR that can upregulate the expression of the specific foreign gene.

The immunogenicity and protective efficacy of the paramyxovirus-vectored vaccine is also affected by the conformation of the expressed foreign antigen. Viral vectors that express the foreign protein in their native conformation induce antigen-specific immune responses which are neutralizing and protective. Sometimes, the native conformation is lost when the foreign antigen is separated from other viral proteins. In this case, coexpression of several viral proteins may be necessary to maintain the native conformation of the foreign antigen. Greater immunogenicity of the foreign glycoprotein is achieved by its incorporation into virus envelope than as soluble protein [93]. In some cases, mutations have been introduced into the foreign antigen to stabilize the structure of the protein [77].

Another factor that can affect the efficacy of a paramyxovirus-vectored vaccine is the repertoire of foreign antigens that is expressed. A vector vaccine typically expresses one or two proteins of the pathogen, whereas in some cases, complete protection may require immune response against several antigens of one pathogen. This is particularly true for complex pathogens with a large number of protective antigens. In these situations, several antigens of the pathogen may need to be expressed by using a combination of vectors to provide protection.

The immunogenicity of a paramyxovirus-vectored vaccine can be enhanced by coexpression of an immunostimulatory molecule such as granulocyte-macrophage colony-stimulating factor, interleukin-2, gamma interferon, or cytochrome C. However, it was found that coexpression of an immunostimulatory molecule not only can increase the immunogenicity of the vector but also can attenuate the vector [6–8].

## 18    Derivatives of Paramyxovirus Vectors

Over the years, several groups have developed innovative strategies to improve the efficacy of paramyxovirus vector vaccines [10, 61]. These approaches broadly address two challenges: (i) attenuation of the wild-type paramyxovirus by reverse genetics for use as potential vectors in humans and animals and (ii) overcoming preexisting anti-vector immunity.

## 18.1    Attenuation of Wild-Type Paramyxovirus for Use as a Vaccine Vector

The most important requirement for a paramyxovirus or any other virus to be used as a vaccine vector is that it must be highly safe in the host where it is intended to be used as a vaccine vector. In general, paramyxoviruses are attenuated in unnatural hosts due to host range restrictions. Furthermore, insertion of a foreign gene for expression often has an attenuating effect by itself. However, some paramyxoviruses are well-known pathogens, and these paramyxoviruses will require attenuation by reverse genetics before use as vaccine vectors in humans and/or animals. In order to use a virulent virus as a vaccine vector, several methods have been used to genetically modify the viral genome to reduce its virulence [10].

The major determinant of NDV virulence is the F protein cleavage site (FPCS). The avirulent NDV strains contain a mono or dibasic amino acid(s) within the FPCS such that the F0 protein can be cleaved into F1 and F2 subunits only by a trypsin-like protease that is present extracellularly in the respiratory and intestinal tracts. Hence, replication of these strains is restricted to the respiratory and intestinal tracts. On the other hand, virulent NDV strains have multibasic amino acids at the FPCS that can be cleaved intracellularly by a ubiquitous furin-like protease, resulting in systemic infections [86, 94]. It was found that modification of FPCS from multibasic to monobasic amino acid attenuated a virulent NDV strain [91]. This approach has been used to attenuate virulent NDV strains for use as live attenuated vaccines in the field. However, it was observed that after few passages in chickens, the mutated monobasic FPCS reverts back to multibasic FPCS making the virus virulent again. Therefore, it remains a challenge to engineer genetically stable live attenuated NDV vaccines by reverse genetics.

Attenuation of HPIVs was achieved by introducing amino acid point mutations by reverse genetics [107] or by replacing the F and

HN genes of HPIV-3 with the corresponding genes of BPIV-3 [102]. Deletion or silencing of nonessential accessory genes responsible for antagonizing interferon has also been used to attenuate NDV [44]. Increasing the length of the intergenic sequence between the HN and L genes led to reduced pathogenicity of NDV [129]. Codon modification through codon pair deoptimization has been used as a method to attenuate paramyxoviruses [73, 124]. By changing codons, the amount of viral protein translated can be modulated without impairing its function. A virulent NDV strain was attenuated by introducing mutations to the receptor binding site of HN protein [99]. Attenuation was also achieved by exchanging internal protein genes between related human and animal or avian viruses [96, 110]. In measles virus, it was found that the 3'UTR of the M gene and the 5'UTR of the F gene, have a cross-regulatory function on F and M gene expressions, respectively. Thus, deletion of these regions reduces viral pathogenesis [114].

## 18.2 Limitations of Paramyxovirus Vectors

Like all other viral vectors, paramyxoviruses also have few limitations:

1. Some of the paramyxoviruses commonly used for vaccine vector are endemic in animal populations. Therefore, preexisting immunity against the virus may reduce the efficacy of the vaccine.
2. Compared to DNA viral vectors, its transgene capacity (~5 kb) is low, which limits the use of the paramyxovirus vector.
3. The expression period of the foreign gene is relatively short (7–14 days), which can limit the duration of immune response.
4. It can be difficult to rescue some paramyxoviruses with a large-size foreign gene.

## 18.3 Overcoming Preexisting Immunity to the Vaccine Vector

Many of the paramyxoviruses that are being developed as vaccine vectors circulate in the human and animal populations, which therefore carry antibodies to the vaccine vectors. Preexisting antibodies to the vector are also induced if a single vector is used multiple times for vaccination. The presence of preexisting antibodies is a hurdle for the vaccine vector. Although preexisting antibodies may not completely block the induction of an immune response by a paramyxovirus vector vaccine [16, 130], they can reduce immune responses and thereby can make a vaccine ineffective. Therefore, for each vector, it is important to determine whether preexisting immunity to it could prevent the vaccine from working.

Preexisting immunity issue can be completely avoided by using a vector based on a virus that is unnatural to the host. For example, humans or animals do not have neutralizing antibodies to avian viral vectors. The issue of preexisting immunity can also be circumvented by using an antigenically unrelated virus as the vaccine vector, for examples, BPIV-3 as a vaccine vector, instead of HPIV3, in humans [102], and APMV-3 as a vaccine vector, instead of NDV, in poultry (our unpublished results). Alternatively, the vaccine vector can be modified to carry a different set of surface antigens. For example, to overcome maternal antibody to NDV, an antigenic chimeric virus was generated by replacing the ectodomains of NDV F and HN proteins with those of APMV-2 [59, 79] or APMV-8 [112]. The antigenic chimeric virus vector was constructed due to inefficient replication of APMV-2 and APMV-8 and incapability of accommodating a large-size foreign gene. Another advantage of antigenically unrelated and antigenic chimeric virus vectors is that they can be used in a prime-boost immunization strategy to enhance cellular and humoral immune responses compared to single immunization with one virus vector [61].

## 19    Specific Paramyxovirus Vectors for Veterinary Vaccine Development

### 19.1    Newcastle Disease Virus

NDV is a member of the genus *Orthoavulavirus* in the family *Paramyxoviridae* [48]. NDV strains vary widely in virulence. They have been classified into three major pathotypes based on the severity of disease produced in chickens: low virulent (lentogenic), moderately virulent (mesogenic), and highly virulent (velogenic). Lentogenic strains such as LaSota and B1 are used widely as live-virus vaccines for more than 60 years with good track records of safety and efficacy.

NDV has several characteristics that make it a suitable vaccine vector for both human and animal uses [58]. Lentogenic NDV strains are highly safe in avian and non-avian species. NDV replicates efficiently in most avian and non-avian species including humans and induces a robust immune response. Contrary to adeno-, herpes-, and poxvirus vectors, NDV encodes only seven proteins, and therefore, there is less competition for the foreign protein for immune response. NDV infects via the intranasal route; consequently, it induces both local and systemic immune responses.

One explanation of the attenuation of NDV in non-avian species is that the NDV V protein, which is produced by RNA editing of the P gene, antagonizes the antiviral effects of IFN. It interacts specifically to avian cell proteins to abrogate IFN response in a species-specific manner. It does not interfere with the IFN response in non-avian cells [92]. Therefore, NDV cannot efficiently prevent innate host defenses of non-avian species.

NDV is an ideal vaccine vector for poultry pathogens because live attenuated NDV vaccines are widely used in the poultry industry. An NDV vector carrying the protective antigen of another avian pathogen can be used as a bivalent vaccine. A bivalent vaccine will be highly economical for poultry farmers. Live attenuated NDV vaccine strains have been evaluated as bivalent vaccines against several poultry pathogens (Table 2).

NDV is also a promising vaccine vector for human use. NDV is highly attenuated in humans due to natural host range restriction. In rare cases, it can infect humans and cause mild conjunctivitis and flu-like symptoms [89]. Another advantage is that most humans do not have preexisting immunity to NDV. Therefore, the entire human population is susceptible to NDV immunization. NDV is known to induce very high type I IFN levels in mammalian cells [39], which contributes to an effective B cell response to this virus and to the foreign antigen [35]. NDV is also being evaluated as an oncolytic agent in humans [29]. The potential of NDV as a vaccine vector for human pathogens has been evaluated in nonhuman primates [9, 23].

The potential of NDV as a vaccine vector for veterinary pathogens has also been evaluated (Table 2). The results suggest that NDV can also be an excellent vaccine vector for veterinary use. One of the major advantages is that NDV is highly safe in most animal species. Another advantage is that most animal species are free of preexisting antibodies to NDV. It replicates well in the respiratory tract of most animal species and induces good antibody response to the foreign antigen.

### 19.2    Avian Paramyxovirus Serotypes 2–20

APMV serotypes are widespread in the wild bird populations. Currently, there are 20 officially recognized APMV serotypes (APMV-1 to APMV-20) [34]. NDV belongs to APMV-1 and is the most characterized member among all APMV serotypes. The potential of other APMV serotypes as vaccine vectors is beginnings to be explored [68, 118, 132, 133]. Recombinant vaccine vectors based on APMV serotypes 2–20 can overcome the activity of anti-NDV antibody present in commercial chickens due to routine vaccination. Recently, it was found that APMV-3 could also be used as a vaccine vector [68, 132, 133] and was able to overcome maternal NDV

**Table 2** Examples of NDV-vectored vaccines against human, animal, and poultry pathogens

| Host | Pathogen | Antigen | NDV strain | Insertion site | Animal model | References |
|---|---|---|---|---|---|---|
| Human | HIV-1 | Gag/Env | B1/LaSota | P/M | Mouse, Guinea pig | [12, 54] |
| | Influenza virus A | HA | B1 | P/M | Mouse | [87] |
| | SARS-CoV | S | LaSota/BC | P/M | Monkey | [23] |
| | EBOV | GP | LaSota | P/M | Monkey | [24] |
| | NoV | VP1 | LaSota | P/M | Mouse | [57] |
| | RSV | F | LaSota | P/M | Mouse | [81] |
| | HPIV-3 | HN | LaSota/BC | P/M | Monkey | [9] |
| | Polio virus | P1,3CD | LaSota | P/M | Guinea pig | [120] |
| | *B. burgdorferi* | BmPA/OspC | LaSota | P/M | Mouse, hamster | [127] |
| Cattle/ sheep | BHV-1 | gD | LaSota | P/M | Calf | [53] |
| | BEFV | G | LaSota | P/M | Calf | [135] |
| | RVFV | Gn/Gc | LaSota | P/M | Lamb | [63, 64] |
| | VSV | G | LaSota | P/M | Mouse | [136] |
| Dog/cat | CDV | HN | LaSota | P/M | Mink | [34] |
| | RV | G | LaSota | P/M | Dog/cat | [33, 134] |
| Pig | NiV | G, F | LaSota | P/M | Pig | [62] |
| Horse | WNV | Pr M/E | LaSota | P/M | Horse | [123] |
| Chicken | HPAIV(H5) | HA | LaSota | P/M | Chicken | [88, 118, 119] |
| | HPAIV(H7) | HA | LaSota | P/M | Chicken | [93] |
| | IBDV | VP2 | LaSota | 3'end | Chicken | [45] |
| | IBV | S1, S2, S | LaSota | P/M | Chicken | [117, 105, 138, 139] |
| | ILTV | gB, gC, gD | LaSota | P/M | Chicken | [48, 140] |
| | Parvovirus | VP3 | Genotype VII | P/M | Goose | [122] |
| Turkey | AMPV | G | LaSota | F/HN | Turkey | [40] |

*HIV-1* human immunodeficiency virus type 1, *SARS-CoV* severe acute respiratory syndrome coronavirus, *EBOV* Ebola virus, *NoV* norovirus, *RSV* respiratory syncytial virus, *HPIV-3* human parainfluenza virus type 3, *BEFV* bovine ephemeral fever virus, *RVFV* Rift Valley fever virus, *VSV* vesicular stomatitis virus, *CDV* canine distemper virus, *RV* rabies virus, *NiV* Nipah virus, *WNV* West Nile virus, *HPAIV (H5)* highly pathogenic avian influenza virus subtype H5, *HPAIV(H7)* highly pathogenic avian influenza virus subtype H7, *IBDV* infectious bursal disease virus, *IBV* infectious bronchitis virus, *ILTV* infectious laryngotracheitis virus, *AMPV* avian metapneumovirus, *BC* Beaudette C

antibody in broiler chicks (our unpublished results). Recombinant APMV-3 vector expressing the Ebola virus glycoprotein induced high levels of mucosal and humoral neutralizing antibodies against Ebola virus in guinea pigs [133]. A recent study showed that a recombinant APMV-3 expressing the H5 hemagglutinin protein protected chickens against highly pathogenic H5N1 avian influenza virus challenge (our unpublished results). One advantage APMV-3 has over NDV as a vaccine vector is that it replicates systemically, thus inducing a robust immune response compared to avirulent NDV vectors, whose replication is restricted only to the respiratory tract. APMV-3 is safe in chickens and turkeys. For these reasons, it seems that APMV-3 has great potential as a vaccine vector. It is most likely that other APMV serotypes will soon be available as vaccine vectors for human and veterinary uses.

## 19.3 Parainfluenza Virus 5

PIV-5 is a member of the genus *Rubulavirus* in the subfamily *Rubulavirinae* in the family *Paramyxoviridae*. PIV-5 was first isolated from monkey cells as a contaminant in 1956 [46]. The origin and natural host of PIV-5 are not clear. PIV-5 does not cause disease in humans and animals except dogs where it is thought to contribute to upper respiratory infections known as "Kennel cough" [22, 82]. PIV-5 has also emerged as a novel vaccine vector for humans and animals [19]. Some of the favorable properties of PIV-5 as a vaccine vector include the ability to infect a large number of mammals without any association with any disease except Kennel cough in dogs, lack of preexisting antibody in most mammalian species, and ability to infect a large number of cell types and also grow to high titers. PIV-5 can accommodate foreign gene sequences of at least 3 kb.

PIV-5 has been evaluated as a vaccine vector against human and avian influenza viruses [76, 85, 116, 142], human respiratory syncytial virus [97, 121], rabies virus [17, 43], and *Mycobacterium tuberculosis* [18]. The protective efficacy of PIV-5 vector vaccine was tested in mice and guinea pigs but has not been tested in other animal models and in the vaccine host species. Therefore, the potential of PIV-5 vector vaccines in the field condition is unknown.

## 19.4 Bovine Parainfluenza Virus 3

BPIV-3 was originally isolated from cattle with shipping fever [100]. Since then, this virus has been recovered from normal cattle and from calves with enzootic pneumonia. The BPIV-3 has worldwide distribution. BPIV-3 is one of the contributing factors in the bovine respiratory disease complex. BPIV-3 is closely related antigenically to HPIV-3. The amino acid sequence identity of the HN and F proteins of BPIV-3 and HPIV-3 are 79% and 75%, respectively. BPIV-3 is restricted in replication in the respiratory tract of rhesus monkeys, chimpanzees, and humans

and is being evaluated as a vaccine against HPIV-3 [21, 49]. Recombinant BPIV-3 viruses have been recovered by reverse genetics for use as human vaccines [36, 102]. BPIV-3 is an ideal vaccine vector for both humans and animals, especially for those pathogens that use the respiratory tract as the portal of entry, such as human and bovine RSVs. A chimeric recombinant BPIV-3 where the F and HN genes were replaced with the F and HN genes of HPIV-3 (B/HPIV-3) was constructed as a vaccine vector for use in humans [102]. However, the potential of recombinant BPIV-3 as a vaccine vector for veterinary pathogens has not been evaluated.

## 20 Conclusion and Future Perspectives

Viral vectors based on recombinant paramyxoviruses hold great promise for the development of effective human and veterinary vaccines. Major advantages of paramyxovirus vaccine vectors are their safety, potency, stability, and the ability to be used in different animal species. Our knowledge of paramyxovirus vaccine vectors is rapidly expanding, and several paramyxovirus-vectored veterinary vaccines are already available in the market. Paramyxovirus vectors will be highly beneficial for prevalent as well as emerging veterinary diseases. These vectors can be used to express structure-based vaccines to induce enhanced immune responses against diseases for which vaccine development has been difficult, such as porcine reproductive and respiratory syndrome virus, and bovine respiratory syncytial virus infections. They will also be valuable as a "one health" concept in protecting both animals and humans. In the future, it may be possible to use paramyxovirus vectors for multivalent and multipathogen vaccines.

To develop a paramyxovirus vectored vaccine, it is important to select the right vaccine vector, because the exact same immunogen can have different immune responses by different vectors. The choice of the viral vector should depend on the susceptibility of the host to the vector, safety considerations, replication properties of the vector in the host, vector genetic stability, preexisting immunity to the vector, the size of the foreign gene that can be accommodated without affecting vector replication, and the type of host tissue that will produce the foreign antigen following infection by the vector.

Currently, NDV-vectored vaccines have shown great potential for poultry, veterinary, and human uses. But there are many other paramyxoviruses, the potential of which as vaccine vectors has not been explored. New reverse genetics systems are being developed for other paramyxoviruses on a constant basis. Some of these viruses may surpass NDV as vaccine vectors.

There are many areas where paramyxovirus vectors require further refinement, which can be achieved through basic research. One area of improvement is to increase the expression level of the foreign antigen because the quantity of antigen may affect the extent of immune response. A second area of improvement is to increase the immunogenicity of the foreign antigen. The immunogenicity of the foreign antigen expressed by a viral vector is often low in comparison with traditional live attenuated virus vaccines. It is important that the foreign antigen expressed by the paramyxovirus vector forms its native conformation. Another area of improvement is the design of paramyxovirus vaccine vectors to overcome host preexisting immunity.

In conclusion, paramyxovirus viral vectors have great potential for development of human, veterinary and poultry vaccines. The safety, potency, and versatility of paramyxovirus vaccine vectors will make them invaluable for future human and veterinary vaccine developments.

## 21    Summary

Paramyxovirus-based vaccine vectors have been utilized in veterinary vaccine development for some years. These viruses stably express a wide variety of heterologous antigens at relatively high levels in many species of animals and have several advantages over other viral vectors making them an efficacious platform for the development
of veterinary vaccines in the future. There are many areas where paramyxovirus vectors require further refinement, which can be achieved through basic research. One area of improvement is to increase the expression level of the foreign antigen because the quantity of antigen may affect the extent of immune response. A second area of improvement is to increase the immunogenicity of the foreign antigen.

**Acknowledgement** The author would like to thank Anandan Paldurai, Berin Varghese, Edris Shirvani, and Mohammed Elbehairy for their assistance for this chapter.

## References

1. Abraham M, Arroyo-Diaz N, Li Z, Zengel J, Sakamoto K, He B. Role of small hydrophobic protein of J paramyxovirus in virulence. J Virol. 2018;92 (20):e00653–18.
2. Ayllon J, García-Sastre A, Martínez-Sobrido L. Rescue of recombinant Newcastle disease virus from cDNA. J Vis Exp. 2013;80:e50830.
3. Baron M, Barrett T. Rescue of rinderpest virus from cloned cDNA. J Virol. 1997;71(2):1265–71.
4. Beaty S, Park A, Won S, Hong P, Lyons M, Vigant F, Freiberg A, TenOever B, Duprex W, Lee B. Efficient and robust paramyxoviridae reverse genetics systems. mSphere. 2017;2(2):e00376.
5. Buchholz U, Finke S, Conzelmann K. Generation of Bovine Respiratory Syncytial Virus (BRSV) from cDNA: BRSV NS2 is not essential for virus replication in tissue culture, and the human RSV leader region acts as a functional BRSV genome promoter. J Virol. 1999;73(1):251–9.
6. Bukreyev A, Whitehead S, Bukreyeva N, Murphy B, Collins P. Interferon expressed by a recombinant respiratory syncytial virus attenuates virus replication in mice without compromising immunogenicity. Proc Natl Acad Sci. 1999;96(5):2367–72.
7. Bukreyev A, Belyakov I, Berzofsky J, Murphy B, Collins P. Granulocyte-macrophage colony-stimulating factor expressed by recombinant respiratory syncytial virus attenuates viral replication and increases the level of pulmonary antigen-presenting cells. J Virol. 2001;75(24):12128–40.
8. Bukreyev A, Skiadopoulos M, McAuliffe J, Murphy B, Collins P, Schmidt A. More antibody with less antigen: can immunogenicity of attenuated live virus vaccines be improved? Proc Natl Acad Sci. 2002;99(26):16987–91.
9. Bukreyev A, Huang Z, Yang L, Elankumaran S, St. Claire M, Murphy B, Samal S, Collins P. Recombinant Newcastle disease virus expressing

a foreign viral antigen is attenuated and highly immunogenic in primates. J Virol. 2005;79(21):13275–84.

10. Bukreyev A, Skiadopoulos M, Murphy B, Collins P. Non segmented negative-strand viruses as vaccine vectors. J Virol. 2006;80(21):10293–306.

11. Calain P, Roux L. The rule of six, a basic feature for efficient replication of Sendai virus defective interfering RNA. J Virol. 1993;67(8):4822–30.

12. Carnero E, Li W, Borderia A, Moltedo B, Moran T, Garcia-Sastre A. Optimization of human immunodeficiency virus Gag expression by Newcastle disease virus vectors for the induction of potent immune responses. J Virol. 2009;83(2):584–97.

13. Chanock R. Association of a new type of cytopathogenic myxovirus with infantile croup. J Exp Med. 1956;104(4):555–76.

14. Chanock R, Parrott R, Cook K, Andrews B, Bell J, Reichelderfer T, Kapikian A, Mastrota F, Huebner R. Newly recognized myxoviruses from children with respiratory disease. N Engl J Med. 1958;258(5):207–13.

15. Chellappa M, Dey S, Gaikwad S, Pathak D, Vakharia V. Rescue of a recombinant Newcastle disease virus strain R2B expressing green fluorescent protein. Virus Genes. 2017;53(3):410–7.

16. Chen Z, Xu P, Salyards G, Harvey S, Rada B, Fu Z, He B. Evaluating a parainfluenza virus 5-based vaccine in a host with pre-existing immunity against parainfluenza virus 5. PLoS One. 2012;7(11):e50144.

17. Chen Z, Zhou M, Gao X, Zhang G, Ren G, Gnanadurai C, Fu Z, He B. A novel rabies vaccine based on a recombinant parainfluenza virus 5 expressing rabies virus glycoprotein. J Virol. 2013;87(6):2986–93.

18. Chen Z, Gupta T, Xu P, Phan S, Pickar A, Yau W, Karls R, Quinn F, Sakamoto K, He B. Efficacy of parainfluenza virus 5 (PIV5)-based tuberculosis vaccines in mice. Vaccine. 2015;33(51):7217–24.

19. Chen Z. Parainfluenza virus 5-vectored vaccines against human and animal infectious diseases. Rev Med Virol. 2017;28(2):e1965.

20. Clarke D, Sidhu M, Johnson J, Udem S. Rescue of mumps virus from cDNA. J Virol. 2000;74(10):4831–8.

21. Coelingh K, Winter C, Murphy B, Rice J, Kimball P, Olmsted R, Collins P. Conserved epitopes on the hemagglutinin-neuraminidase proteins of human and bovine parainfluenza type 3 viruses: nucleotide sequence analysis of variants selected with monoclonal antibodies. J Virol. 1986;60(1):90–6.

22. Cornwell H, McCandlish I, Thompson H, Laird H, Wright N. Isolation of parainfluenza virus SV5 from dogs with respiratory disease. Vet Rec. 1976;98(15):301–2.

23. DiNapoli J, Kotelkin A, Yang L, Elankumaran S, Murphy B, Samal S, Collins P, Bukreyev A. Newcastle disease virus, a host range-restricted virus, as a vaccine for intranasal immunization

against emerging pathogens. Proc Natl Acad Sci. USA 2007;104:9788–9793.

24. DiNapoli J, Yang L, Samal S, Murphy B, Collins P, Bukreyev A. Respiratory tract immunization of non-human primates with a Newcastle disease virus-vectored vaccine candidate against Ebola virus elicits a neutralizing antibody response. Vaccine. 2010;29(1):17–25.

25. Doyle T. A hitherto unrecorded disease of fowls due to a filter-passing virus. J Comparative Pathol Therapeutic. 1927;40:144–69.

26. Elroy-Stein O, Moss B. Cytoplasmic expression system based on constitutive synthesis of bacteriophage T7 RNA polymerase in mammalian cells. Proc Natl Acad Sci. 1990;87(17):6743–7.

27. Engeland C, Bossow S, Hudacek A, Hoyler B, Förster J, Veinalde R, Jäger D, Cattaneo R, Ungerechts G, Springfeld C. A Tupaia paramyxovirus vector system for targeting and transgene expression. J Gen Virol. 2017;98(9):2248–57.

28. Fang Y, Rowland R, Roof M, Lunney J, Christopher-Hennings J, Nelson E. A full-length cDNA infectious clone of north American type 1 porcine reproductive and respiratory syndrome virus: expression of green fluorescent protein in the Nsp2 region. J Virol. 2006;80(23):11447–55.

29. Fournier P, Schirrmacher V. Oncolytic Newcastle disease virus as cutting edge between tumor and host. Biology. 2013;2(3):936–75.

30. Garcin D, Pelet T, Calain P, Roux L, Curran J, Kolakofsky D. A highly recombinogenic system for the recovery of infectious Sendai paramyxovirus from cDNA: generation of a novel copy-back nondefective interfering virus. EMBO J. 1995;14(24):6087–94.

31. Gassen U, Collins F, Duprex W, Rima B. Establishment of a rescue system for canine distemper virus. J Virol. 2000;74(22):10737–44.

32. Ge J, Deng G, Wen Z, Tian G, Wang Y, Shi J, Wang X, Li Y, Hu S, Jiang Y, Yang C, Yu K, Bu Z, Chen H. Newcastle disease virus-based live attenuated vaccine completely protects chickens and mice from lethal challenge of homologous and heterologous H5N1 avian influenza viruses. J Virol. 2007;81(1):150–8.

33. Ge J, Wang X, Tao L, Wen Z, Feng N, Yang S, Xia X, Yang C, Chen H, Bu Z. Newcastle disease virus-vectored rabies vaccine is safe, highly immunogenic, and provides long-lasting protection in dogs and cats. J Virol. 2011;85(16):8241–52.

34. Ge J, Wang X, Tian M, Gao Y, Wen Z, Yu G, Zhou W, Zu S, Bu Z. Recombinant Newcastle disease viral vector expressing hemagglutinin or fusion of canine distemper virus is safe and immunogenic in minks. Vaccine. 2015;33(21):2457–62.

35. Grieves J, Yin Z, Garcia-Sastre A, Mena I, Peeples M, Risman H, Federman H, Sandoval M, Durbin R, Durbin J. A viral-vectored RSV vaccine

induces long-lived humoral immunity in cotton rats. Vaccine. 2018;36(26):3842–52.

36. Haller A, Miller T, Mitiku M, Coelingh K. Expression of the surface glycoproteins of human parainfluenza virus type 3 by bovine parainfluenza virus type 3, a novel attenuated virus vaccine vector. J Virol. 2000;74(24):11626–35.

37. He B, Paterson R, Ward C, Lamb R. Recovery of infectious SV5 from cloned DNA and expression of a foreign gene. Virology. 1997;237(2):249–60.

38. Hoffman M, Banerjee A. An infectious clone of human parainfluenza virus type 3. J Virol. 1997;71 (6):4272–7.

39. Honda K, Sakaguchi S, Nakajima C, Watanabe A, Yanai H, Matsumoto M, Ohteki T, Kaisho T, Takaoka A, Akira S, Seya T, Taniguchi T. Selective contribution of IFN- /signaling to the maturation of dendritic cells induced by double-stranded RNA or viral infection. Proc Natl Acad Sci. 2003;100 (19):10872–7.

40. Hu H, Roth J, Estevez C, Zsak L, Liu B, Yu Q. Generation and evaluation of a recombinant Newcastle disease virus expressing the glycoprotein (G) of avian metapneumovirus subgroup C as a bivalent vaccine in turkeys. Vaccine. 2011;29 (47):8624–33.

41. Hu H, Roth J, Yu Q. Generation of a recombinant Newcastle disease virus expressing two foreign genes for use as a multivalent vaccine and gene therapy vector. Vaccine. 2018;36(32):4846–50.

42. Hu Q, Chen W, Huang K, Baron M, Bu Z. Rescue of recombinant peste des petits ruminants virus: creation of a GFP-expressing virus and application in rapid virus neutralization test. Vet Res. 2012;43(1):48.

43. Huang Y, Chen Z, Huang J, Fu Z, He B. Parainfluenza virus 5 expressing the G protein of rabies virus protects mice after rabies virus infection. J Virol. 2015;89(6):3427–9.

44. Huang Z, Krishnamurthy S, Panda A, Samal S. Newcastle disease virus V protein is associated with viral pathogenesis and functions as an alpha interferon antagonist. J Virol. 2003;77(16):8676–85.

45. Huang Z, Elankumaran S, Yunus A, Samal S. A recombinant Newcastle disease virus (NDV) expressing VP2 protein of infectious Bursal disease virus (IBDV) protects against NDV and IBDV. J Virol. 2004;78(18):10054–63.

46. Hull R, Minner J, Smith J. New viral agents recovered from tissue cultures of monkey kidney cells. Am J Epidemiol. 1956;63(2):204–15.

47. International Committee on Taxonomy of Viruses (ICTV). 2019. *International Committee on Taxonomy of Viruses (ICTV)*. [Online] Available at: https://talk.ictvonline.org/. Accessed 19 Jul 2019.

48. Kanabagatte Basavarajappa M, Kumar S, Khattar S, Gebreluul G, Paldurai A, Samal S. A recombinant Newcastle disease virus (NDV) expressing infectious laryngotracheitis virus (ILTV) surface glycoprotein D protects against highly virulent ILTV and NDV

challenges in chickens. Vaccine. 2014;32 (28):3555–63.

49. Karron R, Makhene M, Gay K, Wilson M, Clements M, Murphy B. Evaluation of a live attenuated bovine parainfluenza type 3 vaccine in two- to six-month-old infants. Pediatr Infect Dis J. 1996;15(8):650–4.

50. Kasel J, Frank A, Keitel W, Taber L, Glezen W. Acquisition of Serum antibodies to specific viral glycoproteins of parainfluenza virus 3 in children. J Virol. 1984;52(3):828–32.

51. Kast W. Cooperation between cytotoxic and helper T lymphocytes in protection against lethal Sendai virus infection. Protection by T cells is MHC- restricted and MHC-regulated; a model for MHC-disease associations. J Exp Med. 1986;164(3):723–38.

52. Kawano M, Kaito M, Kozuka Y, Komada H, Noda N, Nanba K, Tsurudome M, Ito M, Nishio M, Ito Y. Recovery of infectious human parainfluenza type 2 virus from cDNA clones and properties of the defective virus without V-specific cysteine-rich domain. Virology. 2001;284(1):99–112.

53. Khattar S, Collins P, Samal S. Immunization of cattle with recombinant Newcastle disease virus expressing bovine herpesvirus-1 (BHV-1) glycoprotein D induces mucosal and serum antibody responses and provides partial protection against BHV-1. Vaccine. 2010;28(18):3159–70.

54. Khattar S, Manoharan V, Bhattarai B, LaBranche C, Montefiori D, Samal S. Mucosal immunization with Newcastle disease virus vector coexpressing HIV-1 Env and Gag proteins elicits potent serum, mucosal, and cellular immune responses that protect against Vaccinia virus Env and gag challenges. MBio. 2015;6(4):e01005.

55. Kim S, Samal S. Role of untranslated regions in regulation of gene expression, replication, and pathogenicity of Newcastle disease virus expressing Green fluorescent protein. J Virol. 2010;84 (5):2629–34.

56. Kim S, Xiao S, Shive H, Collins P, Samal S. Mutations in the fusion protein cleavage site of avian paramyxovirus serotype 4 confer increased replication and syncytium formation in vitro but not increased replication and pathogenicity in chickens and ducks. PLoS One. 2013;8(1):e50598.

57. Kim S, Chen S, Jiang X, Green K, Samal S. Newcastle disease virus vector producing human norovirus-like particles induces serum, cellular, and mucosal immune responses in mice. J Virol. 2014;88 (17):9718–27.

58. Kim S, Samal S. Newcastle disease virus as a vaccine vector for development of human and veterinary vaccines. Viruses. 2016;8(7):183.

59. Kim S, Paldurai A, Samal S. A novel chimeric Newcastle disease virus vectored vaccine against highly pathogenic avian influenza virus. Virology. 2017;503:31–6.

60. Kim S, Samal S. Reverse genetics for Newcastle disease virus as a vaccine vector. Curr Protoc Microbiol. 2018;48:18.5.1–18.5.12.

61. Kim S, Samal S. Innovation in Newcastle disease virus vectored avian influenza vaccines. Viruses. 2019;11(3):300.

62. Kong D, Wen Z, Su H, Ge J, Chen W, Wang X, Wu C, Yang C, Chen H, Bu Z. Newcastle disease virus-vectored Nipah encephalitis vaccines induce B and T cell responses in mice and long-lasting neutralizing antibodies in pigs. Virology. 2012;432 (2):327–35.

63. Kortekaas J, Dekker A, de Boer S, Weerdmeester K, Vloet R, Wit A, Peeters B, Moormann R. Intramuscular inoculation of calves with an experimental Newcastle disease virus-based vector vaccine elicits neutralizing antibodies against Rift Valley fever virus. Vaccine. 2010a;28(11):2271–6.

64. Kortekaas J, de Boer S, Kant J, Vloet R, Antonis A, Moormann R. Rift Valley fever virus immunity provided by a paramyxovirus vaccine vector. Vaccine. 2010b;28(27):4394–401.

65. Kozak M. An analysis of 5′-noncoding sequences from 699 vertebrate messenger RNAs. Nucleic Acids Res. 1987;15(20):8125–48.

66. Kranveld F. A poultry disease in the Dutch East Indies. Ned Indisch Bl Diergeneeskd. 1926;38:448–50.

67. Krishnamurthy S, Huang Z, Samal S. Recovery of a virulent strain of Newcastle disease virus from cloned cDNA: expression of a foreign gene results in growth retardation and attenuation. Virology. 2000;278 (1):168–82.

68. Kumar S, Nayak B, Collins P, Samal S. Evaluation of the Newcastle disease virus F and HN proteins in protective immunity by using a recombinant avian paramyxovirus type 3 vector in chickens. J Virol. 2011;85(13):6521–34.

69. Kuroya M, Ishida N, Shiratori T. Newborn virus pneumonia (type Sendai) II. Report: the isolation of a new virus possessing hemagglutinin activity. Yokohama Med Bull. 1953;4:217–33.

70. Laing E, Amaya M, Navaratnarajah C, Feng Y, Cattaneo R, Wang L, Broder C. Rescue and characterization of recombinant cedar virus, a non-pathogenic Henipavirus species. Virol J. 2018;15(1):56.

71. Lamb RA, Parks GD. *Paramyxoviridae*: the viruses and their replication. In: Fields BN, Knipe DM, Howley PM, editors. *Fields virology: sixth edition*, vol. 1. Philadelphia: Lippincott, Williams, and Wilkins; 2013. p. 957–95.

72. Le Mercier P, Jacob Y, Tanner K, Tordo N. A novel expression cassette of lyssavirus shows that the distantly related Mokola virus can rescue a defective rabies virus genome. J Virol. 2002;76(4):2024–7.

73. Le Nouen C, Brock L, Luongo C, McCarty T, Yang L, Mehedi M, Wimmer E, Mueller S, Collins P, Buchholz U, DiNapoli J. Attenuation of

human respiratory syncytial virus by genome-scale codon-pair deoptimization. Proc Natl Acad Sci. 2014;111(36):13169–74.

74. Li B, Li X, Lan X, Yin X, Li Z, Yang B, Liu J. Rescue of Newcastle disease virus from cloned cDNA using an RNA polymerase II promoter. Arch Virol. 2011;156(6):979–86.

75. Li J, Hu H, Yu Q, Diel D, Li D, Miller P. Generation and characterization of a recombinant Newcastle disease virus expressing the red fluorescent protein for use in co-infection studies. Virol J. 2012;9(1):227.

76. Li Z, Gabbard J, Mooney A, Chen Z, Tompkins S, He B. Efficacy of parainfluenza virus 5 mutants expressing hemagglutinin from H5N1 influenza a virus in mice. J Virol. 2013;87(17):9604–9.

77. Liang B, Ngwuta J, Herbert R, Swerczek J, Dorward D, Amaro-Carambot E, Mackow N, Kabatova B, Lingemann M, Surman S, Yang L, Chen M, Moin S, Kumar A, McLellan J, Kwong P, Graham B, Schaap-Nutt A, Collins P, Munir S. Packaging and prefusion stabilization separately and additively increase the quantity and quality of respiratory syncytial virus (RSV)-neutralizing antibodies induced by an RSV fusion protein expressed by a parainfluenza virus vector. J Virol. 2016;90(21):10022–38.

78. Liang B, Kabatova B, Kabat J, Dorward D, Liu X, Surman S, Liu X, Moseman A, Buchholz U, Collins P, Munir S. Effects of alterations to the CX3C motif and secreted form of human Respiratory Syncytial Virus (RSV) G protein on immune responses to a parainfluenza virus vector expressing the RSV G protein. J Virol. 2019;93(7):e02043–18.

79. Liu J, Xue L, Hu S, Cheng H, Deng Y, Hu Z, Wang X, Liu X. Chimeric Newcastle disease virus-vectored vaccine protects chickens against H9N2 avian influenza virus in the presence of pre-existing NDV immunity. Arch Virol. 2018;163(12):3365–71.

80. Marsh G, Virtue E, Smith I, Todd S, Arkinstall R, Frazer L, Monaghan P, Smith G, Broder C, Middleton D, Wang L. Recombinant Hendra viruses expressing a reporter gene retain pathogenicity in ferrets. Virol J. 2013;10(1):95.

81. Martinez-Sobrido L, Gitiban N, Fernandez-Sesma A, Cros J, Mertz S, Jewell N, Hammond S, Flano E, Durbin R, Garcia-Sastre A, Durbin J. Protection against respiratory syncytial virus by a recombinant Newcastle disease virus vector. J Virol. 2006;80 (3):1130–9.

82. McCandlish I, Thompson H, Cornwell H, Wright N. A study of dogs with kennel cough. Vet Rec. 1978;102(14):293–301.

83. Merz DC, Scheild A, Choppin PW. Importance of antibodies to the fusion glycoprotein of paramyxoviruses in the prevention of spread of infection. J Exp Med. 1980;151(2):275–88.

84. Molouki A, Peeters B. Rescue of recombinant Newcastle disease virus: current cloning strategies

and RNA polymerase provision systems. Arch Virol. 2016;162(1):1–12.

85. Mooney A, Li Z, Gabbard J, He B, Tompkins S. Recombinant parainfluenza virus 5 vaccine encoding the influenza virus hemagglutinin protects against H5N1 highly pathogenic avian influenza virus infection following intranasal or intramuscular vaccination of BALB/c mice. J Virol. 2013;87 (1):363–71.

86. Nagai Y, Klenk H, Rott R. Proteolytic cleavage of the viral glycoproteins and its significance for the virulence of Newcastle disease virus. Virology. 1976;72 (2):494–508.

87. Nakaya T, Cros J, Park M, Nakaya Y, Zheng H, Sagrera A, Villar E, Garcia-Sastre A, Palese P. Recombinant Newcastle disease virus as a vaccine vector. J Virol. 2001;75(23):11868–73.

88. Nayak B, Rout S, Kumar S, Khalil M, Fouda M, Ahmed L, Earhart K, Perez D, Collins P, Samal S. Immunization of chickens with Newcastle disease virus expressing H5 hemagglutinin protects against highly pathogenic H5N1 avian influenza viruses. PLoS One. 2009;4(8):e6509.

89. Nelson C, Pomeroy B, Schrall K, Park W, Lindeman R. An outbreak of conjunctivitis due to Newcastle Disease Virus (NDV) occurring in poultry workers. Am J Public Health Nations Health. 1952;42 (6):672–8.

90. Newman J, Surman S, Riggs J, Hansen C, Collins P, Murphy B, Skiadopoulos M. Sequence analysis of the Washington/1964 strain of human parainfluenza virus type 1 (HPIV1) and recovery and characterization of wild-type recombinant HPIV1 produced by reverse genetics. Virus Genes. 2002;24(1):77–92.

91. Panda A, Huang Z, Elankumaran S, Rockemann D, Samal S. Role of fusion protein cleavage site in the virulence of Newcastle disease virus. Microb Pathog. 2004;36(1):1–10.

92. Park M, Shaw M, Munoz-Jordan J, Cros J, Nakaya T, Bouvier N, Palese P, Garcia-Sastre A, Basler C. Newcastle Disease Virus (NDV)-based assay demonstrates interferon-antagonist activity for the NDV V protein and the Nipah virus V, W, and C proteins. J Virol. 2003;77(2):1501–11.

93. Park M, Steel J, Garcia-Sastre A, Swayne D, Palese P. Engineered viral vaccine constructs with dual specificity: avian influenza and Newcastle disease. Proc Natl Acad Sci. 2006;103(21):8203–8.

94. Peeters B, De Leeuw O, Koch G, Gielkens A. Rescue of Newcastle disease virus from cloned cDNA: evidence that cleavability of the fusion protein is a major determinant for virulence. J Virol. 1999;73 (6):5001–9.

95. Perrotta A, Been M. The self-cleaving domain from the genomic RNA of hepatitis delta virus: sequence requirements and the effects of denaturant. Nucleic Acids Res. 1990;18(23):6821–7.

96. Pham Q, Biacchesi S, Skiadopoulos M, Murphy B, Collins P, Buchholz U. Chimeric recombinant human metapneumoviruses with the nucleoprotein or phosphoprotein open reading frame replaced by that of avian metapneumovirus exhibit improved growth in vitro and attenuation in vivo. J Virol. 2005;79 (24):15114–22.

97. Phan S, Chen Z, Xu P, Li Z, Gao X, Foster S, Teng M, Tripp R, Sakamoto K, He B. A respiratory syncytial virus (RSV) vaccine based on parainfluenza virus 5 (PIV5). Vaccine. 2014;32(25):3050–7.

98. Radecke F, Spielhofer P, Schneider H, Kaelin K, Huber M, Dötsch C, Christiansen G, Billeter M. Rescue of measles viruses from cloned DNA. EMBO J. 1995;14(23):5773–84.

99. Rangaswamy U, Wang W, Cheng X, McTamney P, Carroll D, Jin H. Newcastle disease virus establishes persistent infection in tumor cells in vitro: contribution of the cleavage site of fusion protein and second sialic acid binding site of hemagglutinin-neuraminidase. J Virol. 2017;91(16):e00770–17.

100. Reisinger R, Heddleston K, Manthei C. A Myxovirus (SF-4) associated with shipping fever of cattle. J Am Med Assoc. 1959;135:147–52.

101. Romer-Oberdorfer A, Buchholz U, Mundt E, Mettenleiter T, Mebatsion T. Generation of recombinant lentogenic Newcastle disease virus from cDNA. J Gen Virol. 1999;80(11):2987–95.

102. Schmidt A, McAuliffe J, Huang A, Surman S, Bailly J, Elkins W, Collins P, Murphy B, Skiadopoulos M. Bovine parainfluenza virus type 3 (BPIV3) fusion and hemagglutinin-neuraminidase glycoproteins make an important contribution to the restricted replication of BPIV3 in primates. J Virol. 2000;74(19):8922–9.

103. Schnell M, Mebatsion T, Conzelmann K. Infectious rabies viruses from cloned cDNA. EMBO J. 1994;13 (18):4195–203.

104. Seligman S, Gould E. Live flavivirus vaccines: reasons for caution. Lancet. 2004;363(9426):2073–5.

105. Shirvani E, Paldurai A, Manoharan V, Varghese B, Samal S. A recombinant Newcastle disease virus (NDV) expressing S protein of infectious bronchitis virus (IBV) protects chickens against IBV and NDV. Sci Rep. 2018;8(1):11951.

106. Shirvani E, Varghese BP, Paldouri A, Samal S. A recombinant avian paramyxovirus type-3 (APMV-3) expressing HA protein of highly pathogenic avian influenza virus (HPAIV) protects chickens against HPAIV. Sci Rep. 2020;10:2221.

107. Skiadopoulos MH, Tao T, Surman S, Collins P, Murphy B. Generation of a parainfluenza virus type 1 vaccine candidate by replacing the HN and F glycoproteins of the live-attenuated PIV3 cp45 vaccine virus with their PIV1 counterparts. Vaccine. 1999;18(5–6):503–10.

108. Skiadopoulos M, Surman S, Durbin A, Collins P, Murphy B. Long nucleotide insertions between the HN and L protein coding regions of human parainfluenza virus type 3 yield viruses with

temperature-sensitive and attenuation phenotypes. Virology. 2000;272(1):225–34.

109. Skiadopoulos M, Surman S, Riggs J, Orvell C, Collins P, Murphy B. Evaluation of the replication and immunogenicity of recombinant human parainfluenza virus type 3 vectors expressing up to three foreign glycoproteins. Virology. 2002;297 (1):136–52.

110. Skiadopoulos M, Schmidt A, Riggs J, Surman S, Elkins W, St. Claire M, Collins P, Murphy B. Determinants of the host range restriction of replication of bovine parainfluenza virus type 3 in rhesus monkeys are polygenic. J Virol. 2003;77(2):1141–8.

111. Skiadopoulos M, Biacchesi S, Buchholz U, Amaro-Carambot E, Surman S, Collins P, Murphy B. Individual contributions of the human metapneumovirus F, G, and SH surface glycoproteins to the induction of neutralizing antibodies and protective immunity. Virology. 2006;345(2):492–501.

112. Steglich C, Grund C, Ramp K, Breithaupt A, Höper D, Keil G, Veits J, Ziller M, Granzow H, Mettenleiter T, Römer-Oberdörfer A. Chimeric Newcastle disease virus protects chickens against avian influenza in the presence of maternally derived NDV immunity. PLoS One. 2013;8(9):e72530.

113. Subbiah M, Khattar S, Collins P, Samal S. Mutations in the fusion protein cleavage site of avian paramyxovirus serotype 2 increase cleavability and syncytium formation but do not increase viral virulence in chickens. J Virol. 2011;85(11):5394–405.

114. Takeda M, Ohno S, Seki F, Nakatsu Y, Tahara M, Yanagi Y. Long untranslated regions of the measles virus M and F genes control virus replication and cytopathogenicity. J Virol. 2005;79(22):14346–54.

115. Thibault P, Watkinson R, Moreira-Soto A, Drexler J, Lee B. Zoonotic potential of emerging paramyxoviruses: knowns and unknowns. Adv Virus Res. 2017;98:1–55.

116. Tompkins S, Lin Y, Leser G, Kramer K, Haas D, Howerth E, Xu J, Kennett M, Durbin R, Durbin J, Tripp R, Lamb R, He B. Recombinant parainfluenza virus 5 (PIV5) expressing the influenza A virus hemagglutinin provides immunity in mice to influenza A virus challenge. Virology. 2007;362(1):139–50.

117. Toro H, Zhao W, Breedlove C, Zhang Z, van Santen V, Yu Q. Infectious bronchitis virus S2 expressed from recombinant virus confers broad protection against challenge. Avian Dis. 2014;58 (1):83–9.

118. Tsunekuni R, Hikono H, Tanikawa T, Kurata R, Nakaya T, Saito T. Recombinant avian paramyxovirus serotypes 2, 6, and 10 as vaccine vectors for highly pathogenic avian influenza in chickens with antibodies against Newcastle disease virus. Avian Dis. 2017;61(3):296–306.

119. Veits J, Wiesner D, Fuchs W, Hoffmann B, Granzow H, Starick E, Mundt E, Schirrmeier H, Mebatsion T, Mettenleiter T, Romer-Oberdorfer A. Newcastle disease virus expressing H5 hemagglutinin gene protects chickens against Newcastle disease and avian influenza. Proc Natl Acad Sci. 2006;103(21):8197–202.

120. Viktorova E, Khattar S, Kouiavskaia D, Laassri M, Zagorodnyaya T, Dragunsky E, Samal S, Chumakov K, Belov G. Newcastle disease virus-based vectored vaccine against poliomyelitis. J Virol. 2018;92(17):e00976–18.

121. Wang D, Phan S, DiStefano D, Citron M, Callahan C, Indrawati L, Dubey S, Heidecker G, Govindarajan D, Liang X, He B, Espeseth A. A single-dose recombinant parainfluenza virus 5-vectored vaccine expressing respiratory syncytial virus (RSV) F or G protein protected cotton rats and African green monkeys from RSV challenge. J Virol. 2017;91 (11):e00066–17.

122. Wang J, Cong Y, Yin R, Feng N, Yang S, Xia X, Xiao Y, Wang W, Liu X, Hu S, Ding C, Yu S, Wang C, Ding Z. Generation and evaluation of a recombinant genotype VII Newcastle disease virus expressing VP3 protein of Goose parvovirus as a bivalent vaccine in goslings. Virus Res. 2015;203:77–83.

123. Wang J, Yang J, Ge J, Hua R, Liu R, Li X, Wang X, Shao Y, Sun E, Wu D, Qin C, Wen Z, Bu Z. Newcastle disease virus-vectored West Nile fever vaccine is immunogenic in mammals and poultry. Virol J. 2016;13(1):109.

124. Wang W, Cheng X, Buske P, Suzich J, Jin H. Attenuate Newcastle disease virus by codon modification of the glycoproteins and phosphoprotein genes. Virology. 2019;528:144–51.

125. Welch S, Chakrabarti A, Wiggleton Guerrero L, Jenks H, Lo M, Nichol S, Spiropoulou C, Albariño C. Development of a reverse genetics system for Sosuga virus allows rapid screening of antiviral compounds. PLoS Negl Trop Dis. 2018;12(3): e0006326.

126. Wyatt L, Moss B, Rozenblatt S. Replication-deficient vaccinia virus encoding bacteriophage T7 RNA polymerase for transient gene expression in mammalian cells. Virology. 1995;210(1):202–5.

127. Xiao S, Kumar M, Yang X, Akkoyunlu M, Collins P, Samal S, Pal U. A host-restricted viral vector for antigen-specific immunization against Lyme disease pathogen. Vaccine. 2011;29(32):5294–303.

128. Xiao S, Khattar S, Subbiah M, Collins P, Samal S. Mutation of the F-protein cleavage site of avian paramyxovirus type 7 results in furin cleavage, fusion promotion, and increased replication in vitro but not increased replication, tissue tropism, or virulence in chickens. J Virol. 2012;86(7):3828–38.

129. Yan Y, Samal S. Role of intergenic sequences in Newcastle disease virus RNA transcription and pathogenesis. J Virol. 2007;82(3):1323–31.

130. Yang L, Sanchez A, Ward J, Murphy B, Collins P, Bukreyev A. A paramyxovirus-vectored intranasal vaccine against Ebola virus is immunogenic in

vector-immune animals. Virology. 2008;377 (2):255–64.

131. Yoneda M, Guillaume V, Ikeda F, Sakuma Y, Sato H, Wild T, Kai C. Establishment of a Nipah virus rescue system. Proc Natl Acad Sci. 2006;103 (44):16508–13.

132. Yoshida A, Samal S. Avian paramyxovirus type-3 as a vaccine vector: identification of a genome location for high level expression of a foreign gene. Front Microbiol. 2017;8:693.

133. Yoshida A, Kim S, Manoharan V, Varghese B, Paldurai A, Samal S. Novel avian paramyxovirus-based vaccine vectors expressing the Ebola virus glycoprotein elicit mucosal and humoral immune responses in Guinea pigs. Sci Rep. 2019;9(1):5520.

134. Yu G, Zu S, Zhou W, Wang X, Shuai L, Wang X, Ge J, Bu Z. Chimeric rabies glycoprotein with a transmembrane domain and cytoplasmic tail from Newcastle disease virus fusion protein incorporates into the Newcastle disease virion at reduced levels. J Vet Sci. 2017;18(1):351–9.

135. Zhang M, Ge J, Wen Z, Chen W, Wang X, Liu R, Bu Z. Characterization of a recombinant Newcastle disease virus expressing the glycoprotein of bovine ephemeral fever virus. Arch Virol. 2016;162 (2):359–67.

136. Zhang M, Ge J, Li X, Chen W, Wang X, Wen Z, Bu Z. Protective efficacy of a recombinant Newcastle disease virus expressing glycoprotein of vesicular stomatitis virus in mice. Virol J. 2017;13(1):31.

137. Zhang X, Liu H, Liu P, Peeters B, Zhao C, Kong X. Recovery of avirulent, thermostable Newcastle disease virus strain NDV4-C from cloned cDNA and stable expression of an inserted foreign gene. Arch Virol. 2013;158(10):2115–20.

138. Zhao H, Peeters BP. Recombinant Newcastle disease virus as a viral vector: effect of genomic location of foreign gene on gene expression and virus replication. J Gen Virol. 2003;84(4):781–8.

139. Zhao R, Sun J, Qi T, Zhao W, Han Z, Yang X, Liu S. Recombinant Newcastle disease virus expressing the infectious bronchitis virus S1 gene protects chickens against Newcastle disease virus and infectious bronchitis virus challenge. Vaccine. 2017;35 (18):2435–42.

140. Zhao W, Spatz S, Zhang Z, Wen G, Garcia M, Zsak L, Yu Q. Newcastle disease virus (NDV) recombinants expressing infectious laryngotracheitis virus (ILTV) glycoproteins gB and gD protect chickens against ILTV and NDV challenges. J Virol. 2014;88(15):8397–406.

141. Zhao W, Zhang Z, Zsak L, Yu Q. P and M gene junction is the optimal insertion site in Newcastle disease virus vaccine vector for foreign gene expression. J Gen Virol. 2015;96(1):40–5.

142. Zhuo L, Mooney A, Gabbard J, Gao X, Xu P, Place R, Hogan R, Tompkins S, He B. Recombinant parainfluenza virus 5 expressing hemagglutinin of influenza A virus H5N1 protected mice against lethal highly pathogenic avian influenza virus H5N1 challenge. J Virol. 2012;87(1):354–62.

# Rhabdoviruses as Vaccine Vectors for Veterinary Pathogens

Gert Zimmer

**Abstract**

Rhabdoviruses are simple RNA viruses, which are open to genetic manipulation. Recombinant vector vaccines based on vesicular stomatitis virus (VSV) or rabies virus (RABV) are capable of inducing strong and protective immune responses in animals and humans as exemplified by the VSV-based Ebola virus vaccine. As several rhabdoviruses are harmful for animals and/or humans, the recombinant vector vaccine derived from them needs to be properly attenuated. Single-cycle vector vaccines and interferon-stimulating viruses represent attractive strategies to achieve attenuation. VSV and RABV are notifiable Office International des Epizooties (OIE)-listed pathogens, and this has impeded their general use in the veterinary field. However, vector vaccines based on different non-notifiable rhabdoviruses may represent an attractive alternative.

**Keywords**

Vesicular stomatitis virus · Rabies virus · Attenuation · Pathogenicity · Interferon · Chimeric virus · Pseudotyping · Single-cycle vector

**Learning Objectives**

After reading this chapter, you should be able to:

1. Explain how VSV represents a promising vector for vaccination of humans and animals
2. Explain how attenuation of rhabdoviral vectors may be achieved by exploiting their sensitivity to interferon
3. Recognize that single-cycle rhabdoviral vectors comply with the highest biosafety standards
4. Recognize that viral vectors based on rhabdoviruses other than VSV and RABV circumvent the problems associated with notifiable pathogens

## 1 Introduction

The family *Rhabdoviridae* belongs to the order *Mononegavirales*, which comprises viruses with a non-segmented, single-stranded, negative-sense RNA genome. Members of the *Rhabdoviridae* reveal a remarkably diverse ecology as they infect vertebrates (mammals, birds, reptiles, amphibians, fish), a wide range of terrestrial and aquatic invertebrates, and plants. The family is currently classified into 18 different genera, of which 4 genera (*Vesiculovirus*, *Ephemerovirus*, *Lyssavirus*, *Tibrovirus*) cause diseases in

G. Zimmer (✉)
Institute of Virology and Immunology (IVI), Mittelhäusern/Bern, Switzerland

Department of Infectious Diseases and Pathobiology (DIP), Vetsuisse Faculty, University of Bern, Bern, Switzerland
e-mail: gert.zimmer@ivi.admin.ch

© Springer Nature Switzerland AG 2021
T. Vanniasinkam et al. (eds.), *Viral Vectors in Veterinary Vaccine Development*,
https://doi.org/10.1007/978-3-030-51927-8_9

mammalian hosts. Members of the genera *Novirhabdovirus*, *Perhabdovirus*, and *Sprivivirus* are found in fish and *Tupavirus* in birds.

Rabies is certainly the most important disease caused by a rhabdovirus and has been known for several thousands of years. RABV is a neurotropic virus, which causes a fatal disease in humans and other mammalian species. VSV is an arthropod-borne virus, which causes a vesicular disease in livestock (horses, cattle, pigs). The disease is rather mild, and mortality is low; however, the symptoms of the disease resemble those of foot-and-mouth disease (FMD). VSV is therefore listed as a notifiable viral pathogen. Several rhabdoviruses cause important diseases in fish.

The genome of rhabdoviruses consists of a linear, single-stranded RNA with a size ranging from 11 kb to 15 kb. The viral genes are usually arranged in a linear, non-overlapping order (Fig. 1). They are flanked by highly conserved transcription start and stop sequences and separated by short intergenic regions. The genome encodes for at least five proteins, the nucleoprotein N, the phosphoprotein P, the matrix protein M, the glycoprotein G, and the large RNA polymerase L (Fig. 1). The N protein binds to the genomic RNA, thereby forming the helical nucleocapsid, which also harbors some copies of the P and L protein, respectively. The nucleocapsid is enveloped by a lipid bilayer membrane, which contains several copies of the transmembrane glycoprotein G, which mediates receptor-binding

and membrane fusion. The M protein lines the inner leaflet of the viral envelope. It plays a pivotal role in virus maturation and is responsible for the typical bullet-shaped morphology of rhabdoviruses (Fig. 1).

The first step of the "life" cycle is mediated by the G protein, which binds to receptor proteins at the host cell surface. Subsequently, the virus is internalized by receptor-mediated endocytosis. In the mature endosome, the G protein undergoes a low pH-induced conformational change and mediates fusion of the viral envelope with the endosomal membrane. Following membrane fusion, the nucleocapsid is released into the cytosol, where transcription and replication of the viral RNA take place.

The L protein and its cofactor P are components of the viral ribonucleoprotein (RNP) complex and perform the primary transcription of the viral RNA genome. The presence of the N protein is also required as the naked RNA is not transcribed. Transcription begins at the 3' end of the RNA genome with the synthesis of a short non-coding leader RNA and proceeds to the 5' end by sequential transcription of the viral genes. The polymerase initiates transcription at conserved transcription start signals that precede each gene and terminates at transcription stop signals after each gene. Since transcription is not always restarted after each stop signal, a typical 3'–5' transcription gradient is formed. The gene products N, P, and L subsequently participate in secondary transcription. When the amounts of N and/or P protein exceed a certain level, the viral RNA polymerase switches to the replication modus in which the start and stop signals are ignored. The resulting full-length positive-sense (antigenomic) RNA is encapsidated by the N protein and subsequently replicated into negative-sense RNPs, which are finally packaged into progeny virus. Both the M and the G proteins contribute to the budding and release of progeny virus from the plasma membrane.

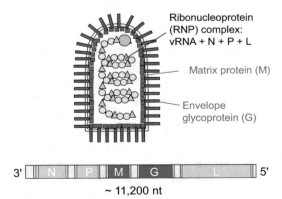

**Fig. 1** Genome organization and encoded proteins of the prototype rhabdovirus VSV

## 2 Construction of the Rhabdoviral Vector

The most widely used reverse genetics system for generation of infectious rhabdoviruses from transfected cDNA takes advantage of the T7-phage RNA polymerase, which allows nucleus-independent cytoplasmic transcription of all components required [1]. The T7 RNA polymerase might be provided by either a recombinant vaccinia virus or transgenic cells constitutively expressing the enzyme. The T7 RNA polymerase-expressing cells are transfected with a recombinant plasmid, which drives T7 RNA polymerase-mediated transcription of the full-length virus genome, resulting in the synthesis of a positive-sense (antigenomic) RNA. A hepatitis delta ribozyme guarantees the formation of the correct 3' end of the positive-sense RNA. In addition to the genome-encoding plasmid, three plasmids containing the viral N, P, and L genes downstream of the T7 promoter are transfected into the cells. Transcription of these genes generates monocistronic mRNAs, which are translated into the viral proteins N, P, and L. Upon binding of the N protein to the antigenomic RNA, the positive-sense RNP complex is formed, which is subsequently replicated by the viral RNA polymerase into negative-sense RNPs. In this way, the viral replication cycle is initiated, which finally leads to the formation and release of progeny virus.

## 3 Propagation-Competent Rhabdoviruses Expressing a Foreign Antigen

Reverse genetics has allowed us to design recombinant rhabdoviruses for the delivery and expression of reporter genes, antigens, cytokines, and other factors in vitro as well as in vivo. For expression of foreign genes, additional transcription units can be inserted into the vector genome. The position of the transgene in the viral genome affects expression levels and stability of the transgene. Due to the transcription gradient, the location of foreign genes close to the 3' end of the

viral RNA genome will result in higher expression levels compared to genes that have been introduced close to the 5' end of the genome. However, depending on number, size, and position, the insertion of foreign genes may cause attenuation of the viral vector leading to slower replication and lower infectious titers.

## 4 Propagation-Competent Chimeric Rhabdoviruses

VSV is known to incorporate foreign viral envelope proteins quite efficiently even if they are derived from viruses that belong to distantly related virus families [2]. If the foreign glycoprotein exhibits receptor-binding and fusion activities, it may functionally substitute for the rhabdovirus G protein, so that propagation-competent chimeric viruses are produced. Consequently, the chimeric virus may show an altered cell tropism compared to the parental rhabdovirus. A well-known chimeric rhabdovirus is VSVΔG/EBOV-GP, which expresses the Ebola virus glycoprotein (EBOV-GP) in place of the VSV G protein [3]. VSVΔG/EBOV-GP turned out to be highly attenuated in even immunocompromised animals [4]. Nevertheless, a single dose was sufficient to induce a protective immune response in nonhuman primates [5]. Chimeric rhabdoviruses may not only serve as vector vaccines but may also be useful for the detection of neutralizing and opsonizing antibodies and for the identification of entry inhibitors.

## 5 Biosafety Aspects of Propagation-Competent Rhabdovirus Vectors

Usually, recombinant RABVs expressing foreign antigens have been used as killed vaccines to comply with biosafety concerns. As far as the less harmful VSV-based vectors are concerned, preference was given to live-attenuated vaccines. The following strategies have been employed in order to attenuate propagation-competent VSV vectors:

1. *Changing the order of genes in the viral genome*
   Gene rearrangements may lead to attenuation of the vector [6].
2. *Modification of the envelope glycoprotein*
   The length of the cytoplasmic domain of the G protein but not a specific sequence is important for VSV morphogenesis. Accordingly, VSV vectors with C-terminally truncated G proteins are attenuated [7].
3. *Replacing the G protein with a foreign viral glycoprotein*
   Foreign viral glycoproteins are less efficiently incorporated into VSV particles than VSV G protein and may also mediate a more restricted cell and tissue tropism than the homotypic envelope protein. For these reasons, chimeric viruses might be attenuated compared to parental VSV and also with respect to the virus from which the foreign glycoprotein is originating. However, there are also notable exceptions from this rule [8].
4. *Modification of viral interferon-antagonistic proteins*
   VSV is highly sensitive to the antiviral effects induced by type I interferon (IFN). To escape the innate immune response, VSV relies on a single antagonistic mechanism. The VSV matrix (M) protein causes a profound host shutoff by efficiently blocking the nucleocytoplasmic transport of cellular mRNAs. In this way, the synthesis of type I interferon (IFN) is suppressed. However, certain mutations in the VSV M protein (e.g., M51R) are known to abrogate host shutoff activity. These mutant viruses trigger the secretion of IFN from infected cells at high levels. The autocrine action of IFN causes a significantly reduced viral transcription/replication leading to lower antigen expression levels. The paracrine action of IFN efficiently inhibits further dissemination of the virus. Nevertheless, M mutant VSV vectors may still be capable of inducing protective immune responses.
5. *Expression of cytokines*
   Vector-driven expression of cytokines such as granulocyte-macrophage colony-stimulating factor (GM-CSF), IFN-λ, IFN-γ, IFN-β, and TNF-α might result in significant attenuation of the viral vector while even inducing enhanced immune responses or increasing efficacy of oncolytic viruses.
6. *Changing codon usage*
   The recoding of even single viral genes may result in attenuation of the vector [9].
7. Use of *ts*-mutants
   Some mutations in viral genes may lead to a temperature-sensitive (*ts*) phenotype [7].

## 6  Propagation-Incompetent Rhabdovirus Vectors

The RNA polymerase complex of rhabdoviruses lacks proof-reading activity with the consequence that a multitude of quasi-species are produced. This increases the risk of reversion of live-attenuated viruses. Propagation-incompetent or single-cycle viruses represent an attractive solution to this problem. For example, rhabdoviruses with a deletion of the G gene have been generated by providing the G protein or a suitable foreign envelope glycoprotein *in trans*. The envelope glycoprotein mediates virus entry into susceptible cells and delivery of the viral genome into the cytosol (Fig. 2). Upon transcription and replication of the viral RNA, all the genes encoded by the genome will be expressed. However, due to the lack of the G protein gene, any infectious progeny virus will not be released (Fig. 2). The lack of virus dissemination adds to the extraordinary biosafety of these vectors, which are not only nonpathogenic but also unable to revert to virulence. Moreover, single-cycle vectors will not shed from vaccinated animals, thus excluding any transmission.

In principle, genes other than G may also be deleted from the virus genome in order to generate a propagation-incompetent virus. However, vector vaccines based on rhabdoviruses that lack either the P, N, or L gene can only perform primary transcription resulting in very low antigen expression levels.

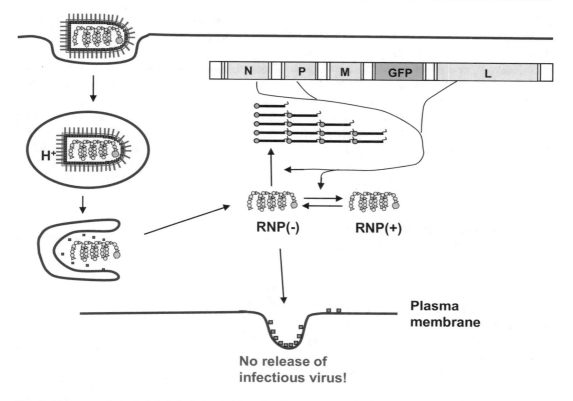

**Fig. 2** *Trans*-complemented rhabdoviruses lacking the *G* gene are replication-competent but do not produce any infectious progeny

## 7 Complementing Defective Rhabdoviruses

Complementing defective viruses are based on at least two propagation-incompetent viruses that contain complementary gene deletions. In co-infected cells, these viruses complement each other and can even be passaged on cells without further complementation. Complementing defective rhabdoviruses have been described for VSV [10] as well as RABV [11] and represent in interesting way of attenuation. In contrast to single-cycle vectors, complementing defective viruses can be propagated on any susceptible cell line, independently of genetically modified helper cells.

## 8 Application of the Viral Vector

### 8.1 *Lyssavirus*-Based Vector Vaccines for Veterinary Pathogens

Recombinant live-attenuated RABVs have been developed for vaccination of wildlife [12]. To improve the biosafety profile of RABV vaccines, propagation-incompetent P gene- or M gene-deleted RABVs have been generated [13, 14]. Apart from that, only a few reports on the use of recombinant RABV-based vector vaccines for protection of livestock or domestic animals are available in the literature. For example, recombinant RABV displaying canine distemper virus (CDV) glycoproteins has been shown to induce protective immunity in ferrets against both RABV and CDV [15]. Similarly, a bivalent vaccine for use in humans and animals against Middle East respiratory syndrome coronavirus (MERS-CoV) and RABV has successfully been evaluated in a receptor-modified mouse model [16]. Both vaccines have been chemically inactivated in order to comply with biosafety concerns. Bivalent RABV-based vaccines for protection of wildlife and humans against both RABV and Ebola virus have been developed as well, either as live-attenuated, propagation-incompetent, or chemically inactivated viruses [17–21]. Finally, a RABV-based Hendra virus vaccine has been generated [22].

## 8.2 *Vesiculovirus*-Based Vector Vaccines for Veterinary Pathogens

Recombinant VSV is a promising vector vaccine for a number of human infectious diseases. However, since VSV is a notifiable pathogen of livestock, VSV-based vector vaccines have rarely been used in the veterinary field. For example, propagation-competent VSV encoding the bovine viral diarrhea virus (BVDV) E2 glycoprotein induced neutralizing antibodies in mice but has not been evaluated in the natural host [23]. Vaccination of chickens and pigs with propagation-incompetent, G gene-deleted VSV vectors encoding the influenza virus antigens HA or NA has been very successful [24–27], whereas vector-driven expression of NP antigen in pigs resulted in enhanced disease upon challenge infection [28]. A single-cycle VSV vector encoding the VP2 outer capsid protein of bluetongue virus conferred complete protection of sheep against challenge infection, indicating that an antigen from a non-enveloped virus, which is not incorporated into VSV particles, can be protective as well [29]. In contrast, the envelope glycoproteins of a porcine arterivirus were poorly immunogenic and did not confer full protection in pigs [30].

Maraba virus, another member of the genus *Vesiculovirus*, which has been isolated from insects, is not linked to any disease in animals or humans. Maraba virus-vectored cancer vaccines represent a safe and novel therapeutic option for cats [31].

## 8.3 *Novirhabdovirus*-Based Vector Vaccines

Recombinant infectious hematopoietic necrosis virus (IHNV) vector vaccines have been used for protection against other fish diseases [32–34] but also for protection of mice from infection with either influenza A virus or West Nile virus [35, 36]. Attenuation of *Novirhabdovirus*-based vector vaccines has been achieved by gene rearrangements or using G gene-deleted single-cycle vectors [37, 38].

## 9 Summary

Rhabdovirus-based vector vaccines have been shown to trigger protective immune responses against a variety of different pathogens. In addition, they represent promising vectors for immunotherapy of cancer. However, VSV, the most commonly used rhabdoviral vector, has rarely been used in livestock or domestic animals. The main reason for using VSV hesitantly is the fact that VSV is a notifiable agent, which causes symptoms resembling those caused by FMDV. Single-cycle, G gene-deleted rhabdoviral vectors represent a promising alternative to live-attenuated vectors as they do not produce any progeny virus, do not cause any disease, and are unable to revert to virulence. Furthermore, they have been shown to induce a robust and protective immune response.

## 10 Future Directions

The high-yield production of propagation-incompetent, G protein-deleted rhabdoviral vectors represents a challenging issue. Complementation of these viruses on cells that have been transiently transfected to express the viral envelope glycoprotein of interest usually results in only low infectious virus titers. Stably transfected helper cells represent an attractive alternative; however, many viral envelope glycoproteins have cytotoxic properties, thus limiting their constitutive expression. A task for the future is the generation of helper cells that allow the expression of viral envelope proteins in a regulated, inducible manner. Preferably, envelope proteins other than VSV G protein might be expressed by helper cells in order to avoid any problems with serological VSV diagnostics. Finally, vector vaccines based on rhabdoviruses other than VSV or RABV might represent an attractive alternative, provided these viruses are not listed as notifiable agents.

# References

1. Schnell MJ, Mebatsion T, Conzelmann KK. Infectious rabies viruses from cloned cDNA. EMBO J. 1994;13 (18):4195–203.
2. Schnell MJ, Buonocore L, Kretzschmar E, Johnson E, Rose JK. Foreign glycoproteins expressed from recombinant vesicular stomatitis viruses are incorporated efficiently into virus particles. Proc Natl Acad Sci U S A. 1996;93(21):11359–65.
3. Garbutt M, Liebscher R, Wahl-Jensen V, Jones S, Moller P, Wagner R, et al. Properties of replication-competent vesicular stomatitis virus vectors expressing glycoproteins of filoviruses and arenaviruses. J Virol. 2004;78(10):5458–65.
4. Geisbert TW, Daddario-Dicaprio KM, Lewis MG, Geisbert JB, Grolla A, Leung A, et al. Vesicular stomatitis virus-based Ebola vaccine is well-tolerated and protects immunocompromised nonhuman primates. PLoS Pathog. 2008;4(11):e1000225.
5. Marzi A, Hanley PW, Haddock E, Martellaro C, Kobinger G, Feldmann H. Efficacy of vesicular stomatitis virus-Ebola virus postexposure treatment in rhesus macaques infected with Ebola virus Makona. J Infect Dis. 2016;214(suppl 3):S360–S6.
6. Wertz GW, Perepelitsa VP, Ball LA. Gene rearrangement attenuates expression and lethality of a nonsegmented negative strand RNA virus. Proc Natl Acad Sci U S A. 1998;95(7):3501–6.
7. Clarke DK, Nasar F, Lee M, Johnson JE, Wright K, Calderon P, et al. Synergistic attenuation of vesicular stomatitis virus by combination of specific G gene truncations and N gene translocations. J Virol. 2007;81(4):2056–64.
8. van den Pol AN, Mao G, Chattopadhyay A, Rose JK, Davis JN. Chikungunya, influenza, nipah, and semliki forest chimeric viruses with vesicular stomatitis virus: actions in the brain. J Virol. 2017;91(6):e02154-16.
9. Wang B, Yang C, Tekes G, Mueller S, Paul A, Whelan SP, et al. Recoding of the vesicular stomatitis virus L gene by computer-aided design provides a live, attenuated vaccine candidate. MBio. 2015;6(2): e00237-15.
10. Muik A, Dold C, Geiss Y, Volk A, Werbizki M, Dietrich U, et al. Semireplication-competent vesicular stomatitis virus as a novel platform for oncolytic virotherapy. J Mol Med (Berl). 2012;90(8):959–70.
11. Hidaka Y, Lim CK, Takayama-Ito M, Park CH, Kimitsuki K, Shiwa N, et al. Segmentation of the rabies virus genome. Virus Res. 2018;252:68–75.
12. Dietzschold ML, Faber M, Mattis JA, Pak KY, Schnell MJ, Dietzschold B. In vitro growth and stability of recombinant rabies viruses designed for vaccination of wildlife. Vaccine. 2004;23(4):518–24.
13. Cenna J, Hunter M, Tan GS, Papaneri AB, Ribka EP, Schnell MJ, et al. Replication-deficient rabies virus-based vaccines are safe and immunogenic in mice and nonhuman primates. J Infect Dis. 2009;200 (8):1251–60.
14. McGettigan JP, David F, Figueiredo MD, Minke J, Mebatsion T, Schnell MJ. Safety and serological response to a matrix gene-deleted rabies virus-based vaccine vector in dogs. Vaccine. 2014;32(15):1716–9.
15. da Fontoura BR, Hudacek A, Sawatsky B, Kramer B, Yin X, Schnell MJ, et al. Inactivated recombinant rabies viruses displaying canine distemper virus glycoproteins induce protective immunity against both pathogens. J Virol. 2017;91(8):e02077-16.
16. Wirblich C, Coleman CM, Kurup D, Abraham TS, Bernbaum JG, Jahrling PB, et al. One-health: a safe, efficient, dual-use vaccine for humans and animals against Middle East respiratory syndrome coronavirus and rabies virus. J Virol. 2017;91(2):e02040-16.
17. Willet M, Kurup D, Papaneri A, Wirblich C, Hooper JW, Kwilas SA, et al. Preclinical development of inactivated rabies virus-based polyvalent vaccine against rabies and filoviruses. J Infect Dis. 2015;212 (Suppl 2):S414–24.
18. Papaneri AB, Bernbaum JG, Blaney JE, Jahrling PB, Schnell MJ, Johnson RF. Controlled viral glycoprotein expression as a safety feature in a bivalent Rabies-Ebola vaccine. Virus Res. 2015;197:54–8.
19. Papaneri AB, Wirblich C, Cann JA, Cooper K, Jahrling PB, Schnell MJ, et al. A replication-deficient rabies virus vaccine expressing Ebola virus glycoprotein is highly attenuated for neurovirulence. Virology. 2012;434(1):18–26.
20. Blaney JE, Marzi A, Willet M, Papaneri AB, Wirblich C, Feldmann F, et al. Antibody quality and protection from lethal Ebola virus challenge in nonhuman primates immunized with rabies virus based bivalent vaccine. PLoS Pathog. 2013;9(5):e1003389.
21. Blaney JE, Wirblich C, Papaneri AB, Johnson RF, Myers CJ, Juelich TL, et al. Inactivated or live-attenuated bivalent vaccines that confer protection against rabies and Ebola viruses. J Virol. 2011;85 (20):10605–16.
22. Kurup D, Wirblich C, Feldmann H, Marzi A, Schnell MJ. Rhabdovirus-based vaccine platforms against henipaviruses. J Virol. 2015;89(1):144–54.
23. Grigera PR, Marzocca MP, Capozzo AV, Buonocore L, Donis RO, Rose JK. Presence of bovine

viral diarrhea virus (BVDV) E2 glycoprotein in VSV recombinant particles and induction of neutralizing BVDV antibodies in mice. Virus Res. 2000;69 (1):3–15.

24. Kalhoro NH, Veits J, Rautenschlein S, Zimmer G. A recombinant vesicular stomatitis virus replicon vaccine protects chickens from highly pathogenic avian influenza virus (H7N1). Vaccine. 2009;27 (8):1174–83.

25. Halbherr SJ, Brostoff T, Tippenhauer M, Locher S, Berger Rentsch M, Zimmer G. Vaccination with recombinant RNA replicon particles protects chickens from H5N1 highly pathogenic avian influenza virus. PLoS One. 2013;8(6):e66059.

26. Halbherr SJ, Ludersdorfer TH, Ricklin M, Locher S, Berger Rentsch M, Summerfield A, et al. Biological and protective properties of immune sera directed to the influenza virus neuraminidase. J Virol. 2015;89 (3):1550–63.

27. Ricklin ME, Vielle NJ, Python S, Brechbuhl D, Zumkehr B, Posthaus H, et al. Partial protection against porcine influenza a virus by a hemagglutinin-expressing virus replicon particle vaccine in the absence of neutralizing antibodies. Front Immunol. 2016;7:253.

28. Ricklin ME, Python S, Vielle NJ, Brechbuhl D, Zumkehr B, Posthaus H, et al. Virus replicon particle vaccines expressing nucleoprotein of influenza A virus mediate enhanced inflammatory responses in pigs. Sci Rep. 2017;7(1):16379.

29. Kochinger S, Renevey N, Hofmann MA, Zimmer G. Vesicular stomatitis virus replicon expressing the VP2 outer capsid protein of bluetongue virus serotype 8 induces complete protection of sheep against challenge infection. Vet Res. 2014;45:64.

30. Eck M, Duran MG, Ricklin ME, Locher S, Sarraseca J, Rodriguez MJ, et al. Virus replicon particles expressing porcine reproductive and respiratory syndrome virus proteins elicit immune priming but do not confer protection from viremia in pigs. Vet Res. 2016;47:33.

31. Hummel J, Bienzle D, Morrison A, Cieplak M, Stephenson K, DeLay J, et al. Maraba virus-vectored cancer vaccines represent a safe and novel therapeutic option for cats. Sci Rep. 2017;7(1):15738.

32. Guo M, Shi W, Wang Y, Wang Y, Chen Y, Li D, et al. Recombinant infectious hematopoietic necrosis virus expressing infectious pancreatic necrosis virus VP2 protein induces immunity against both pathogens. Fish Shellfish Immunol. 2018;78:187–94.

33. Romero A, Figueras A, Thoulouze MI, Bremont M, Novoa B. Recombinant infectious hematopoietic necrosis viruses induce protection for rainbow trout Oncorhynchus mykiss. Dis Aquat Org. 2008;80 (2):123–35.

34. Emmenegger EJ, Biacchesi S, Merour E, Glenn JA, Palmer AD, Bremont M, et al. Virulence of a chimeric recombinant infectious haematopoietic necrosis virus expressing the spring viraemia of carp virus glycoprotein in salmonid and cyprinid fish. J Fish Dis. 2018;41 (1):67–78.

35. Rouxel RN, Merour E, Biacchesi S, Bremont M. Complete protection against influenza virus H1N1 strain A/PR/8/34 challenge in mice immunized with non-adjuvanted novirhabdovirus vaccines. PLoS One. 2016;11(10):e0164245.

36. Nzonza A, Lecollinet S, Chat S, Lowenski S, Merour E, Biacchesi S, et al. A recombinant novirhabdovirus presenting at the surface the E glycoprotein from West Nile Virus (WNV) is immunogenic and provides partial protection against lethal WNV challenge in BALB/c mice. PLoS One. 2014;9(3): e91766.

37. Rouxel RN, Tafalla C, Merour E, Leal E, Biacchesi S, Bremont M. Attenuated infectious hematopoietic necrosis virus with rearranged gene order as potential vaccine. J Virol. 2016;90(23):10857–66.

38. Kim MS, Choi SH, Kim KH. Effect of G gene-deleted recombinant viral hemorrhagic septicemia virus (rVHSV-DeltaG) on the replication of wild type VHSV in a fish cell line and in olive flounder (Paralichthys olivaceus). Fish Shellfish Immunol. 2016;54:598–601.

# Coronaviruses as Vaccine Vectors for Veterinary Pathogens

Ding Xiang Liu, Yan Ling Ng, and To Sing Fung

**Abstract**

Coronaviruses (CoVs, family *Coronaviridae*) are enveloped, plus-stranded RNA viruses that can cause highly contagious upper respiratory diseases in humans and animals with potentially fatal outcomes. Typical symptoms found in chickens infected with infectious bronchitis coronavirus (IBV) include coughing, sneezing, gasping, nasal discharge and tracheal rales. Animal CoVs also cause local epidemics and pandemics with high infection rates, significantly increasing the economic burden on the poultry and livestock industry. With the realization that animal CoVs can be transmitted to humans, these viruses are now considered a global health threat. Improvement in technologies, such as reverse genetics, has conferred the ability to manipulate coronaviral genomes in the development of antiviral intervention and as vaccine vectors against other veterinary pathogens. This chapter summarizes new information on CoV reverse genetics and advances in vaccine development.

**Keywords**

CoV · Vaccine vectors · Bacterial artificial chromosome · In vitro ligation · Vaccinia virus · Infectious bronchitis · Vaccination

**Learning Objectives**

After reading this chapter, you should be able to:

- Explain the basic molecular biology, pathogenesis and replication cycle of coronavirus
- Explain the working mechanisms of the four coronavirus reverse genetics systems and compare and contrast their advantages and limitations
- Use practical examples to demonstrate how coronavirus genomes can be manipulated to serve as vaccine vectors for veterinary pathogens

## 1 Introduction

Coronaviruses (CoVs) are a family of enveloped viruses with non-segmented, single-stranded and positive-sense RNA genomes. Viruses in this family infect and cause diseases in various domesticated and laboratory vertebrates, including cats, dogs, pigs, chickens and mice. In addition, several CoVs are able to infect humans and cause respiratory diseases with mild to severe outcomes. Three recently emergent zoonotic

D. X. Liu (✉) · T. S. Fung
Guangdong Province Key Laboratory Microbial Signals & Disease Co, and Integrative Microbiology Research Centre, South China Agricultural University, Guangzhou, People's Republic of China

Y. L. Ng
School of Biological Sciences, Nanyang Technological University, Singapore, Singapore

© Springer Nature Switzerland AG 2021
T. Vanniasinkam et al. (eds.), *Viral Vectors in Veterinary Vaccine Development*,
https://doi.org/10.1007/978-3-030-51927-8_10

149

CoVs, severe acute respiratory syndrome coronavirus (SARS-CoV), Middle East respiratory syndrome coronavirus (MERS-CoV) and 2019 novel coronavirus (SARS-CoV-2), have crossed the species barrier and infected humans directly or via intermediate hosts, causing lethal diseases in a pandemic scale. In terms of animal CoVs, the high mortality rates in infected young animals and the reduction in product quality and yield have imposed heavy economic burdens on the global livestock industry.

Ever since the identification of the first CoV more than 80 years ago, continuous effort has been made to isolate and cultivate CoVs in tissue and cell cultures with some success, but many CoV field isolates still remain unculturable. Meanwhile, research on the molecular virology of CoV has revealed its highly characteristic replication mechanism, featuring by the synthesis of a nested set of subgenomic mRNA species, shared by other viruses belonging to the order *Nidovirales* (*nidus*, nest). While the replication cycle of CoV is relatively simple occurring exclusively in the cytoplasm, the enormous genome size (27–32 kb) has made the initial attempts to establish reverse genetics systems a daunting task. Nonetheless, several approaches have been developed, and the ability to recover recombinant CoVs harbouring precise deletions and mutations proves to be invaluable in the basic and applied research of CoV.

The large genome size and unique transcription strategy of CoVs make them promising candidates for the development of viral vectors expressing heterologous genes. Studies with CoV replicons, the autonomous replicating RNA molecules, have led to the identification of viral genes and the *cis*-acting elements indispensable for efficient CoV replication. Meanwhile, experiments using reporter genes, such as fluorescent proteins and luciferases, have demonstrated the impacts of genomic localization, sequence composition and transgene size on the expression levels of the heterologous genes and the recovery rate and stability of the recombinant viruses. Several pioneering studies have also explored the feasibility of using CoV-based vectors to construct vaccines and evaluated their efficacies

against virulent strains of CoVs or other viruses. In this chapter, we first review the basic knowledge of CoV biology, followed by a detailed discussion on the four CoV reverse genetic systems. We then summarize current studies on the construction of CoV replicons and CoV vectors and discuss their potential application to serve as vaccine vectors.

## 2 Background

### 2.1 Clinical Importance

Records of diseases attributable to CoVs date back to 1931 in North Dakota, United States, in what was described as "an apparently new respiratory disease of baby chicks" known as infectious bronchitis (IB) [1]. The viral nature of IB was established 2 years later, when Bushnell and Brandly reported a similar disease in chickens [2]. Bacterial and toxin origins were ruled out as the causative agent was filterable. Since then, different CoVs have been discovered and the symptoms they cause extend beyond the respiratory system. In all cases, the viral particles were found to be between 80–150 nm, pleomorphic, membrane-coated and decorated with widely spaced club-shaped surface projections [3–7]. This group of viruses, known as CoVs (*corona* denoting the reminiscent of a solar corona), was officially recognized as a new genus in 1971 [8].

CoV research hit its prime in 2002, when the emergence of the SARS-CoV outbreak resulted in 8096 reported cases worldwide and an approximate 10% mortality rate [9]. The subsequent occurrences of MERS-CoV epidemic in 2012 and SARS-CoV-2 pandemic in 2019 have once again demonstrated that it is highly probable for CoVs circulating amongst bat species and other animals to be introduced into the human population. This prompts the need for more research to elucidate their replication mechanisms to identify potential drug targets and develop vaccines, which can be effective countermeasures against CoVs.

## 2.2 Taxonomy

CoVs are taxonomically classified under the order of the *Nidovirales*, a large group of enveloped, single-stranded, positive-sense RNA genomes which produce a 3′ co-terminal nested set of subgenomic mRNAs during infection [10]. CoVs were traditionally classified into four groups based on phylogenetic clustering. However, extensive changes were made to the order in the 2018 International Committee on Taxonomy of Viruses (ICTV) taxonomy, in which the nidoviruses are reorganized into seven suborders (Fig. 1). The *Coronaviridae* family is now under the suborder of *Cornidovirineae* and further divided into two subfamilies, *Letovirinae* and *Orthocoronavirinae*. The introduction of a new subgenus rank also redistributes the *alpha-*, *beta-*, *delta-* and *gammaCoVs* across 12, 5, 4 and 2 subgenera, respectively. Almost all *alpha-* and *beta*CoVs have mammalian hosts, whereas *gamma-* and *delta*CoVs have been isolated mainly from avian hosts.

## 2.3 Morphology

On negative contrast electron microscopy (EM), CoV virions appear roughly spherical and pleomorphic (Fig. 2), with an average diameter of 80–150 nm [11, 12]. The virion surface is covered with a fringe of 20 nm crown-like surface projections consisting of trimers of spike (S) glycoprotein. In *beta*CoVs, there exists a second type of protein projections on the surface, the homodimeric hemagglutinin-esterase (HE) glycoprotein, which measures about 5–7 nm in length [13, 14].

## 2.4 Genome Structure and Organization

CoVs possess the largest and most complex RNA genome amongst established RNA viruses, with genome sizes typically ranging between 26 and 32 kb. These viruses maintain a well-conserved genome organization, with the essential genes occurring in the order 5′-replicase-S-E-M-N-3′. The genomic RNAs (gRNAs) contain a methylated cap structure at the 5′ termini of its viral genomic and subgenomic RNAs (sgRNAs) [15] and a 3′ terminal polyadenylated tail [16]. The replicase-transcriptase constitutes about two-thirds of the virus genome and is the only gene directly translated from the genomic RNA upon entry into the host cell. Products of their downstream open reading frames (ORFs) are derived from subgenomic mRNAs.

## 2.5 CoV Replication Cycle

### 2.5.1 Attachment and Entry

Viral infection of host cells begins with receptor recognition by CoVs, which is initiated by the binding interaction between the S protein and its cognate receptors (Fig. 3). Host range and tissue tropism are mainly determined by the interaction between the S protein and its receptor. The cellular receptor for *gammaCoVs* is currently unknown, although sialic acids are thought to serve as non-specific attachment factors [17–19].

Following receptor binding, membrane fusion between the viral and host membrane is mediated by conformational changes of the S protein, largely accomplished by the S2 subunit of the S protein. Cleavage of the S protein occurs at two sites within the S protein, firstly at the (S1/S2) site for separating the receptor-binding domain (RBD) and fusion domain of the S protein and secondly at the (S2′) site for exposing the fusion peptide [20]. The insertion of the fusion peptide into the host membrane upon S2′ cleavage then triggers the joining of two heptad repeats (HRs) in S2, forming an antiparallel six-helix bundle [21]. The formation of this bundle permits membrane mixing to eventually deliver the viral genome into the cytoplasm.

### 2.5.2 Replicase Gene Translation and Processing

The next step of the CoV life cycle involves the translation of the replicase gene from the incoming virion genomic RNA. The replicase gene

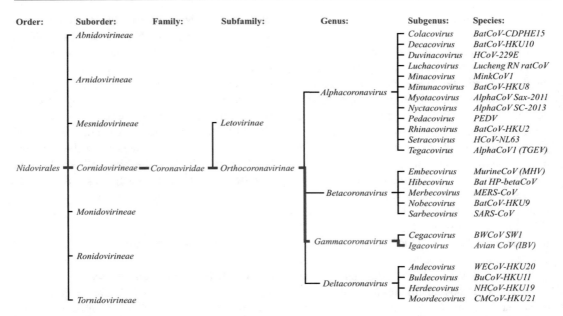

Order: | Suborder: | Family: | Subfamily: | Genus: | Subgenus: | Species:

*Abnidovirineae* — *Colacovirus* — *BatCoV-CDPHE15*
*Decacovirus* — *BatCoV-HKU10*
*Duvinacovirus* — *HCoV-229E*
*Luchacovirus* — *Lucheng RN ratCoV*
*Arnidovirineae* — *Minacovirus* — *MinkCoV1*
*Minunacovirus* — *BatCoV-HKU8*
*Alphacoronavirus* — *Myotacovirus* — *AlphaCoV Sax-2011*
*Nyctacovirus* — *AlphaCoV SC-2013*
*Mesnidovirineae* — *Letovirinae* — *Pedacovirus* — *PEDV*
*Rhinacovirus* — *BatCoV-HKU2*
*Setracovirus* — *HCoV-NL63*
*Tegacovirus* — *AlphaCoV1 (TGEV)*

*Nidovirales* — *Cornidovirineae* — *Coronaviridae* — *Orthocoronavirinae*

*Embecovirus* — *MurineCoV (MHV)*
*Hibecovirus* — *Bat HP-betaCoV*
*Betacoronavirus* — *Merbecovirus* — *MERS-CoV*
*Nobecovirus* — *BatCoV-HKU9*
*Monidovirineae* — *Sarbecovirus* — *SARS-CoV*

*Gammacoronavirus* — *Cegacovirus* — *BWCoV SW1*
*Igacovirus* — *Avian CoV (IBV)*
*Ronidovirineae*

*Andecovirus* — *WECoV-HKU20*
*Buldecovirus* — *BuCoV-HKU11*
*Deltacoronavirus* — *Herdecovirus* — *NHCoV-HKU19*
*Tornidovirineae* — *Moordecovirus* — *CMCoV-HKU21*

**Fig. 1 Current taxonomy of the order *Nidovirales* according to the International Committee on Taxonomy of Viruses (ICTV).** Coronaviruses are classified under the subfamily of *Coronavirinae* in the family *Coronaviridae*

The following virus name abbreviations were used: *BtCoV* bat coronavirus, *PEDV* porcine epidemic diarrhoea virus, *HCoV* human coronavirus, *TGEV* transmissible gastroenteritis virus, *PRCV* porcine respiratory coronavirus, *FCoV* feline coronavirus, *MiCoV* mink coronavirus, *CiCoV* civet severe acute respiratory syndrome CoV, *RatCoV* rat coronavirus, *MHV* murine hepatitis virus, *PHEV* porcine hemagglutinating encephalomyelitis virus, *BCoV* bovine

coronavirus, *CaCoV* canine respiratory coronavirus, *ECoV* equine coronavirus, *DrCoV* dromedary camel coronavirus, *WtDCoV* white-tailed deer coronavirus, *GCoV* giraffe coronavirus, *AnCoV* sable antelope coronavirus, *WBkCoV* waterbuck coronavirus, *SdCoV* sambar deer coronavirus, *RabCoV* rabbit coronavirus, *HeCoV* hedgehog coronavirus, *IBV* infectious bronchitis virus, *CMCoV* common moorhen coronavirus, *WECoV* wigeon coronavirus, *BuCoV* bulbul coronavirus, *ThCoV* thrush coronavirus, *MuCoV* munia coronavirus, *PCoV* porcine coronavirus, *WiCoV* white-eye coronavirus, *NHCoV* night heron coronavirus

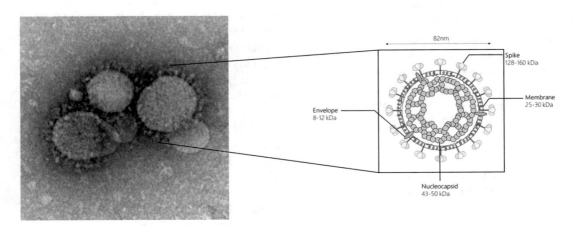

**Fig. 2 Coronavirus morphology.** An electron micrograph image of Middle East respiratory syndrome coronaviruses (MERS-CoV). The structural proteins of coronaviruses that comprise the spike, membrane, nucleocapsid and envelop protein. The molecular weight of each structural protein monomer is as shown. The size of an infectious bronchitis virus virion is 82 nm. (Image source: Cynthia Goldsmith/Maureen Metcalfe/Azaibi Tamin, from https://www.cdc.gov/coronavirus/mers/photos.html)

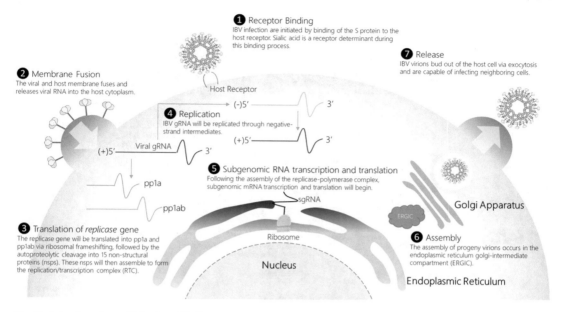

**1** Receptor Binding
IBV infection are initiated by binding of the S protein to the host receptor. Sialic acid is a receptor determinant during this binding process.

**7** Release
IBV virions bud out of the host cell via exocytosis and are capable of infecting neighboring cells.

**2** Membrane Fusion
The viral and host membrane fuses and releases viral RNA into the host cytoplasm.

Host Receptor

(-)5′                      3′

**4** Replication
IBV gRNA will be replicated through negative-strand intermediates.

(+)5′                      3′

Viral gRNA        3′

(+)5′

**5** Subgenomic RNA transcription and translation
Following the assembly of the replicase-polymerase complex, subgenomic mRNA transcription and translation will begin.

pp1a

sgRNA

pp1ab                                           Golgi Apparatus

ERGIC

**3** Translation of *replicase* gene
The replicase gene will be translated into pp1a and pp1ab via ribosomal frameshifting, followed by the autoproteolytic cleavage into 15 non-structural proteins (nsps). These nsps will then assemble to form the replication/transcription complex (RTC).

Ribosome

Nucleus

**6** Assembly
The assembly of progeny virions occurs in the endoplasmic reticulum golgi-intermediate compartment (ERGIC).

Endoplasmic Reticulum

**Fig. 3 Infectious bronchitis virus (IBV) replicase gene and processing scheme of replicase protein products.** (a) Ribosomal frameshifting elements of IBV replicase gene. Pseudoknot stems are indicated as S1 and S2, and loops are indicated as L1 and L3. (b) Translation of the replicase genes ORF1a and ORF1b begins following the release of the viral genome into the host cytoplasm via ribosomal frameshifting, into polyprotein (pp) 1a and 1ab. Pp1a and pp1ab will then be autoproteolytically cleaved at cleavage sites by papain-like protease (PLpro) (in red triangles) and main protease (Mpro) (in orange triangles) into 15 non-structural proteins

encodes two large ORFs, ORF1a and ORF1ab, which share a small overlap and encode two large polyproteins, pp1a and pp1ab, respectively. Polyprotein1ab is translated by a ribosomal frameshifting event (Fig. 4a). As demonstrated by in vitro studies, the incidence of ribosomal frameshifting is as high as 25%. This alternate translation mechanism can help either to regulate the ratio of pp1a and pp1ab proteins or to postpone the translation of products from the 1b coding region until sufficient products of ORF1a are synthesized, thereby creating a suitable environment for RNA replication [22].

Following the production of pp1a and pp1ab, the polyproteins are autoproteolytically processed by its encoded viral proteases to form mature protein products, termed non-structural proteins (nsps) 1–16, except for *gammaCoVs*, which do not encode nsp1 [23] (Fig. 4b). These processed nsps will assemble to form the replication-transcription complex (RTC), creating an environment suitable for RNA synthesis, RNA replication and transcription of subgenomic RNAs.

### 2.5.3 Replication and Transcription

The expression and assembly of the RTC set the stage for viral RNA synthesis, a process which results in the replication of both gRNA and the transcription of multiple sgRNAs. Each sgRNA contains a leader RNA of 70–100 nucleotides (nt) identical to the 5′ end of the genome joined to the body RNA identical to the 3′ portions of the genome. The fusion of the leader and body RNA happens at short motifs on the genome, known as transcriptional regulatory site (TRS).

There is now general consensus that this fusion takes place through discontinuous extension of the negative-stranded RNA [24]. As a result, the negative-stranded sgRNAs in a partial duplex with the positive-stranded gRNA are now used as templates for the synthesis of the corresponding multiple positive-stranded RNAs. In addition to the formation of gRNA and sgRNAs, CoVs can undergo homologous and non-homologous RNA recombination [25, 26]. The ability of these CoVs to recombine may play a prominent role in viral evolution and

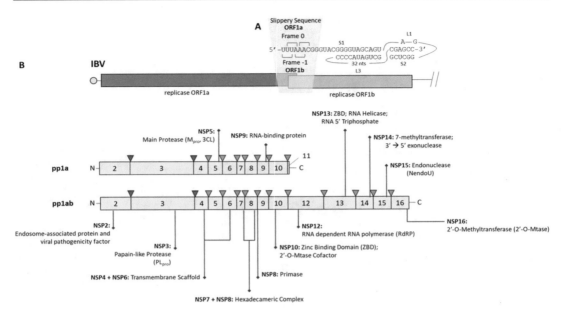

**Fig. 4** **Coronavirus replication cycle**

### 2.5.4   Assembly and Egress

With the replication and transcription of the gRNA and sgRNAs, the next step of the replication cycle is to enable the translation of the structural and accessory proteins, which help to direct the assembly of progeny viruses. The S, membrane (M) and envelope (E) proteins are translated and inserted into the endoplasmic reticulum (ER), where they transit along the secretory pathway into the site of virion assembly – the ER-Golgi intermediate compartment (ERGIC) [27, 28]. Here, the viral genomes encapsulated by the nucleocapsid (N) protein coalesce, in addition with the envelope components, will bud into the ERGIC and translocated to the Golgi to form mature virions [29].

Following assembly and budding, the virions are exported from the infected cells in vesicles and are released by exocytosis. A portion of the S protein, which is not assembled into virions, will translocate to the plasma membrane, where it mediates cell-cell fusion between neighbouring uninfected cells. This leads to the formation of large, multinucleated cells, which promote virus

in preventing the accumulation of deleterious mutations.

spread within an organism without being detected by virus-specific antibodies.

### 2.6   CoV Genetics and Reverse Genetics

In the past, CoV genetics are broadly restricted to the analysis of three types of mutants. The first were naturally arising deletion mutants, which offered clues to some of the phenotypic changes found in various pathogenic strains. The second were defective RNA templates, which depend on replicase proteins provided *in trans* by a helper virus [30–33]. The last were temperature-sensitive (*ts*) mutants isolated from MHV and IBV using chemical mutagenesis and by adaptation to different culture temperatures [34–37]. Reversible and easy to use, *ts* mutants quickly became powerful tools for studying gene function after natural mutants. However, a great deal of efforts must be expended to produce a comprehensive collection of mutants representing the possible complementation groups of cistrons encoded in the CoV genome, consequently limiting their use in CoV genetics.

Reverse genetics for CoVs were developed as early as the 1990s, when the large size of the CoV genome and expression of specific CoV cDNA sequences in bacterial cloning systems became major impediments in the generation of recombinant CoVs using classical genetic techniques [38]. Here, we review four approaches developed for building CoV infectious cDNAs using IBV, FIPV and TGEV as models and how each of these systems has been extended to the generation of CoV replicon RNAs and CoV-based vaccine vectors.

# 3 Construction of the CoV Vectors

## 3.1 Reverse Genetics Systems for CoV

### 3.1.1 Homologous RNA Recombination

Targeted RNA recombination was the first reverse genetics method developed when it was not clear if the construction of full-length infectious cDNA clones would ever be feasible [39]. This system was originally designed for MHV, which harnesses on the intrinsic property of the CoV replication machinery to recombine RNA molecules. It involves generation of a chimeric RNA donor bearing the desired mutations being transfected into cells, which have been infected with a parental virus presenting certain characteristics (i.e. *ts* or host range-based selection) that can be selected against [40–42]. Recombinant CoVs generated by targeted RNA recombination can then be isolated by counterselection of the parental virus and purified.

Despite the existence of RNA recombination since the 1990s, attempts at recombining IBV RNA molecules have proved to be unsuccessful until recently, when van Beurden and colleagues established a targeted RNA recombination reverse genetics system on IBV H52 (Fig. 5) [43]. Using this approach, the IBV donor plasmid, pIBV, was constructed as stepwise ligation of fragments derived from five plasmids. Each of these five plasmids contains gene fragments of various IBV structural and accessory genes. To construct a recombinant chimeric murinized (m) IBV intermediate (mIBV) donor plasmid, the ectodomain of MHV A59 spike gene was amplified by PCR and ligated into a plasmid to produce p-MHV-S [44–48]. Targeted RNA recombination of IBV proceeds in two steps. Construction of the recombinant chimeric mIBV was the first step. In the next step, the recombinant IBV (rIBV) was generated by exchanging the IBV S ectodomain back into the mIBV genome.

With this method, manipulating the 3′ one-third of the CoV genome, which consists of the structural and accessory genes as well as the 3′ UTR, has been successful in FIPV [49], TGEV [50] and most recently, in IBV [43]. However, despite its value, there are clear drawbacks in this system. For technical reasons, this method cannot be expanded to the 5′ two-thirds of the genome, where the replicase is located due to the requirement of these gene products for virus passage. As such, to get around the barriers presented by the huge size of the replicase gene and the instability of key regions propagated in bacterial clones, three other methods were developed to overcome these challenges, namely, construction of the full-length cDNA clones in bacterial artificial chromosomes (BACs), in vitro ligation and propagation of full-length cDNAs in vaccinia virus-based vectors.

### 3.1.2 Bacterial Artificial Chromosomes (BACs)

The BAC system is based on *Escherichia coli* and its single-copy F vector, which follows a strictly controlled replication to produce only one or two copies per cell. Amongst bacterial cloning systems, BAC has been chosen for reverse genetics as it has been shown to be capable of maintaining large DNA fragments from various genomic sources with high structural stability even after generations of serial growth [51]. The first full-length cDNA-based reverse genetics system for CoVs using the BAC approach was developed for TGEV in 2000 (Fig. 6) [52, 53]. In this system, the full-length cDNA copy of the TGEV genome was assembled in a synthetic low-copy-number BAC pBeloBAC11, downstream of a

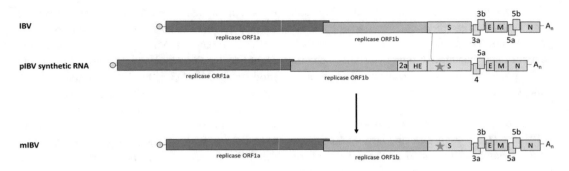

**Fig. 5 Targeted RNA recombination.** An interspecies chimeric murinized IBV containing an MHV S domain (mIBV) is generated via a single recombination event between the IBV genomic RNA with the synthetic donor plasmid (p-IBV synthetic RNA) bearing a mutation (star) in the spike (S)

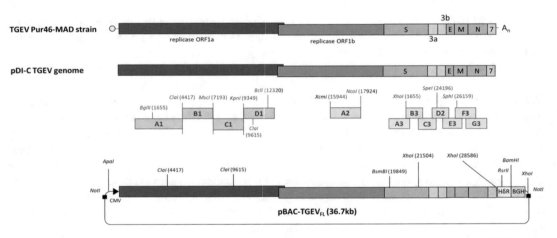

**Fig. 6 Assembly of full-length TGEV cDNA clone using bacterial artificial chromosome (BAC) method.** To construct the complete cDNA of the TGEV genome, a plasmid containing a cDNA encoding TGEV-derived DI (pDI-C) was used. As pDI-C RNA contains three deletions across the genome, a set of cDNAs encoding the fragments of the missing regions were generated by RT-PCR (highlighted in yellow boxes). The final TGEV cDNA was flanked by a CMV promoter at the 5′ end and a 24 bp poly(A) tail at the 3′ end, followed by the hepatitis delta virus ribozyme (HδR) and bovine GH termination (BGH) and polyadenylation sequences (Adapted from [52])

cytomegalovirus (CMV) promoter [54]. Downstream of the 3′ end of the genomic RNA are also a poly(A) tail, the hepatitis delta virus (HDV) ribozyme and a bovine growth hormone (BGH) termination and polyadenylation sequence to ensure that the synthetic RNAs produced contain the genuine 3′ end sequence of the genome. The infection is initiated from the transcription of the transfected cDNA by using the host RNA polymerase II [55]. This method is advantageous in avoiding the potential limitations of in vitro capping and transcription of genomic RNA. More

recently, modified BAC approaches have been used to generate the full-length cDNA clone of SARS-CoV Frankfurt-1 under the control of a T7 RNA polymerase promoter instead of the CMV promoter, to prevent mRNA splicing after transcription of the full-length cDNA by host RNA polymerase II in the nucleus [56].

BAC clones can also be modified into E. coli cells by homologous recombination, using the red recombination system and homing endonuclease I-SceI, for counter-selection [57]. This method employs      a      sequence      bearing      the      desired modifications and allows an accurate and efficient introduction of modifications to a BAC clone in a "scarless" fashion.

The BAC system poses several advantages. First, manipulation of the BAC clones follows standard cloning protocols, such as selecting suitable restriction sites and ligating cDNA fragments into the vector, making it relatively easy and essentially similar to molecular manipulation of a conventional plasmid [51]. Likewise, the high stability of the exogenous DNA sequences provided by the BACs permitted infinite cDNA production in the cell. Together with the high transfection efficiency of BACs in mammalian cells, this system quickly becomes an attractive tool for CoV reverse genetics [58] and has been successfully employed to study the role of various viral proteins in replication and pathogenesis as well as for the production of genetically attenuated viruses that are potential vaccine candidates for SARS-CoV [59–61] and MERS-CoV [62].

### 3.1.3 In Vitro Ligation of Full-Length Genomic cDNA

The third system constructs a full-length genomic cDNA through the systematic assembly of smaller cloned cDNA fragments via in vitro ligation [44]. Using this approach, each smaller clone was ligated in a directed manner using asymmetric restriction sites and then transfected to permissive host cells. The full-length cDNA was in vitro transcribed to generate a capped, full-length RNA transcript, from which infectious viruses can be rescued after co-transfection with a capped N gene RNA transcript. This method has been successfully used to recover clones of many CoV, including TGEV [44], MERS-CoV [63], MHV [46], SARS-CoV [47] and IBV [64–66].

This in vitro cDNA assembly approach is proven to be simple and straight forward, as it allows rapid mutagenesis of independent cDNA fragments in parallel using conventional techniques and is compatible with other reverse genetics methods, such as BAC and vaccinia vectors. However, the initial application of this approach met two potential caveats: one was the

generation of premature T7 transcription termination signals during in vitro transcription of the full-length viral RNA, and another was the lack of restriction enzymes that provide a unique overhang that does not randomly self-assemble. To overcome the first problem, several mutations were inserted into the genome to avoid potential T7 transcription termination signals. To overcome the second caveat, a variation of the approach was initially developed to engineer the MHV infectious cDNA by utilizing type IIs restriction enzymes, such as *BsaI* and *SapI* at the ends of each cDNA fragments [46]. These enzymes recognize asymmetrical sites and leave behind 1–4 nt overhangs that could be ligated in vitro using standard protocols without changing the viral sequence. Through cleavage and ligation of these fragments, the added restriction sites are removed, leaving the exact viral sequence at the junction, allowing one to generate cDNA clones without introducing mutations to the viral genome sequence. Additionally, it is possible to introduce mutations to the CoV genome at any position by designing primers, which incorporate a type IIs restriction site and mutation of interest to the viral genome [46, 67].

By using this strategy, the construction of a full-length cDNA clone derived from a Vero cell-adapted IBV Beaudette strain was obtained by in vitro assembly of five cDNA fragments spanning the entire IBV genome (Fig. 7) [64]. Each of these fragments was generated by RT-PCR using specific primers that introduced *BsmBI* or *BsaI* into the 5′ and 3′ ends of the fragments. These cDNA fragments were digested and systematically and unidirectionally assembled into a full-length cDNA by in vitro ligation. Following in vitro transcription, the genome-length transcripts were electroporated into Vero cells and used to recover the infectious viruses [63, 68].

This approach has been used with success at identifying and evaluating promising zoonotic vaccine candidate strains for SARS-CoV [69] and developing a recombination safe vaccine platform by rewiring the viral TRSs [70].

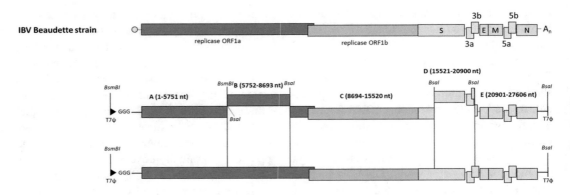

**Fig. 7 Assembly of an IBV full-length cDNA clone by in vitro ligation.** A full-length cDNA of Vero cell-adapted IBV Beaudette strain was assembled by in vitro ligation of five contiguous cDNA fragments (A to E) spanning the entire viral genome, which were flanked by native or engineered BsmBI and BsaI restriction sites. The assembled full-length cDNA contained a T7 RNA polymerase promoter (T7) at the 5′ end and poly(A) tail at the 3′ end, allowing for the in vitro transcription of a full-length, capped polyadenylated transcripts. The viral genes and relevant restriction sites are indicated (Adapted from [64])

### 3.1.4 Vaccinia Virus-Based Vectors

The last system involves the use of vaccinia virus cloning vector to generate full-length CoV genome [71, 72]. This system represents a basic approach to CoV genetics and was first reported for the generation of human CoV 229E (HCoV-229E) and, soon after, extended to FIPV [73, 74], IBV [45] and MHV-A59 [75]. Vaccinia virus vectors were attractive reverse genetics tools for several reasons. First, vaccinia virus has the capacity to accommodate at least 25 kb of foreign DNA sequence, allowing recombinant vaccinia viruses to replicate in cell culture without compromising on viral titer when compared to non-recombinant virus [76]. Second, vaccinia virus vectors have been designed for foreign DNA insertion by in vitro ligation, negating the need to generate plasmid intermediates to carry the entire cDNA insert [77]. Finally, the cloned cDNA insert is open to mutagenesis through vaccinia virus-mediated homologous recombination [78, 79].

The generation of a reverse genetics system for IBV will be described to illustrate the vaccinia virus-based system (Fig. 8) [45]. In principle, the procedure can be divided into two parts: generation of a full-length IBV cDNA and the recovery of the infectious rIBV. Firstly, three plasmids (pFRAG-1, pFRAG-2 and pFRAG-3) which contain the contiguous regions of the IBV genome were constructed for the final assembly of the full-length cDNA. The assembly of the full-length cDNA clone, which encompasses the entire IBV genome downstream of the T7 promoter, was achieved via a two-step in vitro ligation method. Secondly, the in vitro ligation products, comprising the full-length IBV cDNA with dephosphorylated Bsp102I ends, were ligated to the *NotI* arm present in the genomic DNA of the vaccinia virus vector vNotI/tk. The ligation products were then transfected into CK cells, which were previously infected with a fowlpox virus (FPV) expressing T7 polymerase as a helper to recover the recombinant IBV [80]. A second plasmid, pCi-Nuc under the control of both the CMV and T7 RNA polymerase promoter, was also co-transfected to express IBV N [81]. FPV was selected as a helper virus, because of the abortive FPV infection in the mammalian cells together with the rarity of recombination events between two poxviruses [71, 82].

Two recombination methods have also been developed based on the sequential use of *E. coli* guanine-phosphoribosyl transferase gene (gpt) as both a positive and negative selection marker

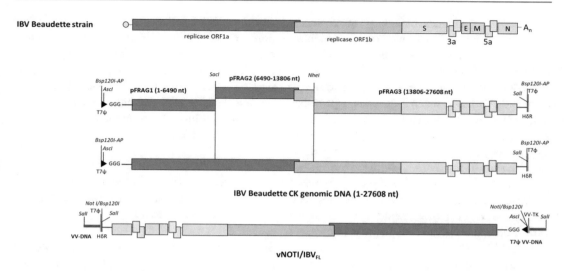

**Fig. 8** **Assembly of an IBV full-length cDNA clone in the vaccinia virus genome.** Three contiguous cDNA fragments (pFRAG1 to pFRAG3) spanning the entire viral genome of IBV Beaudette CK strain was cloned in vaccinia virus DNA (V) by in vitro ligation. pFRAG1 contained a T7 RNA polymerase promoter (T7) at the 5' end and pFRAG3 fragment a 28 nt poly(A) tail at the 3' end. The assembly of the full-length IBV cDNA and its insertion into the vaccinia virus genome is shown with the final orientation of the cDNA in the vaccinia genome. (Adapted from [45]). *AP* dephosphorylated ends, *T7ψ* T7 promoter, *T7φ* T7 terminator, *HδR* hepatitis delta antigenome ribozyme

[72, 83]. In the first method, the CoV region of interest in the recombinant vaccinia virus was substituted with the gpt gene in a plasmid, and the gpt gene was flanked by CoV sequences to facilitate a double recombination event. The second method, known as transient dominant selection [79, 83], requires the modified CoV cDNA region to be inserted into a plasmid containing the gpt selective marker under the control of a vaccinia virus promoter. The complete sequence of the plasmid (including the CoV genome) was then transiently integrated into the vaccinia virus by homologous recombination with a single crossover event. The recombinant vaccinia virus containing the modified CoV cDNA region was then gpt-positively selected.

Collectively, it is possible to genetically modify CoV genomes at desirable positions using these approaches. Recombinant CoVs with gene inactivation, deletions or attenuation modifications can now be generated to study the role of specific gene products in viral replication and pathogenesis. Most importantly, attenuated viruses can be potential vaccine candidates, and modified CoV genomes have been developed as eukaryotic, multigene expression vectors [84].

## 3.2 Construction of CoV Replicons

In line with the development of viral vectors based on CoVs, previous studies have also explored the possibilities of constructing CoV replicons. Replicons are autonomous replicating RNAs encoding all viral proteins and *cis*-acting elements required for RNA replication and are considered as a safe alternative to full-length viral genomes as they lack structural genes to produce infectious virus particles. In order to construct CoV replicons, it is important to identify viral genes and sequence elements that are indispensable for efficient CoV replication. By analogy to other positive-stranded RNA viruses, it was assumed that the replicase gene and *cis*-acting elements at the 5' and 3' termini would be sufficed for efficient CoV replication and transcription. However, reverse genetics analyses revealed a role of CoV N proteins for efficient CoV replication [85, 86]. Therefore, the basic units of a CoV replicon include (i) the replicase gene, (ii) the 5′ and 3′ genomic termini containing the *cis*-acting elements required for replication and transcription and (iii) the nucleocapsid gene.

CoV replicons have been constructed for many animal CoVs using reverse genetics approaches [85, 87] and are currently being used as a non-infectious system to analyze CoV replication and transcription [88–90]. This system is particularly useful if the corresponding virus grows poorly in tissue culture or if the pathogenicity of the virus is a matter of concern. Several replicon versions have also been reported to include a variety of reporter genes such as green fluorescent protein (GFP), firefly luciferase (FLuc) or Renilla luciferase (RLuc) alone or in combination with antibiotic resistance genes [91–93]. In the case of SARS-CoV, these replicons are particularly useful in the identification of viral and host factors involved in CoV RNA synthesis [94] and antiviral drug testing [89].

With reverse genetics, the generation of CoV replicons can be derived from replication-competent, propagation-defective viruses, as in the case for TGEV which lacked an E gene [87, 95] as well as MERS-CoV [62]. The rationale behind this approach is to generate CoV replicons through deletion of one or more structural genes and to express the missing structural gene(s) *in trans*. In a study reporting the construction of a replicon for HCoV-229E, all the structural genes (including N) and accessory genes were removed, while three reporter genes (chloramphenicol acetyltransferase (CAT), FLuc and GFP) were incorporated, each located downstream of an HCoV-229E TRS [84]. Expression of the reporter genes could be detected in BHK-21 cells transfected with the replicon, but the packaging of the replicon genome into a virus-like particle (VLP) required simultaneous transfection of the replicon RNA, a helper HCoV-229E full-length genomic RNA, and the mRNA for the N gene. However, the availability of VLPs produced in this study was low, and a substantial amount of helper virus was found in VLP stocks [84].

In order to improve the biosafety standard of such vectors, safety guards must be introduced to work in a BSL2 containment facility. Reconstitution of wild-type constructs using vectors containing rearranged structured genes [96] or rewiring of the CoV transcription circuit [70] may be used. In this line, SARS-CoV replicons contain changed TRS elements, which resulted in an incompatibility of recombinant and wild-type TRS elements [70].

## 3.3 Expression of Heterologous Genes by Recombinant CoV or CoV Replicons

Several approaches to incorporate heterologous genes into the genomes or replicons have been tested for various animal coronaviruses. These include (1) insertion of a transcriptional cassette containing a TRS upstream of the gene of interest, (2) replacement of the coding sequence for one or more accessory genes with that of a heterologous gene, (3) expression of heterologous genes fused with coronavirus structural gene, and (4) insertion of heterologous genes between non-structural proteins encoded by the replicase gene.

Using reporter genes, such as EGFP and FLuc, the expression efficiency of a heterologous gene and the genetic stability of various recombinant viruses were investigated using IBV as a vector. When the EGFP preceded by a TRS from IBV ORF5 was inserted after the M or N gene or when EGFP was fused to the C-terminal region of the S gene, infectious recombinant viruses expressing EGFP could be recovered [97]. However, these recombinant viruses were only genetically stable up to passage 5, after which the inserted EGFP gene was lost [97]. On the other hand, when the accessory genes 3a and 3b were replaced with FLuc gene, the recombinant IBV exhibited stable expression of luciferase activity up to passage 15 [97]. In sharp contrast, the insertion of TRS-FLuc cassette after the M or N gene resulted in highly unstable viruses. Using the same system, two viral proteins (SARS-CoV ORF6 and DENV1 core) and a host protein (eIF3f) were also expressed [97]. While the genomic location is a critical determinant for successful heterologous gene insertion in the CoV vector, the sequence composition and size of the inserted gene also have major impacts on the expression level and stability of the recombinant virus.

It is also possible to express heterologous genes by modifying the replicase gene. This has been demonstrated for HCoV-229E-based replicon RNA, where a selectable marker, the neomycin resistance (neo) gene, has been inserted between nsp1 and nsp2 [88]. Since the C-terminus of nsp1 and the N-terminus of nsp2 are released by proteolytic processing in the wild-type context, a "2A-like" autoprocessing peptide has been used to liberate a slightly modified nsp1 C-terminus. The neo gene was cloned downstream of the 2A-like element followed by an IRES that drives the translation of nsp2-nsp16. This selection cassette, composed of the 2A-like element, the neo gene and the IRES element, has been used to generate stable replicon-containing cell lines using conventional G418 selection after transfection of the replicon RNA into eukaryotic cells. An attractive application for these non-infectious replicon cell lines is the identification and evaluation of CoV replicase inhibitors [88, 89].

The nsp2 is dispensable for the replication of MHV and SARS-CoV [98]. In one study, the coding sequence for nsp2 was deleted in the MHV genome, and a reporter gene (GFP or FLuc) was fused to nsp3 at the N-terminus, resulting in the recovery of recombinant virus MHV-Δ2-GFP/FFL3 [99]. Meanwhile, the coding sequence for GFP or FLuc was fused at the N-terminus of nsp2, and a cleavage site was introduced to allow proteolytic cleavage between nsp1 and GFP/FFL-nsp2, resulting in the recovery of recombinant virus MHV-GFP/FFL2 [96]. Compared with wild type control, replication of MHV-Δ2-GFP/FFL3 was delayed and the virus titer reduced by 1-$\log_{10}$, whereas replication of MHV-GFP/FFL2 was not significantly affected. MHV-GFP/FFL2 was genetically stable up to passage 5 [96]. Subsequent studies have also used the same approach to express nsp2 proteins fused with APEX2 ascorbate peroxidase for electron microscopy or BirA$_{R118G}$ biotin ligase for proximity labelling studies [100].

## 4 Application of Recombinant CoVs as Vaccine Vectors

### 4.1 Recombinant CoVs as Vaccine Vectors

The observation that accessory genes are dispensable for coronavirus replication allows for the replacement of these genes by heterologous genes. In principle, CoV vectors have room for the insertion of large heterologous genes and expression of multiple genes is feasible [84]. CoV reverse genetic systems are therefore used to assess the efficacy of CoV-based expression vectors for producing large amounts of heterologous proteins or as immunogenic vectors in the context of vaccine development and immunotherapy. The extraordinary large genome and unique transcription strategy make CoVs promising candidates for the development of multigene expression vectors [101, 102].

Based on the following three facts, CoVs would be promising vectors for vaccine development [103]. First, some of CoV accessory genes are amenable to deletion and can be used to produce attenuated viruses. Second, CoV host tropism can be altered by manipulating the S protein. Finally, heterologous genes can be expressed by inserting into CoV genomes with appropriate CoV transcription signals.

### 4.2 CoV Vaccines Against Virulent Strains by the Replacement of S Gene

S protein is the major determinant of host and tissue tropisms for coronaviruses. Using the vaccinia virus approach, Godeke and colleagues identified 64 residues comprising the endodomain and transmembrane domain of the MHV S protein to be critical in packaging into MHV VLPs [104]. This information was then used to generate recombinant viruses expressing a chimeric S protein made up of the C-terminal 64 residues of MHV fused to the ectodomain of FIPV S protein [105]. The recombinant virus, fMHV, lost the ability to infect murine cells but can now infect feline cells, demonstrating that the chimeric spike protein conferred a switch in species tropism. Similar studies have been conducted on TGEV and IBV, providing the evidence that the cell tropism of these viruses is correlated with the S protein expressed by the recombinant viruses [50, 106–108].

In one earlier attempt, the ectodomain of the S protein of an apathogenic Beaudette strain of IBV (Beau-R) was replaced with that from the pathogenic M41 strain to produce a recombinant IBV BeauR-M41(S) [106]. The replacement changed the cell tropism of the virus in vitro, but no significant differences in pathogenicity were observed between the recombinant BeauR-M41 (S) and the apathogenic parent Beau-R. Remarkably, BeauR-M41(S) induced much greater protection (77%) against challenge with M41 compared with Beau-R (11%). Therefore, the S gene exchange between apathogenic and pathogenic strains may be a new direction in IBV vaccine development.

In a more recent study, the S ectodomain of a Vero cell-adapted Beaudette strain of IBV (p65) was replaced with that from the H120 vaccine strain [109]. The resulting recombinant IBV retained the ability to replicate in Vero cells and induced the production of specific antibodies for S, M and N proteins in the immunized chickens. It also induced protection (80%) against challenge with the virulent M41 strain. In a separate study, the entire S gene of H120 strain was replaced with that of the cell-adapted p65 strain [110]. The recombinant R-H120-p65(S) virus thus acquired the ability to grow in Vero cells, while in vivo pathogenicity remained similar to the parental H120 strain. The R-H120-p65 (S) also induced protection (80%) against challenge with the virulent M41 strain. These studies demonstrated that by partial or complete replacement of the S gene, recombinant coronaviruses can be generated that exhibit desirable neutralizing epitopes and/or acquire cell culturability, providing new insights for the development of novel recombinant vaccines.

## 4.3 CoV as Vaccine Vectors Expressing Heterologous Viral Proteins

The potential to express heterologous genes in CoV-based vaccine vectors has been well demonstrated in studies using TGEV [111]. To increase the cloning capacity, ORF3a and ORF3b were deleted in the parental virus (rPUR-MAD-SC11), resulting in the generation of the rTGEV-Δ3 vector. The heterologous GFP gene, followed by TRS of 3a, was then inserted to replace the deleted ORF3a and ORF3b, yielding rTGEV-Δ3-TRS3a-GFP. Both rTGEV-Δ3 and rTGEV-Δ3-TRS3a-GFP showed similar replication kinetics to those of the parental virus in cell culture, although slightly attenuated in infected animals. Importantly, rTGEV-Δ3-TRS3a-GFP stably expressed GFP ($>40$ μg/$10^6$ cells) for up to 20 passages. Immunization of pregnant sows with recombinant virus demonstrated that GFP-specific antibodies were detected in the immunized sows and their progeny, as well as in the colostrum from day 1 of lactation. This study was amongst one of the early attempts to demonstrate that promising potential of using CoV as vaccine vectors.

Using the same system, engineered TGEV vectors expressing antigens from porcine reproductive and respiratory syndrome virus (PRRSV) were constructed. Currently, live PRRSV vaccines only provide partial protection against clinical disease and sometimes revert to virulence, while killed PRRSV vaccines are generally less effective. Previous studies have shown that GP5 and M proteins are involved, respectively, in the induction of neutralizing antibodies and cellular immune response during PRRSV infection [112]. To express PRRSV GP5 and M, a bicistronic expression cassette was adopted to replace the ORF3a/ORF3b expression cassette in TGEV. Both proteins were efficiently expressed in the infected cells and tissues from infected piglets. When 1-week-old piglets were immunized with this recombinant TGEV (rTGEV), antibodies specific for PRRSV GP5 and M were produced, but the immune response was limited against challenge with a virulent European PRRSV strain [109]. This is likely due to the low level of neutralizing antibodies induced by the rTGEV vector.

Presumably, due to the toxicity of GP5, rTGEV vectors expressing PRRSV GP5 and M were not fully stable, and GP5 expression was lost after 8–10 passage in cell culture. In order to promote induction of neutralizing antibodies, a point mutation was introduced into a glycosylation site near the epitope critical for neutralization, resulting in the generation of rTGEV-GP5-N46S-M virus. Piglets immunized with killed or live virus of this rTGEV vector produced a higher level of anti-GP5 and neutralizing antibodies and exhibited less severe disease upon challenge, although the immune response was not strong enough for full protection.

Because the full-length GP5 protein was not stably expressed, rTGEV expressing a truncated version of GP5 consisting of its N-terminal domain without the signal peptide was constructed [113]. Additionally, rTGEVs expressing a smaller domain containing neutralizing epitopes of the minor envelope protein GP3 and GP4 were also created. In yet another rTGEV, the PRRSV M protein was used as a scaffold to express the GP3 neutralizing epitope in the N-terminal region. The truncated PRRSV GP3, GP4 and GP5 fragments were stable up to passage 16 of the rTGEV vectors. These four rTGEVs together with one that expresses PRRSV M protein were used to immunize 12-days-old piglets. Compared with the non-immunized group inoculated with empty rTGEV vector, the immunized group exhibited less severe clinical symptoms and lower levels of lung inflammation when challenged with the virulent PRRSV Olot91-like strain. PRRSV titre was also slightly lower in the immunized group with a higher humoral response against GP5, suggesting that the rTGEV vectors expressing PRRSV antigens indeed provided partial protection against PRRSV infection.

## 5    Future Directions and Other Potential Applications

The high mortality posed to livestock and domestic animals by animal CoVs and the lack of specific antivirals and vaccines have greatly motivated investigators to understand this family of viruses at the molecular level, in order to decipher its interactions with the host and to develop strategies to prevent and control CoV infections. To this end, structural and molecular genetics analyses of the CoV genome have been enabled by reverse genetics. Previous caveats faced in the development of infectious cDNA clones, such as the large size of the CoV genome and the instability of its cDNA sequences in bacterial systems have now triumphed over, thanks to new creative approaches in four reverse genetics systems. These reverse genetics approaches based on homologous recombination, full-length cDNA clones in BACs, in vitro ligation and vaccinia virus vectors made CoV full-length infectious clones available for the study of CoV replication, virus-host interactions and pathogenesis, as well as for antiviral drug screening. It would accelerate the development of vaccines and may be used as a vaccine vector for veterinary pathogens without the need for manipulating infectious viruses. Nevertheless, it should be kept in mind that reverse genetics has its own limitations. For instance, full elucidation of viral gene functions that are essential for RNA synthesis by reverse genetics has proved to be difficult. For example, research aimed at investigating the functions of replicase genes by introducing mutations/insertion at certain positions of the viral genome, rendering no recovery of infectious virus. To overcome these pitfalls, individual gene functions may be studied using bioinformatics or structural means. Knowledge deduced from these studies may give insights into the CoV RNA synthesis while exploiting the full potential of CoV reverse genetics.

An important prerequisite for viral vaccine vectors is the delivery efficiency of genetic material to specific target cells, such as targeting of viral vaccine vectors to antigen-presenting cells

(APCs). In the case of MHV, its cognate receptor carcinoembryonic antigen-related cell adhesion molecule 1 (CEACAM1) is expressed on murine DCs, and given the abundance of CEACAM1 on the DC surface, MHV-based VLPs containing MHV-based vector RNAs may be used to transduce murine DCs [114]. Along with well-established immunological techniques in inbred and transgenic mice, recombinant MHV vectors in the murine model may guide the development of CoV vaccine vectors and pave the way for CoV-based vaccine in livestock and domestic animals [114, 115].

Self-replicative mRNA vaccines, which have been developed with certain RNA viruses, would offer an alternative strategy. For instance, recombinant alphavirus replicon particles are created exclusively from the structural proteins of the donor alphavirus, but the genomic RNAs contained in these particles are chimeric. In this case, the structural proteins of alphavirus are replaced by those from heterologous viruses. Using a similar strategy, a PEDV vaccine was developed using the Venezuelan equine encephalitis virus replicons expressing the PEDV S gene [116]. With a better functional understanding of the CoV replicase gene, a similar approach may be viable to establish mRNA vaccine vectors based on CoV.

## 6    Summary

The large genome size and unique transcription strategy of CoVs make them promising candidates for the development of vaccine vectors. Other characteristics that make them ideal for use as vaccine vectors include the ability to manipulate their genome leading to the generation of attenuated viruses and change their cell tropism as well as the ability to use them to create multigene expressing vectors. Further understanding of genes such as the replicase gene is required to further develop CoV as an efficacious veterinary vaccine vector.

## References

1. Schalk AF, Hawn MC. An apparently new respiratory disease of baby chicks. J Am Vet Med Assoc. 1931;78:413–6.
2. Bushnell LD, Brandly CA. Laryngotracheitis in chicks. Poult Sci. 1933;12(1):55–60.
3. Almeida JD, Tyrrell DA. The morphology of three previously uncharacterized human respiratory viruses that grow in organ culture. J Gen Virol. 1967;1(2):175–8.
4. Hamre D, Procknow JJ. A new virus isolated from the human respiratory tract. Proc Soc Exp Biol Med. 1966;121(1):190–3.
5. McIntosh K, Becker WB, Chanock RM. Growth in suckling-mouse brain of "IBV-like" viruses from patients with upper respiratory tract disease. Proc Natl Acad Sci U S A. 1967;58(6):2268–73.
6. McIntosh K, Dees JH, Becker WB, Kapikian AZ, Chanock RM. Recovery in tracheal organ cultures of novel viruses from patients with respiratory disease. Proc Natl Acad Sci U S A. 1967;57(4):933–40.
7. Witte KH, Tajima M, Easterday BC. Morphologic characteristics and nucleic acid type of transmissible gastroenteritis virus of pigs. Arch Gesamte Virusforsch. 1968;23(1):53–70.
8. Tyrrell DA, Almeida JD, Cunningham CH, Dowdle WR, Hofstad MS, McIntosh K, et al. Coronaviridae Intervirol. 1975;5(1–2):76–82.
9. Gu J, Korteweg C. Pathology and pathogenesis of severe acute respiratory syndrome. Am J Pathol. 2007;170(4):1136–47.
10. Lefkowitz EJ, Dempsey DM, Hendrickson RC, Orton RJ, Siddell SG, Smith DB. Virus taxonomy: the database of the international committee on taxonomy of viruses (ICTV). Nucleic Acids Res. 2018;46(D1):D708–D17.
11. McIntosh K. Coronaviruses: a comparative review. Curr Top Microbiol Immunol. 1974;63:85–129.
12. Masters PS. The molecular biology of coronaviruses. Adv Virus Res. 2006;66:193–292.
13. Kienzle TE, Abraham S, Hogue BG, Brian DA. Structure and orientation of expressed bovine coronavirus hemagglutinin-esterase protein. J Virol. 1990;64(4):1834–8.
14. Yokomori K, La Monica N, Makino S, Shieh CK, Lai MM. Biosynthesis, structure, and biological activities of envelope protein gp65 of murine coronavirus. Virology. 1989;173(2):683–91.
15. Chen Y, Guo D. Molecular mechanisms of coronavirus RNA capping and methylation. Virol Sin. 2016;31(1):3–11.
16. Yang D, Leibowitz JL. The structure and functions of coronavirus genomic 3′ and 5′ ends. Virus Res. 2015;206:120–33.
17. Winter C, Schwegmann-Wessels C, Cavanagh D, Neumann U, Herrler G. Sialic acid is a receptor determinant for infection of cells by avian infectious bronchitis virus. J Gen Virol. 2006;87(Pt 5):1209–16.

18. Winter C, Herrler G, Neumann U. Infection of the tracheal epithelium by infectious bronchitis virus is sialic acid dependent. Microbes Infect. 2008;10 (4):367–73.

19. Abd El Rahman S, El-Kenawy AA, Neumann U, Herrler G, Winter C. Comparative analysis of the sialic acid binding activity and the tropism for the respiratory epithelium of four different strains of avian infectious bronchitis virus. Avian Pathol. 2009;38(1):41–5.

20. Yamada Y, Liu DX. Proteolytic activation of the spike protein at a novel RRRR/S motif is implicated in furin-dependent entry, syncytium formation, and infectivity of coronavirus infectious bronchitis virus in cultured cells. J Virol. 2009;83(17):8744–58.

21. Bosch BJ, van der Zee R, de Haan CA, Rottier PJ. The coronavirus spike protein is a class I virus fusion protein: structural and functional characterization of the fusion core complex. J Virol. 2003;77 (16):8801–11.

22. Araki K, Gangappa S, Dillehay DL, Rouse BT, Larsen CP, Ahmed R. Pathogenic virus-specific T cells cause disease during treatment with the calcineurin inhibitor FK506: implications for transplantation. J Exp Med. 2010;207(11):2355–67.

23. Lim KP, Ng LF, Liu DX. Identification of a novel cleavage activity of the first papain-like proteinase domain encoded by open reading frame 1a of the coronavirus avian infectious bronchitis virus and characterization of the cleavage products. J Virol. 2000;74(4):1674–85.

24. Sawicki SG, Sawicki DL, Siddell SG. A contemporary view of coronavirus transcription. J Virol. 2007;81(1):20–9.

25. Lai MM, Baric RS, Makino S, Keck JG, Egbert J, Leibowitz JL, et al. Recombination between nonsegmented RNA genomes of murine coronaviruses. J Virol. 1985;56(2):449–56.

26. Keck JG, Makino S, Soe LH, Fleming JO, Stohlman SA, Lai MM. RNA recombination of coronavirus. Adv Exp Med Biol. 1987;218:99–107.

27. Tooze J, Tooze S, Warren G. Replication of coronavirus MHV-A59 in sac- cells: determination of the first site of budding of progeny virions. Eur J Cell Biol. 1984;33(2):281–93.

28. Krijnse-Locker J, Ericsson M, Rottier PJ, Griffiths G. Characterization of the budding compartment of mouse hepatitis virus: evidence that transport from the RER to the Golgi complex requires only one vesicular transport step. J Cell Biol. 1994;124 (1–2):55–70.

29. de Haan CA, Rottier PJ. Molecular interactions in the assembly of coronaviruses. Adv Virus Res. 2005;64:165–230.

30. Izeta A, Smerdou C, Alonso S, Penzes Z, Mendez A, Plana-Duran J, et al. Replication and packaging of transmissible gastroenteritis coronavirus-derived synthetic minigenomes. J Virol. 1999;73 (2):1535–45.

31. Narayanan K, Makino S. Cooperation of an RNA packaging signal and a viral envelope protein in coronavirus RNA packaging. J Virol. 2001;75 (19):9059–67.

32. Repass JF, Makino S. Importance of the positive-strand RNA secondary structure of a murine coronavirus defective interfering RNA internal replication signal in positive-strand RNA synthesis. J Virol. 1998;72(10):7926–33.

33. Williams GD, Chang RY, Brian DA. A phylogenetically conserved hairpin-type 3′ untranslated region pseudoknot functions in coronavirus RNA replication. J Virol. 1999;73(10):8349–55.

34. Sturman LS, Eastwood C, Frana MF, Duchala C, Baker F, Ricard CS, et al. Temperature-sensitive mutants of MHV-A59. Adv Exp Med Biol. 1987;218:159–68.

35. Martin JP, Koehren F, Rannou JJ, Kirn A. Temperature-sensitive mutants of mouse hepatitis virus type 3 (MHV-3): isolation, biochemical and genetic characterization. Arch Virol. 1988;100 (3–4):147–60.

36. Shen S, Liu DX. Characterization of temperature-sensitive (ts) mutants of coronavirus infectious bronchitis virus (IBV). Adv Exp Med Biol. 2001;494:557–62.

37. Stobart CC, Lee AS, Lu X, Denison MR. Temperature-sensitive mutants and revertants in the coronavirus nonstructural protein 5 protease (3CLpro) define residues involved in long-distance communication and regulation of protease activity. J Virol. 2012;86(9):4801–10.

38. Masters PS. Reverse genetics of the largest RNA viruses. Adv Virus Res. 1999;53:245–64.

39. Masters PS, Rottier PJ. Coronavirus reverse genetics by targeted RNA recombination. Curr Top Microbiol Immunol. 2005;287:133–59.

40. Makino S, Keck JG, Stohlman SA, Lai MM. High-frequency RNA recombination of murine coronaviruses. J Virol. 1986;57(3):729–37.

41. Baric RS, Fu K, Schaad MC, Stohlman SA. Establishing a genetic recombination map for murine coronavirus strain A59 complementation groups. Virology. 1990;177(2):646–56.

42. Kusters JG, Jager EJ, Niesters HG, van der Zeijst BA. Sequence evidence for RNA recombination in field isolates of avian coronavirus infectious bronchitis virus. Vaccine. 1990;8(6):605–8.

43. van Beurden SJ, Berends AJ, Kramer-Kuhl A, Spekreijse D, Chenard G, Philipp HC, et al. A reverse genetics system for avian coronavirus infectious bronchitis virus based on targeted RNA recombination. Virol J. 2017;14(1):109.

44. Yount B, Curtis KM, Baric RS. Strategy for systematic assembly of large RNA and DNA genomes: transmissible gastroenteritis virus model. J Virol. 2000;74(22):10600–11.

45. Casais R, Thiel V, Siddell SG, Cavanagh D, Britton P. Reverse genetics system for the avian coronavirus

infectious bronchitis virus. J Virol. 2001;75 (24):12359–69.

46. Yount B, Denison MR, Weiss SR, Baric RS. Systematic assembly of a full-length infectious cDNA of mouse hepatitis virus strain A59. J Virol. 2002;76(21):11065–78.

47. Yount B, Curtis KM, Fritz EA, Hensley LE, Jahrling PB, Prentice E, et al. Reverse genetics with a full-length infectious cDNA of severe acute respiratory syndrome coronavirus. Proc Natl Acad Sci U S A. 2003;100(22):12995–3000.

48. Baric RS, Sims AC. Development of mouse hepatitis virus and SARS-CoV infectious cDNA constructs. Curr Top Microbiol Immunol. 2005;287:229–52.

49. Haijema BJ, Volders H, Rottier PJ. Switching species tropism: an effective way to manipulate the feline coronavirus genome. J Virol. 2003;77(8):4528–38.

50. Sanchez CM, Izeta A, Sanchez-Morgado JM, Alonso S, Sola I, Balasch M, et al. Targeted recombination demonstrates that the spike gene of transmissible gastroenteritis coronavirus is a determinant of its enteric tropism and virulence. J Virol. 1999;73 (9):7607–18.

51. Shizuya H, Birren B, Kim UJ, Mancino V, Slepak T, Tachiiri Y, et al. Cloning and stable maintenance of 300-kilobase-pair fragments of human DNA in Escherichia coli using an F-factor-based vector. Proc Natl Acad Sci U S A. 1992;89(18):8794–7.

52. Almazan F, Gonzalez JM, Penzes Z, Izeta A, Calvo E, Plana-Duran J, et al. Engineering the largest RNA virus genome as an infectious bacterial artificial chromosome. Proc Natl Acad Sci U S A. 2000;97 (10):5516–21.

53. Gonzalez JM, Penzes Z, Almazan F, Calvo E, Enjuanes L. Stabilization of a full-length infectious cDNA clone of transmissible gastroenteritis coronavirus by insertion of an intron. J Virol. 2002;76 (9):4655–61.

54. Wang K, Boysen C, Shizuya H, Simon MI, Hood L. Complete nucleotide sequence of two generations of a bacterial artificial chromosome cloning vector. BioTechniques. 1997;23(6):992–4.

55. Dubensky TW Jr, Driver DA, Polo JM, Belli BA, Latham EM, Ibanez CE, et al. Sindbis virus DNA-based expression vectors: utility for in vitro and in vivo gene transfer. J Virol. 1996;70 (1):508–19.

56. Pfefferle S, Krahling V, Ditt V, Grywna K, Muhlberger E, Drosten C. Reverse genetic characterization of the natural genomic deletion in SARS-coronavirus strain Frankfurt-1 open reading frame 7b reveals an attenuating function of the 7b protein in-vitro and in-vivo. Virol J. 2009;6:131.

57. Tischer BK, Smith GA, Osterrieder N. En passant mutagenesis: a two step markerless red recombination system. Methods Mol Biol. 2010;634:421–30.

58. Montigny WJ, Phelps SF, Illenye S, Heintz NH. Parameters influencing high-efficiency transfection of bacterial artificial chromosomes into cultured

mammalian cells. BioTechniques. 2003;35 (4):796–807.

59. DeDiego ML, Alvarez E, Almazan F, Rejas MT, Lamirande E, Roberts A, et al. A severe acute respiratory syndrome coronavirus that lacks the E gene is attenuated in vitro and in vivo. J Virol. 2007;81 (4):1701–13.

60. Enjuanes L, Dediego ML, Alvarez E, Deming D, Sheahan T, Baric R. Vaccines to prevent severe acute respiratory syndrome coronavirus-induced disease. Virus Res. 2008;133(1):45–62.

61. Fett C, DeDiego ML, Regla-Nava JA, Enjuanes L, Perlman S. Complete protection against severe acute respiratory syndrome coronavirus-mediated lethal respiratory disease in aged mice by immunization with a mouse-adapted virus lacking E protein. J Virol. 2013;87(12):6551–9.

62. Almazan F, DeDiego ML, Sola I, Zuniga S, Nieto-Torres JL, Marquez-Jurado S, et al. Engineering a replication-competent, propagation-defective Middle East respiratory syndrome coronavirus as a vaccine candidate. MBio. 2013;4(5):e00650–13.

63. Scobey T, Yount BL, Sims AC, Donaldson EF, Agnihothram SS, Menachery VD, et al. Reverse genetics with a full-length infectious cDNA of the Middle East respiratory syndrome coronavirus. Proc Natl Acad Sci U S A. 2013;110(40):16157–62.

64. Fang S, Chen B, Tay FP, Ng BS, Liu DX. An arginine-to-proline mutation in a domain with undefined functions within the helicase protein (Nsp13) is lethal to the coronavirus infectious bronchitis virus in cultured cells. Virology. 2007;358(1):136–47.

65. Youn S, Leibowitz JL, Collisson EW. In vitro assembled, recombinant infectious bronchitis viruses demonstrate that the 5a open reading frame is not essential for replication. Virology. 2005;332 (1):206–15.

66. Tan YW, Fang S, Fan H, Lescar J, Liu DX. Amino acid residues critical for RNA-binding in the N-terminal domain of the nucleocapsid protein are essential determinants for the infectivity of coronavirus in cultured cells. Nucleic Acids Res. 2006;34 (17):4816–25.

67. Donaldson EF, Graham RL, Sims AC, Denison MR, Baric RS. Analysis of murine hepatitis virus strain A59 temperature-sensitive mutant TS-LA6 suggests that nsp10 plays a critical role in polyprotein processing. J Virol. 2007;81(13):7086–98.

68. Becker MM, Graham RL, Donaldson EF, Rockx B, Sims AC, Sheahan T, et al. Synthetic recombinant bat SARS-like coronavirus is infectious in cultured cells and in mice. Proc Natl Acad Sci U S A. 2008;105 (50):19944–9.

69. Deming D, Sheahan T, Heise M, Yount B, Davis N, Sims A, et al. Vaccine efficacy in senescent mice challenged with recombinant SARS-CoV bearing epidemic and zoonotic spike variants. PLoS Med. 2006;3(12):e525.

70. Yount B, Roberts RS, Lindesmith L, Baric RS. Rewiring the severe acute respiratory syndrome coronavirus (SARS-CoV) transcription circuit: engineering a recombination-resistant genome. Proc Natl Acad Sci U S A. 2006;103(33):12546–51.

71. Thiel V, Herold J, Schelle B, Siddell SG. Infectious RNA transcribed in vitro from a cDNA copy of the human coronavirus genome cloned in vaccinia virus. J Gen Virol. 2001;82(Pt 6):1273–81.

72. Eriksson KK, Makia D, Thiel V. Generation of recombinant coronaviruses using vaccinia virus as the cloning vector and stable cell lines containing coronaviral replicon RNAs. Methods Mol Biol. 2008;454:237–54.

73. Tekes G, Hofmann-Lehmann R, Stallkamp I, Thiel V, Thiel HJ. Genome organization and reverse genetic analysis of a type I feline coronavirus. J Virol. 2008;82(4):1851–9.

74. Tekes G, Spies D, Bank-Wolf B, Thiel V, Thiel HJ. A reverse genetics approach to study feline infectious peritonitis. J Virol. 2012;86(12):6994–8.

75. Coley SE, Lavi E, Sawicki SG, Fu L, Schelle B, Karl N, et al. Recombinant mouse hepatitis virus strain A59 from cloned, full-length cDNA replicates to high titers in vitro and is fully pathogenic in vivo. J Virol. 2005;79(5):3097–106.

76. Smith GL, Moss B. Infectious poxvirus vectors have capacity for at least 25 000 base pairs of foreign DNA. Gene. 1983;25(1):21–8.

77. Merchlinsky M, Moss B. Introduction of foreign DNA into the vaccinia virus genome by in vitro ligation: recombination-independent selectable cloning vectors. Virology. 1992;190(1):522–6.

78. Ball LA. High-frequency homologous recombination in vaccinia virus DNA. J Virol. 1987;61(6):1788–95.

79. Britton P, Evans S, Dove B, Davies M, Casais R, Cavanagh D. Generation of a recombinant avian coronavirus infectious bronchitis virus using transient dominant selection. J Virol Methods. 2005;123 (2):203–11.

80. Scheiflinger F, Dorner F, Falkner FG. Construction of chimeric vaccinia viruses by molecular cloning and packaging. Proc Natl Acad Sci U S A. 1992;89 (21):9977–81.

81. Hiscox JA, Wurm T, Wilson L, Britton P, Cavanagh D, Brooks G. The coronavirus infectious bronchitis virus nucleoprotein localizes to the nucleolus. J Virol. 2001;75(1):506–12.

82. Thiel V, Herold J, Schelle B, Siddell SG. Viral replicase gene products suffice for coronavirus discontinuous transcription. J Virol. 2001;75(14):6676–81.

83. Keep SM, Bickerton E, Britton P. Transient dominant selection for the modification and generation of recombinant infectious bronchitis coronaviruses. Methods Mol Biol. 2015;1282:115–33.

84. Thiel V, Karl N, Schelle B, Disterer P, Klagge I, Siddell SG. Multigene RNA vector based on coronavirus transcription. J Virol. 2003;77(18):9790–8.

85. Almazan F, Galan C, Enjuanes L. The nucleoprotein is required for efficient coronavirus genome replication. J Virol. 2004;78(22):12683–8.

86. Schelle B, Karl N, Ludewig B, Siddell SG, Thiel V. Selective replication of coronavirus genomes that express nucleocapsid protein. J Virol. 2005;79 (11):6620–30.

87. Curtis KM, Yount B, Baric RS. Heterologous gene expression from transmissible gastroenteritis virus replicon particles. J Virol. 2002;76(3):1422–34.

88. Hertzig T, Scandella E, Schelle B, Ziebuhr J, Siddell SG, Ludewig B, et al. Rapid identification of coronavirus replicase inhibitors using a selectable replicon RNA. J Gen Virol. 2004;85(Pt 6):1717–25.

89. Chen L, Gui C, Luo X, Yang Q, Gunther S, Scandella E, et al. Cinanserin is an inhibitor of the 3C-like proteinase of severe acute respiratory syndrome coronavirus and strongly reduces virus replication in vitro. J Virol. 2005;79(11):7095–103.

90. Almazan F, Dediego ML, Galan C, Escors D, Alvarez E, Ortego J, et al. Construction of a severe acute respiratory syndrome coronavirus infectious cDNA clone and a replicon to study coronavirus RNA synthesis. J Virol. 2006;80(21):10900–6.

91. Ahn DG, Lee W, Choi JK, Kim SJ, Plant EP, Almazan F, et al. Interference of ribosomal frameshifting by antisense peptide nucleic acids suppresses SARS coronavirus replication. Antivir Res. 2011;91(1):1–10.

92. Tanaka T, Kamitani W, DeDiego ML, Enjuanes L, Matsuura Y. Severe acute respiratory syndrome coronavirus nsp1 facilitates efficient propagation in cells through a specific translational shutoff of host mRNA. J Virol. 2012;86(20):11128–37.

93. Pan J, Peng X, Gao Y, Li Z, Lu X, Chen Y, et al. Genome-wide analysis of protein-protein interactions and involvement of viral proteins in SARS-CoV replication. PLoS One. 2008;3(10):e3299.

94. Moreno JL, Zuniga S, Enjuanes L, Sola I. Identification of a coronavirus transcription enhancer. J Virol. 2008;82(8):3882–93.

95. Ortego J, Escors D, Laude H, Enjuanes L. Generation of a replication-competent, propagation-deficient virus vector based on the transmissible gastroenteritis coronavirus genome. J Virol. 2002;76(22):11518–29.

96. de Haan CA, Volders H, Koetzner CA, Masters PS, Rottier PJ. Coronaviruses maintain viability despite dramatic rearrangements of the strictly conserved genome organization. J Virol. 2002;76 (24):12491–502.

97. Shen H, Fang SG, Chen B, Chen G, Tay FP, Liu DX. Towards construction of viral vectors based on avian coronavirus infectious bronchitis virus for gene delivery and vaccine development. J Virol Methods. 2009;160(1–2):48–56.

98. Graham RL, Sims AC, Brockway SM, Baric RS, Denison MR. The nsp2 replicase proteins of murine hepatitis virus and severe acute respiratory syndrome

coronavirus are dispensable for viral replication. J Virol. 2005;79(21):13399–411.

99. Freeman MC, Graham RL, Lu X, Peek CT, Denison MR. Coronavirus replicase-reporter fusions provide quantitative analysis of replication and replication complex formation. J Virol. 2014;88(10):5319–27.

100. V'Kovski P, Gerber M, Kelly J, Pfaender S, Ebert N, Braga Lagache S, et al. Determination of host proteins composing the microenvironment of coronavirus replicase complexes by proximity-labeling. elife. 2019;8

101. Rice CM. Examples of expression systems based on animal RNA viruses: alphaviruses and influenza virus. Curr Opin Biotechnol. 1992;3(5):523–32.

102. Enjuanes L, Sola I, Almazan F, Ortego J, Izeta A, Gonzalez JM, et al. Coronavirus derived expression systems. J Biotechnol. 2001;88(3):183–204.

103. de Haan CA, Haijema BJ, Boss D, Heuts FW, Rottier PJ. Coronaviruses as vectors: stability of foreign gene expression. J Virol. 2005;79(20):12742–51.

104. Godeke GJ, de Haan CA, Rossen JW, Vennema H, Rottier PJ. Assembly of spikes into coronavirus particles is mediated by the carboxy-terminal domain of the spike protein. J Virol. 2000;74(3):1566–71.

105. Kuo L, Godeke GJ, Raamsman MJ, Masters PS, Rottier PJ. Retargeting of coronavirus by substitution of the spike glycoprotein ectodomain: crossing the host cell species barrier. J Virol. 2000;74 (3):1393–406.

106. Hodgson T, Casais R, Dove B, Britton P, Cavanagh D. Recombinant infectious bronchitis coronavirus Beaudette with the spike protein gene of the pathogenic M41 strain remains attenuated but induces protective immunity. J Virol. 2004;78(24):13804–11.

107. Hodgson T, Britton P, Cavanagh D. Neither the RNA nor the proteins of open reading frames 3a and 3b of the coronavirus infectious bronchitis virus are essential for replication. J Virol. 2006;80(1):296–305.

108. Tarpey I, Orbell SJ, Britton P, Casais R, Hodgson T, Lin F, et al. Safety and efficacy of an infectious bronchitis virus used for chicken embryo vaccination. Vaccine. 2006;24(47–48):6830–8.

109. Wei YQ, Guo HC, Dong H, Wang HM, Xu J, Sun DH, et al. Development and characterization of a recombinant infectious bronchitis virus expressing the ectodomain region of S1 gene of H120 strain. Appl Microbiol Biotechnol. 2014;98(4):1727–35.

110. Zhou Y, Yang X, Wang H, Zeng F, Zhang Z, Zhang A, et al. The establishment and characteristics of cell-adapted IBV strain H120. Arch Virol. 2016;161(11):3179–87.

111. Sola I, Alonso S, Zuniga S, Balasch M, Plana-DuranJ, Enjuanes L. Engineering the transmissible gastroenteritis virus genome as an expression vector inducing lactogenic immunity. J Virol. 2003;77 (7):4357–69.

112. Cruz JL, Zuniga S, Becares M, Sola I, Ceriani JE, Juanola S, et al. Vectored vaccines to protect against PRRSV. Virus Res. 2010;154(1–2):150–60.

113. Becares M, Sanchez CM, Sola I, Enjuanes L, Zuniga S. Antigenic structures stably expressed by recombinant       TGEV-derived       vectors.       Virology. 2014;464–465:274–86.

114. Eriksson KK, Makia D, Maier R, Cervantes L, Ludewig B, Thiel V. Efficient transduction of dendritic cells using coronavirus-based vectors. Adv Exp Med Biol. 2006;581:203–6.

115. Thiel V, Siddell SG. Reverse genetics of coronaviruses using vaccinia virus vectors. Curr Top Microbiol Immunol. 2005;287:199–227.

116. Kim H, Lee YK, Kang SC, Han BK, Choi KM. Recent vaccine technology in industrial animals. Clin Exp Vaccine Res. 2016;5(1):12–8.

# Alphavirus-Based Vaccines

Kenneth Lundstrom

### Abstract

Alphavirus vectors based on Semliki Forest virus, Sindbis virus, and Venezuelan equine encephalitis virus have been widely applied for vaccine development. Naked RNA replicons, recombinant viral particles, and layered DNA vectors have been subjected to immunization in preclinical animal models with antigens for viral targets and tumor antigens. Moreover, a limited number of clinical trials have been conducted in humans. Vaccination with alphavirus vectors has demonstrated efficient immune responses and has showed protection against challenges with lethal doses of virus and tumor cells, respectively. Moreover, vaccines have been developed against alphaviruses such as Chikungunya virus, which have caused epidemics.

### Keywords

Alphavirus replicon vectors · Immunization · Viral vaccines · Cancer vaccines · Protection

**Learning Objectives**

After reading this chapter, you should be able to:

- Define and explain how self-replicating alphavirus vectors generate large quantities of recombinant RNA molecules providing efficient translation of recombinant proteins

- Explain how alphavirus vectors can be delivered as RNA replicons, recombinant viral particles, or DNA/RNA layered plasmid vectors and are considered safe for use in preclinical animals studies and human clinical trials

- Characterize the immune responses generated by alphavirus vectors (i.e., strong cellular and humoral immune responses)

- Explain how vaccination with alphavirus vectors provides protection against challenges with lethal doses of infectious agents and tumor cells

## 1 Introduction

Alphaviruses belong to the family *Togaviridae* consisting of a positive-sense single-stranded RNA (ssRNA) genome encapsulated in a capsid protein covered by a membrane envelope structure [74]. Alphaviruses are geographically present as the Old World viruses including Semliki Forest virus (SFV) and Sindbis virus (SIN) in Europe, Asia, Africa, and Australia, while the origin of the New World viruses such as Venezuelan equine encephalitis virus (VEE) and eastern equine encephalitis virus (EEE) is in North and South America [22]. Arthropod-borne alphaviruses are commonly spread by mosquitoes [74] although SIN and EEE are also found in mites and lice [9, 70]. Alphaviruses use birds as primary vertebrate

K. Lundstrom (✉)
PanTherapeutics, Lutry, Switzerland

© Springer Nature Switzerland AG 2021
T. Vanniasinkam et al. (eds.), *Viral Vectors in Veterinary Vaccine Development*,
https://doi.org/10.1007/978-3-030-51927-8_11

hosts supported by findings of a South American strain of EEE in the blood of migrating birds in the Mississippi delta [10]. Several alphaviruses can act as pathogens in both domestic animals and humans [57, 59, 61]. In this context, EEE and western equine encephalitis virus (WEE) infections have caused fever, anorexia, depression, and clinical signs of encephalomyelitis even leading to fatal encephalitis in humans in North and South America [4]. Furthermore, VEE has caused epidemics in both horses and humans in South America [82]. Similarly, both SFV and SIN have been associated with fever epidemics in Africa [50, 54], and natural SIN variants have caused painful polyarthritis in Northern Europe [55]. More recently, Chikungunya virus (CHIK) has been responsible for outbreaks in the Republic of Congo [33], the island of Reunion [31], and Brazil [62]. Due to the potential pathogenicity of alphaviruses, avirulent laboratory strains with no association with disease have been applied for the engineering of expression systems [24].

## 2 Alphavirus Structure, Genome, and Life Cycle

The alphavirus icosahedral nucleocapsid surrounded by glycoproteins in an icosahedral lattice has been confirmed by X-ray crystallography and cryo-electron microscopy for SFV and SIN [12, 25, 56]. The alphavirus envelope contains the E1 and E2 membrane proteins although translated as the precursor PE (SIN) and p62 (SFV) polyprotein, respectively, where the cleaved off SFV E3 remains associated with mature viral particles [41].

The 11.7 kb alphavirus ssRNA genome contains the nonstructural protein (nsP1–4) genes responsible for RNA replication activity and the structural protein (capsid-E2-E2-6K-E1) genes [44]. During RNA replication, a minus-strand copy serves as a template for full-length genomic RNA production and transcription of subgenomic RNA responsible for the translation of structural proteins [73].

Related to the life cycle of alphaviruses, there are significant differences to persistent lifelong infection and presence in arthropod host and the acute short-term infection in vertebrates leading to apoptosis and rapid cell death [74]. The broad host range of both invertebrates (mosquitoes, other hematophagous insects) and vertebrates (mammals, birds, amphibians, reptiles) relates to the various host cell receptors targeted by alphaviruses [11]. For instance, neurovirulent and avirulent SIN strains do not compete for the same receptor [72], and several host cell receptors such as MHC, HLA-A, and HLA-B have been suggested for SFV [28]. Furthermore, studies on monoclonal antibodies revealed that SIN target laminin receptors [79]. Additionally, it has been demonstrated that two laboratory strains of SFV and SIN can target heparan receptors [8]. Once a host cell receptor has been identified, the alphavirus envelope fuses through a highly conserved hydrophobic domain of the E1 glycoprotein with the cellular membrane, releasing the nucleocapsid into the cytoplasm [35]. The entrance occurs through endocytosis in clathrin-coated vesicles followed by transfer to endosomes due to conformational reorganization of E1-E2 heterodimers caused by a low pH [26]. Studies have also indicated that alphaviruses can infect cells by direct fusion to the cell surface [27, 29]. The nucleocapsid disassembly releases the RNA genome for the initial replication of a minus-strand copy of full-length and 26S subgenomic RNA [84]. The replicase complex formed by individual nsP1–4 proteins cleaved from the polyprotein nsP1234 provides efficient RNA replication and translation [11, 73]. Assembly of nucleocapsids occurs by recognition of the packaging signal in the SIN nsP1 [34] and SFV nsP2 [85], respectively, whereas while serologically related to SIN, the packaging signal of Aura virus is located in the 26S region [67]. The alphavirus envelope proteins are folded in the endoplasmic reticulum (ER) and transported to the Golgi complex, and viral particle assembly occurs on the plasma membrane followed by budding of mature alphavirus particles [15, 42, 74]. The released virus particles are then capable of infecting a broad range of mammalian and nonmammalian cells.

# 3    Alphavirus Expression Vector Systems

The most common alphavirus expression systems are based on SFV [38], SIN [88], and VEE [14]. More recently, vector systems have been developed for CHIK [78] and the SIN-like XJ-160 virus [91]. Moreover, the naturally oncolytic M1 alphavirus has been engineered for cancer therapy [39].

Basically, three types of expression systems have been engineered [89] (Fig. 1). All three expression systems rely on the alphavirus replicon, based on the replicase complex composed of the nsP1–4 proteins, which generates an extreme number of RNA copies immediately available for translation in the cytoplasm. The system based on alphavirus expression and helper vectors generates **replication-deficient recombinant replicon particles**, which are capable of delivering transgenes for expression in infected host cells without generating any further viral progeny. In contrast, **replication-competent recombinant replicon particles** comprise the full-length alphavirus genome plus the gene of interest downstream of a second subgenomic promoter. Infection of host cells results in high levels of transgene expression, but also production of new viral infectious particles. In the third expression system, a mammalian host cell-compatible eukaryotic RNA polymerase II type promoter such as CMV has been inserted upstream of the replicon genes, creating the **DNA/RNA layer replicon** system for direct plasmid DNA transfer [16]. In addition to the abovementioned expression systems, alphavirus vectors can be delivered in the form of naked RNA replicons [40].

Plenty of attention has been dedicated to the engineering of modified improved alphavirus vectors. For instance, point mutations in the nsP2 and nsP4 regions of SFV [49] and SIN [2] have provided reduced cell cytotoxicity resulting in prolonged and enhanced transgene expression. Moreover, alphavirus vector systems based on avirulent strains such as the SFV A7(74) can also provide higher levels and extension of the duration of transgene expression both in vitro

[19] and in vivo [79]. Another approach has been to insert the translation enhancement signal of the SFV capsid protein in the expression vector, which generated 5–20-fold higher expression levels [63]. In attempts to improve biosafety and reduce the production of wild-type alphavirus particles, point mutations were introduced in the p62 in the SFV helper vector rendering only conditionally infectious SFV particles [7]. Furthermore, SFV [71] and SIN [68] split helper systems accommodating the capsid and envelope protein genes on separate helper vectors reduced the wild-type virus production to theoretical levels.

During the last 25 years, alphavirus vectors have frequently been applied for the expression of topologically different recombinant proteins in mammalian cell lines [38], primary cells [18], and in vivo [48]. In particular, integral membrane proteins expressed from SFV vectors have been subjected to structural and functional studies in support of drug discovery [45]. Related to vaccines, numerous immunization studies with alphavirus-based vectors have elicited strong cellular and humoral responses in animal models and also provided protection against challenges with lethal doses of infectious agents and tumor cells [47].

# 4    Application of Alphavirus Vectors

Alphavirus vectors have been frequently applied for immunization studies in animal models targeting a large number of infectious diseases and various cancers (Table 1). One advantage of alphavirus-based immunization is the flexibility of using recombinant replicon particles, naked RNA replicons, or DNA/RNA layered plasmid vectors as delivery vehicles. There is no clear indication of which delivery approach is best, and it seems that the rank order varies from one case to another. Most of the immunizations conducted so far for both preclinical and clinical applications relate to targeting human disease. However, this review focuses on veterinary vaccines as described in Table 1 and below.

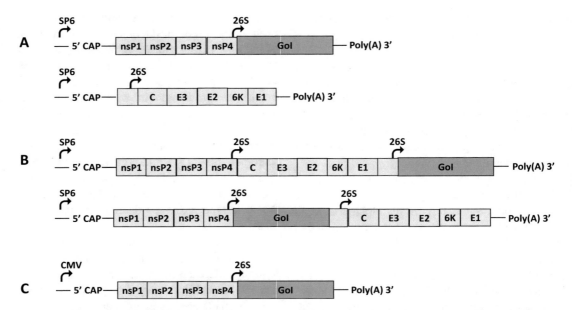

**Fig. 1 Alphavirus expression systems. (a) Replication-deficient particles.** The expression vector comprises the replicase genes (nsP1–4) and the gene of interest (GoI) introduced downstream of the 26S subgenomic promoter. The helper vector provides the alphavirus structural protein genes. In vitro transcribed RNA from the SP6 RNA polymerase promoter followed by co-electroporation of RNA from both expression and helper vectors resulting in replication-deficient "suicide" particles. **(b) Replication-competent particles.** The GoI can be inserted downstream of either the structural or nonstructural genes for in vitro transcription of RNA generating replication-competent viral progeny. **(c) DNA/RNA layer replicon vector.** Replacement of the SP6 RNA polymerase promoter with a CMV promoter allows direct DNA transfection for recombinant protein expression

In the context of influenza A virus Hong Kong isolate, the hemagglutinin (HA) gene was expressed from a VEE vector and immunization with VEE viral replicon particles (VRPs) was evaluated in chicken embryos and young chicks [69]. Immunization *in ovo* or at 1 day of age provided partial protection against challenges with lethal doses of influenza A/HK/156/97 virus. However, a single dose of VEE VRPs was sufficient to give complete protection to 2-week-old birds. In another study, 1-day-old chicks immunized with SFV-LacZ particles elicited significantly higher β-gal antibody levels than animals immunized with SFV-based DNA replicons or conventional plasmid DNA vectors [58]. Moreover, immunization with SFV particles expressing the VP2 protein or the VP2/VP4/VP3 polyprotein of infectious bursal disease virus (IBDV) generated specific antibodies in all animals but neutralizing antibodies in only some birds. Recombinant SFV particles expressing the prME and NS1 proteins of louping ill virus (LIV) were subjected to immunization of sheep, which resulted in complete protection against subcutaneous challenges with LIV and partial protection after intranasal LIV administration [52]. In another study, the major envelope proteins ($G_L$ and M) of equine arteritis virus (EAV) were expressed from VEE vectors for evaluation in horses [6]. Immunization with VEE particles expressing both EAV $G_L$ and M as a heterodimer elicited neutralizing antibodies and only mild equine viral arteritis (EVA) after intranasal or intrauterine challenges with the virulent EAV KY84 strain. In contrast, immunization with only VEE-$G_L$ or VEE-M did not protect the animals from severe EVA.

Related to simian immunodeficiency virus (SIV), a triple-vector approach based on intradermal priming with DNA, followed by two subcutaneous injections with recombinant SFV particles and a final intramuscular administration

**Table 1** Examples of alphavirus-based veterinary vaccines

| Disease | Vector | Target | Response | References |
|---|---|---|---|---|
| Influenza | VEE-VRPs | Influenza HA | Protection against challenges in chicks | [69] |
| IBDV | SFV-VRPs | VP2/VP4/VP3 | Specific antibodies in birds | [58] |
| LIV | SFV-VRPs | prME, NS1 | Protection against LIV in sheep | [52] |
| EVA | VEE-VRPs | EAV $G_L$, M | Protection against EAV in horses | [6] |
| SIV | SFV-VRPs + DNA + MVA | Gag, pol, tat, rev., nef, env | Protection against SIV in monkeys | [51] |
| PRV | SIN-DNA | gC, gD | Survival of pigs after PRV challenge | [17] |
| SVDV | SFV-DNA | 1BCD | Protection against SVDV in pigs | [75] |
| DHV-1 | SFV-DNA | VP1 | Protection against DHV-1 in ducklings | [20] |
| CSFV | SFV-DNA | E2 | Protection against CSFV in rabbits and pigs | [36] |
| | SFV-DNA | E2 | Reduced dose, no clinical symptoms | [37] |
| | SFV-DNA | E2 | rAdV boost enhanced titers | [76] |
| | rAdV-SFV | E2 | Improved protection with BG adjuvant | [87] |
| HCC | SFV-VRPs | IL-12 | Tumor regression in woodchucks | [64] |
| STs | SFV-RCPs | EGFP | No adverse events in dogs | [5] |
| FMD | SFV-VRPs | FMDV P1-2A FMDV 3Cpro | Protection against FMDV in cattle | [23] |
| BVDV | SFV-VRPs | E2 | Some protection against BVDV in calves | [43] |
| DTMUV | SFV-DNA | DTMUV E | Protection against DTMUV in ducklings | [77] |
| IAV-S | VEE-VRPs | HA | Protection against challenges in pigs | [1] |
| HPAI | VEE-VRPs | HA | Protection against challenges in birds* | [32] |
| | | | Conditional drug approval in the United States | |
| PRRSV | BV-SFV | Gp5, M | Specific antibodies | [86] |
| HCV | XJ-160 VRPs | E1, E2 | Humoral and cellular immune responses | [91] |
| Bladder CA | M1-VRPs | M1 | Tumor regression, prolonged survival | [30] |
| VEE | IRES-VEE | VEE | Protection against challenges in mice | [65] |
| | | | Protection against challenges in macaques | [66] |

*Prime-boost regimen with recombinant turkey herpes virus; *BVDV* bovine viral diarrhea virus, *BV-SFV* baculovirus-pseudotyped SFV replicon, *FMDV 3Cpro* FMDV 3C protease, *DHV-1* duck hepatitis virus-1, *EAV* equine arteritis virus, *EVA* equine viral arteritis, *FMD* foot-and-mouth disease, *FMDV* foot-and-mouth disease virus, *HA* hemagglutinin, *HCC* hepatocellular carcinoma, *HCV* hepatitis C virus, *HPAI H5* avian influenza virus, *IAV-S* influenza A virus in swine, *IBDV* infectious bursal disease virus, *IL-12* interleukin-12, *IRES* internal ribosomal entry site, *LIV* louping ill virus, *MVA* modified vaccinia virus Ankara, *PRRSV* porcine reproductive and respiratory syndrome virus, *PRV* pseudorabies virus, *SFV* Semliki Forest virus, *RCPs* replication-competent particles, *SIV* simian immunodeficiency virus, *ST* spontaneous tumors, *SVDV* swine vesicular disease virus, *VEE* Venezuelan equine encephalitis virus, *VRPs* viral replicon particles

of a recombinant modified vaccinia virus Ankara (rMVA) strain, was established [51]. Each vector expressed the Gag, Pol, Tat, Rev, Nef, and Env proteins of SIV. Vaccination of cynomolgus monkeys elicited T-helper proliferative responses and challenges with pathogenic SIVmac251 resulted in full protection in three of four animals.

In another approach, a prime-boost vaccination strategy against the lethal pseudorabies virus (PRV) was investigated in pigs [17]. A SIN-based DNA vector expressing the PRV glycoproteins gC and gD was applied for immunization of pigs followed by boosting with an Orf (ORFV) expressing gC and gD. The vaccination induced strong humoral and cellular-like PRV-specific immune responses, and immunized pigs survived lethal challenges with PRV. Related to swine vesicular disease virus (SVDV), the 1BCD gene was expressed from an SFV DNA vector [75]. Three intramuscular injections in guinea pigs and swine elicited anti-SVDV antibodies and neutralizing antibodies. Moreover, half of the animals were protected against challenges with SVDV. In another study, SFV DNA replicon-based expression of the duck hepatitis virus type 1 (DHV-1) VP1 gene was applied for immunization in ducklings [20]. Two intramuscular injections at 14 days intervals elicited anti-DHV-1 antibodies and protected ducklings against challenges with wild-type DHV-1.

Several vaccine studies have targeted classical swine fever virus (CSFV) due to significant losses CSFV causes for the pig industry [36]. Immunization of rabbits and pigs with an SFV DNA replicon expressing the CSFV E2 glycoprotein (pSFV1CS-E2) elicited CSFV-specific neutralizing antibodies and provided protection against challenges with lethal doses of CSFV. Moreover, application on lower doses and fewer inoculations (twice) with 100 µg of pSFV1CS-E2 generated high titers of specific neutralizing antibodies against CSFV and demonstrated no clinical symptoms after challenges with the virulent CSFV Shimen strain [37]. In another study, a prime-boost strategy in pigs with the pSFV1CS-E2 DNA vaccine and recombinant adenovirus (rAdV-E2) elicited significantly higher titers of CSFV-specific neutralizing antibodies in comparison with double immunizations with rAdV-E2 [76]. No clinical symptoms or viremia was detected in vaccinated pigs challenged with the virulent CSFV Shimen strain. In another approach, the adenovirus-delivered SFV replicon (rAdV-SFV-E2) was evaluated in pigs together with a *Salmonella enteritidis*-derived bacterial ghost adjuvant (BG) [87]. Two intramuscular injections of $10^5$ median tissue culture infective doses ($TCID_{50}$) of rAdV-SFV-E2 combined with $10^{10}$ colony forming units (CFU) of BG, $10^5$ or $10^6$ $TCID_{50}$ rAdV-SFV-E2 alone, or $10^{10}$ CFU BG alone were administered to pigs and

challenged with the highly virulent CSFV Shimen strain. Pigs immunized with $10^5$ $TCID_{50}$ rAdV-SFV-E2 plus BG or $10^6$ $TCID_{50}$ rAdV-SFV-E2 alone were completely protected against lethal challenges with CSFV. In contrast, only partial or no protection was obtained for immunizations with $10^5$ $TCID_{50}$ rAdV-SFV-E2 or BG alone.

In another approach, application of the SFV-enh vector containing the capsid translation enhancement signal [63] for the expression of interleukin-12 (IL-12) in intratumorally injected woodchucks with implanted hepatocellular carcinoma (HCC) xenografts demonstrated high levels of cytokine secretion into serum [64]. Moreover, a dose-dependent partial tumor remission was observed in five of six immunized woodchucks with reductions in tumor volume up to 80%. In preparation for veterinary applications of alphavirus replicon vectors, the nonpathogenic replication-competent SFV VA7-EGFP vector was assessed for canine tumor cell lines and in laboratory beagle dogs [5]. The SFV-VA7-EGFP demonstrated the replication and killing of Abrams and D17 canine tumor cell lines. Moreover, the safety was evaluated in two adult beagle dogs after a single intravenous injection of $2 \times 10^5$ pfu of SFV-VA7-EGFP. Neither adverse events nor infective viruses were detected.

Foot-and-mouth disease (FMD) presents one of the most economically problematic global infectious diseases for the veterinary field [23]. SFV vectors expressing the foot-and-mouth disease virus (FMDV) capsid protein precursor (P1-2A) alone or together with the FMDV 3C protease (3Cpro) have been evaluated as vaccines in cattle. Co-expression of FMDV P1-2A and 3Cpro generated empty capsid particles in infected cells. Although anti-FMDV antibodies were elicited in cattle after immunization with SFV-FMDV vectors alone, no protection against FMDV challenges was obtained. In contrast, primary vaccination with SFV-FMDV followed by boosting with empty FMDV capsid particles provided protection against FMDV challenges. The reverse immunization strategy of priming with empty FMDV capsid particles followed by SFV-FMDV was significantly less efficient. Another important viral pathogen in cattle is bovine viral diarrhea virus (BVDV), which has been addressed for vaccine development by immunization with VEE particles expressing the BVDV E2 glycoprotein [43]. Immunization of calves elicited neutralizing antibodies and high-dose vaccinations provided some protection against challenges with BVDV and significantly reduced leukopenia caused by viral infection.

Related to the Chinese poultry industry, the duck Tembusu virus (DTMUV) causes huge economic losses and has therefore been the target of vaccine development. An SFV DNA vector expressing the DTMUV E glycoprotein was applied for immunization of ducklings [77]. Intramuscular administration elicited robust humoral and cellular immune responses in ducklings providing protection for all vaccinated ducklings against challenges with the virulent DTMUV AH-F10 strain. Another important pathogen comprises the influenza A virus in swine (IAV-S) for which immunization studies with VEE expressing the IAV-S HA gene have been conducted [1]. Immunization with monovalent and bivalent VEE-HA formulations provided efficient protection against challenges with IAV-S viruses with matched and mismatched HA, although in one mismatched HA challenge group, protection was reduced. Another influenza virus-related target relates to the epidemic in poultry in North America in 2014–2015, which caused the death of more than 47 million poultry due to the exposure to the highly pathogenic clade 2.3.4.4 H5 avian influenza (HPAI) virus [32]. Three different vaccines are based on inactivated reverse genetic virus encoding a clade 2.3.4.4 H5 HA gene (rgH5), recombinant turkey herpes virus encoding a clade 2.2. H5 HA (rHVT-AI), and recombinant replication-deficient alphavirus particles expressing the 2.3.4.4 H5 HA (RP-H5). All vaccines increased the survival rates in young birds, although immunization with rHVT-AI or RP-H5 alone did not provide 100% protection. However, a prime-boost regimen with rHVT-A1 and RP-H5 provided complete protection against challenges with lethal doses of H5N2 HPAI. Based on these results, the United States Department of Agriculture (USDA) granted

conditional approval for the use of recombinant vaccines in turkeys.

In an interesting approach, a pseudotyped baculovirus vector was engineered to contain a hybrid cytomegalovirus (CMV) promoter and an SFV replicon for co-expression of the GP5 and M proteins of porcine reproductive and respiratory syndrome virus (PRRSV) (BV-SFV-5m6) [86]. Compared to a CMV-driven pseudotyped baculovirus (BV-CMV-5m6) immunization of BALB/c demonstrated a stronger immune response for the baculovirus vector containing the SFV replicon eliciting GP5-specific antibodies and neutralizing antibodies.

In addition to the commonly used alphavirus vectors based on SFV, SIN, and VEE, the SIN-like XJ-160 has been applied for immunization studies [91]. In this context, the XJ-160 overexpressing the hepatitis C virus (HCV) glycoproteins E1 and E2 was subjected to a prime-boost regimen comprising intramuscular vaccination with XJ-E1E2 combined with Freund's incomplete adjuvant. The outcome was strong humoral and cellular immune responses against HCV E1 and E2 glycoproteins. Moreover, the oncolytic M1 alphavirus has demonstrated selective targeting and killing of zinc-finger antiviral protein (ZAP)-deficient cancer cells in vitro, in vivo, and ex vivo [39]. As ZAP is commonly deficient in human cancers, M1 alphaviruses provide the opportunity to promote personalized cancer therapy. In another study, it was demonstrated that M1 alphaviruses selectively kill bladder cancer cells without causing damage to normal cells [30]. This was confirmed by tail vein injections of orthotopic mice showing significant inhibition of tumor growth and prolonged survival time. Moreover, the antitumor effect of M1 was stronger than first-line treatment with cisplatin.

Finally, as alphaviruses such as VEE are responsible for epidemics in both horses and humans, they have been targets for vaccine development. In this context, an IRES-based VEE vaccine candidate was demonstrated to provide complete protection against challenges with a lethal IE subtype VEE in mice [65]. In another study, the IRES-VEE vaccine elicited robust neutralizing antibodies and offered protection against febrile disease in cynomolgus macaques [66].

## 5 Summary

In summary, numerous studies have demonstrated that alphavirus vectors can elicit strong antibody responses in a variety of immunized animals including rodents, birds, pigs, sheep, and primates. In several cases, protection against challenges with lethal doses of infectious agents and tumor cells has been obtained. Several clinical trials have been conducted to develop vaccines against, for example, HIV [83] and breast, colorectal, and pancreatic cancer [53]. Similarly, alphavirus vectors have found numerous applications in the field of veterinary vaccines as described above and listed in Table 1. The promising outcome of a study on avian influenza virus resulted in the USDA granting a conditional approval for the tested vaccine in turkeys [32].

The self-amplifying nature of the alphavirus RNA replicon generates large quantities of cytoplasmic RNA, which can be immediately translated in the cytoplasm. It also ensures high biosafety as the RNA cannot be integrated into the host genome. Moreover, the flexibility of delivery methods including RNA replicons, layered DNA/RNA vectors, and recombinant replication-deficient and replication-competent viral particles is a great asset. However, as for all drug and vaccine development, there are possibilities for improvement. In this context, different modes of vector targeting have been considered. For example, it was demonstrated that a single point mutation in the SIN E2 glycoprotein was responsible for providing dendritic cell (DC) tropism [21]. In this context, SIN replicon particles were shown to infect skin-resident mouse DCs in vivo. Moreover, SIN particles expressing HIV Gag elicited strong Gag-specific T-cell responses in vitro and in vivo. In another targeting approach, the human papillomavirus type 16 (HPV-16) E7 was evaluated for immunogenicity by engineering three constructs comprising a cytosolic/nuclear protein (E7), a secreted protein (Sig/E7), and an endosomal/lysosomal compartment-targeting protein (Sig/E7/LAMP-1) (lysosome-associated membrane protein 1) [13]. The Sig/E7/LAMP-1 fusion generated the highest E7-specific T-cell-mediated immune responses and antitumor effects. Moreover, it was demonstrated that SIN-transfected apoptotic cells were taken up by bone marrow-derived DCs. Alphavirus vectors have also been targeted to tumors by encapsulation of SFV particles in liposomes [46]. Intraperitoneal administration of encapsulated SFV-LacZ particles demonstrated the accumulation of β-galactosidase expression in tumors of mice with LNCaP xenografts. Moreover, intravenous administration of encapsulated SFV-IL-12 particles (LipoVIL12) in melanoma and kidney carcinoma patients in a phase I clinical trial resulted in 5–10-fold increase in IL-12 plasma levels. The treatment triggered no adverse events and due to the protection of viral particles being recognized by the host immune system, allowed repeated LipoVIL12 administration. In another study, the oncolytic M1 alphavirus was encapsulated in liposomes (M-LPO) in attempts to reduce the immunogenicity and immune clearance in vivo [80]. Intravenous administration of M-LPO demonstrated clearly reduced production of M1-neutralizing antibodies compared to delivery of naked M1 virus, suggesting reduced immunogenicity and improved anticancer therapy.

One approach that will serve future vaccine development relates to the establishment of a vaccine platform based on the utilization of VEE particles for the expression of recombinant viral proteins demonstrated for noroviruses and coronaviruses as model systems [3]. Employment of the attenuated VEE strain 3526 allowed packaging of recombinant particles under biosafety level 2 (BSL2) for rapid generation of candidate vaccines against emerging infectious agents. Another issue which needs to be addressed to improve the process for alphavirus-based vaccines relates to the engineering of an efficient packaging cell line. Efforts made in the late 1990s generated a panel of packaging cell lines for SIN and SFV although providing moderate titers in the range of $10^7$ infectious units (IU)/ml [60]. More recently, a BHK packaging cell line for XJ-160 also applicable to SFV has been engineered [90]. However, the titers were relatively low, in the range of 5–9 x $10^6$ IU/ml, which stresses the need for more efficient packaging systems.

Overall, the published findings and current progress in alphavirus-based vaccine development including the capacity and flexibility for rapid and high-level transient antigen expression combined with targeting antigen-presenting cells and tumors, suggest that alphaviruses may play an important role in future gene therapy and vaccine approaches.

# References

1. Abente E, Rajao D, Gauger P, Vincent A. Alphavirus-vectored hemagglutinin subunit vaccine provides partial protection against heterologous challenge in pigs. Vaccine. 2019;37:1533–9.
2. Agapov E, Frolov I, Lindenbach B, et al. Noncythopathic Sindbis virus RNA vectors for heterologous gene expression. Proc Natl Acad Sci USA. 1998;95:12989–94.
3. Agnihothram S, Menachery V, Yount B, Lindesmith L, et al. Development of a broadly accessible Venezuelan equine encephalitis virus replicon particle vaccine platform. J Virol. 2018;92:e0027–18.
4. Arechiga-Ceballos N, Aguilar-Setien A. Alphaviral equine encephalomyelitis (Eastern, Western and Venezuelan). Rev Sci Tech. 2015;34:491–501.
5. Autio K, Ruotsalainen J, Anttila M, Niittykoski M, et al. Attenuated Semliki Forest virus for cancer treatment in dogs: safety assessment in two laboratory beagles. BMC Vet Res. 2015;11:170.
6. Balasuriya U, Heidner H, Davis N, et al. Alphavirus replicon particles expressing the two major envelope proteins of equine arteritis virus induce high level protection against challenge with virulent virus in vaccinated horses. Vaccine. 2002;20:1609–17.
7. Berglund P, Sjöberg M, Garoff H, et al. Semliki Forest virus expression system: production of conditionally infectious recombinant particles. Biotechnol NY. 1993;11:916–20.
8. Byrnes A, Griffin D. Binding of Sindbis virus to cell surface heparan sulfate. J Virol. 1998;72:7349–56.
9. Calisher C, Karabatsos N. Arbovirus serogroups: definition and geographic distribution. In: Monath T, editor. The arboviruses: epidemiology and ecology. Baca Raton, Florida: CRC Press, Inc; 1988. p. 19–57.
10. Calisher C, Maness K, Lord R, et al. Identification of two south American strains of eastern equine encephalomyelitis virus from migrant birds captured on the Mississippi delta. Am J Epidemiol. 1971;94:172–8.
11. Chamberlain R. Epidemiology of arthropod-borne togaviruses: the role of arthropods as hosts and vectors and of vertebrate hosts in natural transmission cycles. In: Schlesinger RW, editor. The Togaviruses: biology, structure, replication. New York: Academic Press, Inc; 1980. p. 175–227.
12. Cheng R, Kuhn R, Olson N, et al. Three-dimensional structure of an enveloped alphavirus with T=4 icosahedral symmetry. Cell. 1995;80:621–30.
13. Cheng W, Hung C, Hsu K, Chai C, et al. Enhancement of sindbis virus self-replicating RNA vaccine potency by targeting antigen to endosomal/lysosomal compartments. Hum Gene Ther. 2001;12:235–52.
14. Davis N, Brown K, Johnston R. In vitro synthesis of infectious Venezuelan equine encephalitis virus RNA from a cDNA clone: analysis of a viable deletion mutant. Virology. 1989;171:189–204.
15. de Curtis I, Simons K. Dissection of Semliki Forest virus glycoprotein delivery from the trans-Golgi network to the cell surface in permeabilized BHK cells. Proc Natl Acad Sci USA. 1988;85:8052–6.
16. DiCiommo D, Bremner R. Rapid, high level protein production using DNA-based Semliki Forest virus vectors. J Biochem Chem. 1998;273:18060–6.
17. Dory D, Fischer T, Béven V, et al. Prime-boost immunization using DNA vaccine and recombinant Orf virus protects pigs against pseudorabies virus (herpes suid 1). Vaccine. 2006;24:6256–63.
18. Ehrengruber MU, Lundstrom K, Schweitzer C, et al. Recombinant Semliki Forest virus and Sindbis virus efficiently infect neurons in hippocampal slice cultures. Proc Natl Acad Sci USA. 1999;96:7041–6.
19. Ehrengruber M, Renggli M, Raineteau O, et al. Semliki Forest virus A7(74) transduces hippocampal neurons and glial cells in a temperature-dependent dual manner. J Neurovirol. 2003;9:16–28.
20. Fu Y, Chen Z, Liu G. Protective immune responses in ducklings induced by a suicidal DNA vaccine of the VP1 gene of duck hepatitis virus type 1. Vet Microbiol. 2012;160:314–8.
21. Gardner J, Frolov I, Perri S, Ji Y, et al. Infection of human dendritic cells by sindbis virus replicon vector is determined by a single amino acid substitution in the E2 glycoprotein. J Virol. 2009;74:11849–57.
22. Garmashova N, Gorchakov R, Volkova E, et al. The old and New World alphaviruses use different virus-specific proteins for induction of transcriptional shut-off. J Virol. 2007;81:2472–84.
23. Gullberg M, Lohse L, Botner A, McInerney G, et al. A prime-boost vaccination strategy in cattle to prevent foot-and-mouth disease using a "single-cycle" alphavirus vector and empty capsid particles. PLoS One. 2016;11:e0157435.
24. Hanson R, Sulkin S, Buescher E, et al. Arbovirus infections of laboratory workers. Science. 1967;158:1283–6.

25. Harrison S, Schlesinger S, Schlesinger M, et al. Crystallization of Sindbis virus and its nucleocapsid. J Mol Biol. 1992;226:277–80.

26. Helenius A. Semliki Forest virus penetration from endosomes. Biol Cell. 1984;51:181–6.

27. Helenius A. Virus entry: looking Back and moving forward. J Mol Biol. 2018;430:1853–62.

28. Helenius A, Morein B, Fries E, et al. Human (HLA-A and -B) and murine (H2-K and -D) histocompatibility antigens are cell surface receptors for Semliki Forest virus. Proc Natl Acad Sci USA. 1978;75:3846–50.

29. Helenius A, Kartenbeck J, Simons K, et al. On the entry of Semliki Forest virus into BHK-21 cells. J Cell Biol. 1980;84:404–20.

30. Hu C, Liu Y, Lin Y, Liang J, et al. Intravenous injections of the oncolytic virus M1 as a novel therapy for muscle-invasive bladder cancer. Cell Death Dis. 2018;15:274.

31. Jansen K. The 2005-2007 Chikungunya epidemic in Reunion: ambiguous etiologies, memories, and meaning-making. Med Anthropol. 2013;32: 174–89.

32. Kapczynski D, Sylte M, Killian M, Torchetti M, et al. Protection of commercial turkeys following inactivated or recombinant H5 vaccine application against the 2015U.S. H5N2 clade 2,3,4,4 highly pathogenic avian influenza virus. Vet Immunol Immunopathol. 2017;191:74–9.

33. Kelvin A. Outbreak of Chikungunya in the republic of Congo and the global picture. J Infect Dev Ctries. 2011;5:441–4.

34. Kim D, Firth A, Atasheva S, et al. Conservation of a packaging signal and the viral genome RNA packaging mechanism in alphavirus evolution. J Virol. 2011;85:8022–36.

35. Kondor-Koch C, Burke B, Garoff H. Expression of Semliki Forest virus proteins from cloned complementary DNA. I. the fusion activity of the spike glycoprotein. J Cell Biol. 1983;97:644–51.

36. Li N, Qiu H, Zhao J, et al. A Semliki Forest virus replicon vectored DNA vaccine expressing the E2 glycoprotein of classical swine fever virus protects pigs from lethal challenge. Vaccine. 2007a;25:2907–712.

37. Li N, Zhao J, Zhao H, et al. Protection of pigs from lethal challenge by a DNA vaccine based on an alphavirus replicon expressing the E2 glycoprotein of classical swine fever virus. J Virol Methods. 2007b;144:73–8.

38. Liljestrom P, Garoff H. A new generation of animal cell expression vectors based on the Semliki Forest virus replicon. Biotechnol NY. 1991;9:1356–61.

39. Lin Y, Zhang H, Liang J, et al. Identification and characterization of alphavirus M1 as a selective oncolytic virus targeting ZAP-defective human cancers. Proc Natl Acad Sci USA. 2014;111: F4504–12.

40. Ljungberg K, Liljestrom P. Self-replicating alphavirus RNA vaccines. Expert Rev Vaccines. 2015;14:177–94.

41. Lobigs M, Zhao H, Garoff H. Function of Semliki Forest virus E3 peptide in virus assembly: replacement of E3 with an artificial signal peptide abolishes spike heterodimerization and surface expression of E1. J Virol. 1990;64:4346–55.

42. Lopez S, Yao J-S, Kuhn RJ, et al. Nucleocapsid-glycoprotein interactions required for alphavirus assembly. J Virol. 1994;68:1316–23.

43. Loy J, Gander J, Mogler M, Vander Veen R, et al. Development and evaluation of a replicon particle vaccine expressing the E2 glycoprotein of bovine viral diarrhea virus (BVDV) in cattle. Virol J. 2013;10:35.

44. Lundstrom K. Latest development in viral vectors for gene therapy. Trends Biotechnol. 2003a;21:117–1122.

45. Lundstrom K. Semliki Forest virus vectors for rapid and high-level expression of integral membrane proteins. Biochim Biophys Acta. 2003b;1610:90–6.

46. Lundstrom K. Biology and application of alphaviruses in gene therapy. Gene Ther. 2005;12:S92–7.

47. Lundstrom K. Alphavirus-based vaccines. Viruses. 2014;6:2392–415.

48. Lundstrom K, Richards J, Pink J, et al. Efficient *in vivo* expression of a reporter gene in rat brain after injection of recombinant replication-deficient Semliki Forest virus. Gene Ther Mol Biol. 1999;3:15–23.

49. Lundstrom K, Rotmann D, Hermann D. Novel mutant Semliki Forest virus vectors: gene expression and localization studies in neuronal cells. Histochem Cell Biol. 2000;115:83–91.

50. Mathiot C, Grimaud G, Garry P, et al. An outbreak of human Semliki Forest virus infection in Central African Republic. Am J Trop Med Hyg. 1990;42:386–93.

51. Michelini Z, Negri D, Baroncelli S, et al. T-cell-mediated protective efficacy of a systemic vaccine approach in cynomolgus monkeys after SIV mucosal challenge. J Med Primatol. 2004;33:251–61.

52. Morris-Downes M, Sheahan B, Fleeton M, et al. A recombinant Semliki Forest virus particle vaccine encoding the prME and NS1 proteins of louping ill virus is effective in a sheep challenge model. Vaccine. 2001;19:3877–84.

53. Morse M, Hobelka A, Osada T, Berglund P, et al. An alphavirus vector overcomes the presence of neutralizing antibodies and elevated numbers of Tregs to induce immune responses in humans with advanced cancer. J Clin Investig. 2010;120:3234–41.

54. Niklasson B. Sindbis and Sindbis-like viruses. In: Monath TP, editor. The arboviruses: epidemiology and ecology. Boca Raton: CRC Press Inc; 1988. p. 167–76.

55. Niklasson B, Aspmark A, LeDuc J, et al. Association of a Sindbis-like virus with Ockelbo disease in Sweden. Am J Trop Med Hyg. 1984;33:1212–7.

56. Paredes A, Brown D, Rothnagel R, et al. Three-dimensional structure of a membrane-containing virus. Proc Natl Acad Sci USA. 1993;90:9095–9.

57. Peters C, Dalrymple J. Alphaviruses. In: Fields BN, Knipe DM, editors. Virology. New York: Raven Press; 1990. p. 713–61.

58. Phenix K, Wark K, Luke C, et al. Recombinant Semliki Forest virus vector exhibits potential for avian virus vaccine development. Vaccine. 2001;19:3116–23.

59. Phillips D, Murray J, Asakov J, et al. Clinical and subclinical Barmah Forest virus infection in Queensland. Med J Aust. 1990;152:463–6.

60. Polo JM, Belli BA, Driver BA, et al. Stable alphavirus packaging cell line for Sindbis virus and Semliki Forest virus-derived vectors. Proc. Natl. Acad. Sci. USA 1999;96:4598–603.

61. Rennels M. Arthropod-borne virus infections of the central nervous system. Neurol Clin. 1984;2:241–54.

62. Rodrigues Faria N, Lourenco J, Marques de Cerqueira E, et al. Epidemiology of Chikungunya virus in Bahia, Brazil, 2014–2015. PLoS Curr. 2016; Feb 1; 8. pii: ecurrents.outbreaks. c97507e3e48efb946401755d468c28b2

63. Rodriguez-Madoz J, Prieto J, Smerdou C. Semliki forest virus vectors engineered to express higher IL-12 levels induce efficient elimination of murine colon adenocarcinomas. Mol Ther. 2005;12:153–63.

64. Rodriguez-Madoz J, Liu K, Quetglas J, Ruiz-Guillen-M, et al. Semliki forest virus expressing interleukin-12 induces antiviral and antitumoral responses in woodchucks with chronic viral hepatitis and hepatocellular carcinoma. J Virol. 2009;83:12266–78.

65. Rossi S, Guerbois M, Gorchakov R, Plante K, et al. IRES-based Venezuelan equine encephalitis vaccine candidate elicits protective immunity in mice. Virology. 2013;437:81–8.

66. Rossi S, Russell-Lodrique K, Killeen S, Wang S, et al. IRES-containing VEEV vaccine protects Cynomolgus macaques from IE Venezuelan equine encephalitis virus aerosol challenge. PLoS Negl Trop Dis. 2015;9: e0003797.

67. Rumenapf T, Strauss E, Strauss H. The subgenomic mRNA of Aura alphavirus is packaged into virions. J Virol. 1994;68:56–62.

68. Schlesinger S. Alphavirus vectors: development and potential therapeutic applications. Expert Opin Biol Ther. 2001;1:177–91.

69. Schultz-Cherry S, Dybing J, Davis N, et al. Influenza virus (A/HK/156/97) hemagglutinin expressed by an alphavirus replicon system protects chickens against lethal infection with Hong Kong-origin H5N1 viruses. Virology. 2000;278:55–9.

70. Scott T, Weaver S. Eastern equine encephalomyelitis virus: epidemiology and evolution of mosquito transmission. Adv Virus Res. 1989;37:277–328.

71. Smerdou C, Liljestrom P. Two-helper RNA system for production of recombinant Semliki Forest virus particles. J Virol. 1999;73:1092–8.

72. Smith A, Tignor G. Host cell receptors for two strains of Sindbis virus. Arch Virol. 1980;66:11–26.

73. Strauss E, Strauss J. Structure and replication of the alphavirus genome. In: Schlesinger S, Schlesinger M, editors. The Togaviridae and Flaviviridae. New York: Plenum Publishing Corp; 1986. p. 35–90.

74. Strauss J, Strauss E. The alphaviruses; gene expression, replication and evolution. Microbiol Rev. 1994;58:491–562.

75. Sun S, Liu X, Guo H, et al. Protective immune responses in Guinea pigs and swine induced by a suicidal DNA vaccine of the capsid gene of swine vesicular disease virus. J Gen Virol. 2007;88:842–8.

76. Sun Y, Li N, Li H, et al. Enhanced immunity against classical swine fever in pigs induced by prime-boost immunization using an alphavirus replicon-vectored DNA vaccine and a recombinant adenovirus. Vet Immunol Immunopathol. 2010;137:20–7.

77. Tang J, Bi Z, Ding M, Yin D, et al. Immunization with a suicidal DNA vaccine expressing the E glycoprotein protects ducklings against duck Tembusu virus. Virol J. 2018;15:140.

78. Utt A, Quirin T, Saul S, et al. Versatile trans-replication systems for Chikungunya virus allow functional analysis and tagging of every replicase protein. PLoS One. 2016;11:e0151616.

79. Vähä-Koskela M, Tuittila M, Nygardas P, et al. A novel neurotrophic expression vector based on the avirulent A7(74) strain of Semliki Forest virus. J Neurovirol. 2003;9:1–15.

80. Wang K-S, Kuhn R, Strauss E, et al. High-affinity laminin receptor is a receptor for Sindbis virus in mammalian cells. J Virol. 1992;66:4992–5001.

81. Wang Y, Huang H, Zou H, Tian X, et al. Liposome encapsulation of oncolytic virus M1 to reduce immunogenicity and immune clearance in vivo. Mol Pharmacol. 2019;16:779–85.

82. Weaver S, Salas R, Rico-Hesse R, et al. Re-emergence of epidemic Venezuelan equine encephalomyelitis in South America. VEE Study Group. Lancet. 1996;348:436–40.

83. Wecker M, Gilbert P, Russell N, Hural J, et al. Phase I safety and immunogenicity evaluations of an alphavirus replicon HIV-1 subtype C gag vaccine in healthy HIV-1-uninfected adults. Clin Vaccine Immunol. 2012;19:1651–60.

84. Wengler G, Wengler G. Identification of a transfer of viral core protein to cellular ribosomes during the early stages of alphavirus infection. Virology. 1984;134:435–42.

85. White C, Thomson M, Dimmock N. Deletion analysis of a defective interfering Semliki Forest virus RNA genome defines a region in the nsP2 sequence that is required for efficient packaging of the genome into virus particles. J Virol. 1998;72:4320–6.

86. Wu Q, Xu F, Fang L, Xu J, et al. Enhanced immunogenicity induced by an alphavirus replicon-based pseudotyped baculovirus vaccine against porcine reproductive and respiratory syndrome virus. Virol Methods. 2013;187:251–8.

87. Xia S, Lei J, Du M, et al. Enhanced protective immunity of the chimeric vector-based vaccine rAdV-SFV-

E2 against classical swine fever in pigs by a Salmonella bacterial ghost adjuvant. Vet Res. 2016;47:64.

88. Xiong C, Levis R, Shen P, et al. Sindbis virus: an efficient, broad host range vector for gene expression in animal cells. Science. 1989;243:1188–91.

89. Zajakina A, Spunde K, Lundstrom K. Application of alphavirus vectors for immunomodulation in cancer therapy. Curr Pharm Des. 2017;23:4906–32.

90. Zhu W, Liang G. Selection and characterization of packaging cell lines for XJ-160 virus. Intervirology. 2009;52:100–6.

91. Zhu W, Fu J, Lu J, et al. Induction of humoral and cellular immune responses against hepatitis C virus by vaccination with replicon particles derived from Sindbis-like virus XJ-160. Arch Virol. 2013;158: 1013–9.

# Part IV

# Application of Viral Vector Vaccines, Challenges and Future Directions

# Manufacturing and Control of Viral Vectored Vaccines: Challenges

Zahia Hannas, Joanna Sook Mun Tan, Yang Zhang, Frederic Lhermitte, Catherine Cleuziat, Lauri Motes-Kreimeyer, Philippe Dhoms, and Michel Bublot

**Abstract**

The manufacturing of veterinary viral vector vaccines follows the same principles of production of live veterinary vaccines. It includes upstream and downstream production of the active ingredient, as well as formulation, freeze-drying and/or packaging. The process development is key to allow robust production of high-quality, potent, safe and stable batches of vaccine at an acceptable cost. Potency test is generally based on virus titre, but additional specific analysis may be required to evaluate the potency of vector vaccines. Production processes of HVT, poxvirus and adenovirus vectors are summarized. Once the process and quality control are established for a particular vector vaccine, they can be reused for future products based on the same vector speeding up the process development and outline of production acceptance by the regulatory authorities. Process improvements include the use of continuous cell lines in suspension and in medium without substances of animal origin or chemically defined, using single-use technology, automatization, digitalization, data management and artificial intelligence.

**Keywords**

Seed lot · Good manufacturing practices · HVT · Poxvirus · Viral vector · Adenovirus

## Abbreviations

| | |
|---|---|
| AI | active ingredient |
| CEF | chicken embryo fibroblast |
| cGMP | current good manufacturing practice |
| COG | cost of good |
| EPC | end of production cell bank |
| FMD | foot and mouth disease |
| hAd5 | human adenovirus 5 |
| HVT | herpesvirus of turkey |
| MCB | master cell bank |
| MDV | Marek's disease virus |
| MES | Manufacturing execution system |
| MOI | multiplicity of infection |
| MSV | master seed virus |
| QbD | quality by design |

Z. Hannas · J. S. M. Tan · Y. Zhang · F. Lhermitte ·
C. Cleuziat · P. Dhoms · M. Bublot (✉)
Boehringer Ingelheim R&D, 813, cours du 3ème
Millénaire, Saint Priest, France
e-mail: Zahia.HANNAS@boehringer-ingelheim.com;
joanna.tan@boehringer-ingelheim.com; yang_2.zhang.
ext@boehringer-ingelheim.com; Frederic.
LHERMITTE@boehringer-ingelheim.com; Catherine.CL
EUZIAT@boehringer-ingelheim.com; Philippe.
DHOMS@boehringer-ingelheim.com; michel.bublot@
boehringer-ingelheim.com

L. Motes-Kreimeyer
Boehringer Ingelheim R&D, Athens, GA, USA
e-mail: lauri.motes-kreimeyer@boehringer-ingelheim.
com

© Springer Nature Switzerland AG 2021
T. Vanniasinkam et al. (eds.), *Viral Vectors in Veterinary Vaccine Development*,
https://doi.org/10.1007/978-3-030-51927-8_12

QC      quality control
SPF     specific pathogen-free
SUT     single-use technology
WCB     working cell bank
WSV     working seed virus

**Learning Objectives**
After reading this chapter, you should be able to:

- Recognize that the manufacturing of viral vector vaccines follows the same principles of live vaccine manufacturing
- Relate that the quality of viral vector vaccines is the quality of the process
- Explain how the same process may be applied to several vector vaccines using the same vector
- Recognize that a key challenge is to develop a process allowing the production of a vaccine with sufficient quality, safety and efficacy and within an acceptable cost of good (COG)

# 1      Introduction

Viral vectored vaccines are live viruses that carry at least one gene coding for a protein able to induce a protective immune response against a pathogen agent. This "protective protein" will be expressed in vivo, in the infected cells of the vaccinated animal. The manufacturing of viral vectored vaccines is aiming to produce and preserve an adequate amount of this infectious virus. It does not differ fundamentally from that of live conventional vaccines. They both follow the same principles which are well summarized in the Chapter 1.1.8 of the OIE Terrestrial Manual (http://www.oie.int/fileadmin/Home/eng/Health_standards/tahm/1.01.08_VACCINE_PRODUCTION.pdf). The current chapter includes an overview in the process development and current production practices in veterinary vaccine manufacturing as well as specificities for a few chosen vector examples, including avipoxvirus, herpesvirus of turkey (HVT) and human adenovirus production. This chapter is also covering key challenges when developing and commercializing veterinary vaccines as well as future improvements in vaccine production. It is based on both authors experience and literature review.

# 2      Vector Vaccine Manufacturing Process Development

The process definition and optimization are critical steps in the development of vector vaccines leading to market authorization. Table 1 gathers different steps of process development. The cell substrate (i.e. the cells used to manufacture the viral vector) needs first to be selected. It could be primary cells or cell lines and, among the latter, classical cell lines or genetically modified cell lines. Cells could be grown in a monolayer (roller bottles (Fig. 1), cell factory) or in a bioreactor (on micro-carriers, fixed-bed or suspension culture; Fig. 2). When working with cell lines, master and working cell banks (MCB and WCB, respectively) will have to be established during development to ensure a reliable and consistent supply of cells. The MCB and, in a lesser extent, the WCB (depending on the local regulatory guidance) will be fully controlled and characterized both in vitro and in vivo (mainly for avian targets) prior to use for production (see Table 1 for more details). Both MCB and WCB will be stored in vapour or liquid phase of liquid nitrogen in at least one secure place (two for MCB) since the loss of MCB will mean the loss of the product (as based on a seed lot system). The MCB and WCB production and testing is a long and costing process.

Similar to MCB, a master seed virus (MSV) will have to be produced with the chosen vector vaccine candidate. Extensive testing will have to be done on this MSV to prove its purity and identity. In addition, many safety studies will be done with the MSV to prove that it has no residual pathogenicity, that it does not revert to virulence, that its tropism, shedding and eventual spreading are not changed compared to the parental virus. The genetic and phenotypic (transgene expression) stability will have to be assessed after passages. Similar to MCB, MSV is securely

**Table 1** Different steps in the development of production process and of quality control of veterinary vector vaccines

| General step | Includes |
|---|---|
| Cells | Selection of cells for virus production |
| | Selection of media |
| | Definition of optimal conditions for cell production |
| | Production of master cell bank (MCB), working cell bank (WCB) and end of production cell bank (EPC) usually corresponding to MCB + 20 passages |
| | Testing of the MCB, WCB and EPC: |
| | Identity testing |
| | Sterility/mycoplasmas |
| | Extraneous agents (depending on the guidelines) |
| | Karyotype (MCB, EPC) |
| | Tumorigenicity (MCB) |
| | Electronic microscopy (MCB, WCB) |
| Virus seed | Selection of best viral vector candidate |
| | Production of master seed virus (MSV) |
| | Testing of the MSV: |
| | Identity testing |
| | Sterility/mycoplasmas |
| | Extraneous agents |
| | Genetic structure |
| | Phenotypic and genotypic stability |
| | Safety and in vivo biological properties (including reversion to virulence, tissue tropism, shed and spread, safety for target and non-target species, recombination potential) |
| Raw materials | Definition of specifications and source for all product ingredients with particular attention to substances of animal origin. |
| Virus seed lot process development and working seed virus production and control | Definition and optimization of process |
| | Working seed virus (WSV) production and storage WSV control |
| Vaccine process development and production of R&D batch(es) | Definition and optimization of: |
| | Upstream active ingredient (AI) production |
| | Downstream active ingredient (AI) production |
| | Formulation |
| | Freeze-drying |
| | Filling |
| | Packaging |
| | Production, quality control, and stability follow-up of vaccine batch(es) for development studies |
| Development and validation of analytical tools | Development and validation of analytical tools |
| | To test MSV and WSV |
| | To test active ingredient |
| | To test vaccine batches |
| | To evaluate stability |
| Transfer of process from R&D to manufacturing and production of pre-license serials | Definition of the process capability |
| | Definition of detailed outline of production |
| | Transfer and validation of production process |
| | Transfer and validation of quality control tools |
| | Production, QC, and stability follow-up of three consecutive pre-license serial batches for validation |

**Fig. 1** Production of viral vaccines in monolayers in roller bottles

stored under defined conditions usually at $-70\,°C$ or $-40\,°C$, in aliquots of uniform composition. One or more of these aliquots will be used to produce working seed virus (WSV) that will then be used for the production of the viral vector, as the active ingredient of the vaccine.

The optimal medium allowing the cell growth and/or viral production will have to be defined. All the raw materials used for the preparation of these seed lot systems (cell and virus) and the active ingredient should be qualified and avoid wherever possible components of animal origin.

The definition and optimization of the upstream and downstream process for active ingredient production is time-consuming and requires testing of many different culture conditions. Most of the live viral vectored vaccines are presented in a freeze-dried form and must be kept under refrigerated conditions. The lyophilisation stabiliser will have to be adequately defined as well as the optimal freeze-drying conditions. This step is critical to assure good stability and so long shelf life of such live vaccines. There are also vector vaccines presented in liquid form (for instance, canarypox vector for horses or vaccinia vector for wildlife vaccination (see below)) or as frozen live infected cells (for instance, herpesvirus of turkey (HVT;

Meleagrid herpesvirus 1) vector for chickens; see below). The type of container (usually glass vials), labelling and secondary packaging will need to be defined too.

In parallel to the development of the process, analytical tools will have to be developed to support process development and to test the seed lots, the active ingredient and the final vaccine. One key test is the potency test that will have to be carefully validated. For live viral vaccines, the potency test is often based on the viral infectious titre since it usually correlates with the protection. Efficacy studies will have to establish the minimum protective dose. The minimum release dose will have to be set up to assure that the vaccine at the end of the shelf life still contains the minimum dose required for protection. For vaccines where vector itself is a vaccine, such as HVT vector vaccines, a viral titre for both the vector and the insert may have to be measured. This is classically done by titration of the vector in a plaque assay and of the insert by staining the plaques with a monoclonal antibody recognizing the expressed foreign gene product(s). This test may be used as an identity test as well; alternative identity tests can be developed using specific molecular tools such as polymerase chain reaction.

**Fig. 2** Production of viral vaccines in stainless steel bioreactor

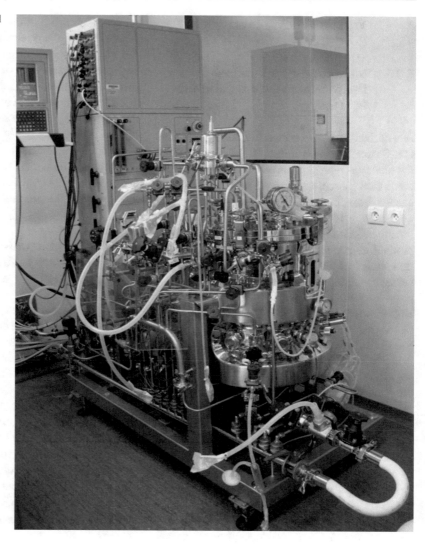

Once process outline, control testing, active ingredient (AI) and vaccine specifications are defined, activities will have to be transferred to production and quality control (QC) departments. The personnel will have to be trained and qualified on the production process and analytical tools. The process will be validated in the manufacturing department by producing successfully three consecutive batches (called pre-license serials in some countries). These batches are typically used in the field clinical

trials implemented during the development of new vaccines.

More and more the process development tends to reach the "Quality by Design" (QbD) principles. QbD identifies and scientifically understands characteristics during the whole process from raw materials to the release of the final product that are critical to the quality of the vaccine for the patient and setting up the right controls, quality assurance and risk management at critical production steps.

# 3    Vector Vaccine Industrial Production

## 3.1    Manufacturing Vaccines is the Monitoring of a Complex Process

A veterinary vector vaccine is a biological product. Basically, it is a medicinal product of which the quality is reproducible over the time through a controlled process of production designed to ensure its robustness. In some countries, the application of current good manufacturing practices (cGMP) is mandatory. The vaccine is a product that results of a chain of several production steps (see Table 2) involving biological processes where live material is handled and physico-chemical processes where specifically engineered equipment is used to perform different tasks (for instance, mixing, heating, cooling, transferring, freeze-drying and filling). It requires from the production personnel the control of each step to ensure its reliability. It requires also that the quality control personnel performs the appropriate tests to ensure the product complies with the quality specifications at each step and for the finished product, basically, its purity, its safety and its efficacy.

The reliability of vaccine manufacturing can be defined as the probability of the finished product to be released. The manufacturing process is as a chain of steps from the raw materials enforcement to the finished product dispensing (see Table 2).

The reliability of the whole process $R_{final}$ is the product of the reliability of each process steps $R_1$; $R_2$; ...; $R_n$. As a result, the failure of the reliability in at least one process step will impact the whole process reliability. Furthermore, when considering the complexity of a combo vaccine that brings about several antigens, the probability of the finished product release is the product of the probability of each antigen release as each antigen is an independent variable of the control process with its own testing. As a result, the failure in the release of at least one antigen regarding its own quality specifications in the combo vaccine will

prevent the release of the whole finished product. It would imply very serious economic losses for the vaccine manufacturing companies. As a consequence, the whole set of vaccine manufacturing process will be designed, developed and validated at the industrial scale in order to ensure the maximum reliability and the maximum probability of release at each production step and for each antigenic component. That is the reason why, most of the time, the following sentence is considered in the vaccine industry: "The quality of a vaccine is the quality of the process".

## 3.2    Manufacturing Vaccines Is Built on Barriers Aimed to Prevent the Entropy and the Butterfly Effect

Following the second principle of thermodynamics, entropy is the measure of the disorder. In a given closed system, the transformation of material always implies the increase of the entropy. Entropy is also the measure of uncertainty in a given system. The less the information is available regarding an event, the less its known probability of occurrence, the highest is the entropy. Entropy is unavoidable in a given manufacturing system that is transformed over time: machine ageing and attrition, personnel turnover, supplier missing, raw material changes, knowledge loss and so on. When considering the concept of entropy with the vaccine manufacturing processing, the natural trends to loss information and process robustness over the time is critical to be controlled and prevented. That's the reason why a whole set of barriers against entropy are designed, developed and transferred to the industrial scale. For instance:

- The raw materials, the seeds and the product at each step of manufacturing are controlled through a defined set of reference test techniques.
- The reference test techniques are in compliance with various regulatory guidelines.
- The genetic and/or phenotypic identity of the different seeds, cells and virus are

**Table 2** Different steps in the manufacturing and quality control of an industrial batch of vector vaccine

| General step | Includes |
|---|---|
| Media | Raw materials sourcing and testing |
| | Media preparation and sterilization |
| Cell cultures | Cell culture initiation from the thawed WCB |
| | Cell culture expansion |
| Active ingredient (AI) production upstream and downstream processes | Upstream process: |
| |   Cell infection and incubation |
| |   Virus harvest |
| | Downstream process (variable): |
| |   Cell disintegration |
| |   Clarification (centrifugation, filtration) |
| |   Concentration |
| |   Storage |
| Vaccine formulation | Mixing of one or more AI with freeze-drying stabilizer |
| | Filling in vials |
| | Freeze-drying |
| | Capping of vials |
| Labelling | Labelling |
| | Secondary packaging |
| Quality control (AI and vaccine) | Physico-chemical (such as pH) and visual aspects |
| | Titration (potency test) |
| | Identity test |
| | Sterility and mycoplasma testing |
| | Extraneous agent testing |
| Documentation, quality assurance and vaccine release | Detailed records of all steps and personnel |
| | Dossier constitution with production and QC raw data |
| | Data audit |
| | Release of vaccine by a qualified person (QP if cGMP are applied) based on the specifications in the dossier |

characterized as accurately as possible and checked at the critical steps of manufacturing.

- The manufacturing steps of each antigen are rigorously detailed in the manufacturing instructions.
- The manufacturing operations for each batch are finely tracked in the batch records.
- The raw material suppliers are audited for the quality of their control and for the robustness of their own raw material quality and their manufacturing processes.
- The manufacturing and the quality control personnel are trained on a regular basis in order to maintain their professional skills.
- The knowledge of product development and production is managed through an expert network.

The butterfly effect is known as a popular metaphor for sensitive dependence on initial conditions of a given system conceived by the mathematician E. Lorentz in 1972. It was demonstrated by the chaos theory that, for a non-linear system depending on several variables, the impact of a small change in the value of one of the system parameters could be very critical. Considering the vaccine manufacturing process system, it is well-known that it is typically a non-linear system with a tremendous number of variables. As a consequence, a minor change in the quality of a raw material or in the process could lead to a non-satisfactory finished product regarding its quality, safety and efficacy. This kind of change could imply very long troubleshooting, trying to find the event root causes and serious economic losses for a vaccine manufacturer through destruction of unsatisfactory products and market back-orders. As quoted above, "the quality of the vaccine is the quality of the process". That's the reason why a whole set of barriers against the butterfly effect are designed, developed and transferred to the industrial scale. For instance:

– The determination of the critical parameters defined as the parameters where a minor change of value could imply a non-satisfactory product
– The risk analysis based on literature and on the expert experiences
– The determination of the robustness of the process in its limits of control set-up during the development for each critical parameter
– The set-up and the systematic use of control charts following the rules of the statistical control of the process
– The validation of the control techniques and at each renewal of the control reagents
– The product quality reviews on a regular base

All of these ways of risk mitigation constitute barriers to the butterfly effect and people are committed in their implementation and rigorous follow-up. It will avoid the consequences that such events have on the vaccine manufacturing companies like an unsafe or under-efficacious batch that may lead to stopping the production or the withdrawal of the batch or even product from the market upon demand by the regulatory authorities.

## 3.3 Manufacturing Vaccines Relies on People

The commitment of the team (including people from development, production, quality control and quality assurance) is required for ensuring the quality of the vaccines placed on the market. Although artificial intelligence may support the process, it will not replace the knowledge, the talent and the shrewdness of the people involved in the vaccine manufacturing industry. People are the warrants that the quality of the process makes

the quality of the vaccine. But all these skills have to be maintained through training. Basically, training has to be set up and performed on the workplace for the core knowledge. Qualification grids help to plan and to track the training implementation. A tutorial process is set up to ensure that the newcomers are trained by skilled long-knowing employees considered as experts in their domain, themselves trained for running training. This allows setting a system of routine human control in the workshops where the team can prevent or repair immediately the failure or the human error of one.

The organization of the manufacturing resources is most of the time under the monitoring of a manufacturing execution system (MES) in order to rationalize the operation sequences and operation execution times. The most modern MES today captures the production data and guides the operation sequence roll-out by direct digital computing system where the workers are helped for keeping the manufacturing instructions.

These measures for risk mitigation regarding the human and organizational factors are today more and more favoured by the authorities in their regulatory guidance and during inspections that are performed on a regular basis in order to ensure the compliance of the vaccines produced and distributed on the markets.

## 4 Manufacturing of Poxvirus Vectors

Attenuated poxviruses are excellent vectors for prophylactic and therapeutic vaccines. Vaccinia, fowlpox and canarypox viruses have successfully been developed as veterinary vaccine vectors for wildlife, poultry and companion animals (horse, dog, cat and ferret) [1–10]. The acquired expertise on recombinant fowlpox and canarypox vector vaccines for multiple species facilitates the rapid generation, development and manufacturing of new candidates. At the industrial level, the same manufacturing process is used for the various avipoxvirus vector vaccines ensuring batch-to-batch consistency [3]. As an example, the manufacturing process of the canarypox vector will be described.

Seeds and active ingredients are produced in sterile disposable plastic flasks and/or roller bottles (Fig. 1) seeded with primary chicken embryo fibroblast (CEF). CEF cells are produced by enzymatic digestion of chicken embryo from specific pathogen-free (SPF) hens. Viral inoculation is performed on CEF monolayer, and the viral culture is incubated a few days until a clear cytopathogenic effect is visible; the number of days depends in part on the multiplicity of infection (MOI) used at viral inoculation. The harvested cell lysates may be centrifuged and sonicated to favour virus release from cell debris before clarification and filtration. The addition of a stabilizer allows more stability of AI during storage at $-40\ °C$ or colder. Once the AI batch is titrated and controlled, it can be formulated and possibly freeze-dried into sterile vials before capping, labelling and packaging. Some canarypox-based vaccines are presented as a liquid viral suspension and may contain carbomer, a polymer primarily made from acrylic acid as an adjuvant. For the vaccinia vector developed to vaccinate wildlife against rabies, the vaccine is in a liquid form in a bag packaged in baits that are released in the field [10]. Each batch/serial of the final product is submitted to QC tests before release. The potency test is based on the infectious viral titre; the titration may include the detection of the foreign gene product using specific monoclonal antibody and immunofluorescence.

In contrast to the fowlpox vector which is fully replicative in chickens, the canarypox vector is non-replicative in mammals. After vaccination, the canarypox virus will enter into mammalian cells and start its replication cycle during which the foreign gene will be expressed; however, the replication cycle is not complete; no infectious virus will be generated from this infected cell. The minimum protective dose (and, consequently, the minimum release dose) of a canarypox vector in mammals is therefore much higher than that of a fowlpox in poultry. It is therefore important to optimise the canarypox virus production in order to reach a high titre at an acceptable cost. Improvement of the process

may include the use of a permissive continuous cell line, bioreactor and chemically defined media [11–13].

## 5    Manufacturing of Herpesvirus Vectors

The main virus that has been developed as a vector for poultry is the *Meleagrid herpesvirus 1* (Marek's disease virus serotype3), also known as herpesvirus of turkey (HVT). The HVT virus is non-oncogenic and is used as a vaccine against Marek's disease since the 1970s [14]. It was developed as a vector for different avian diseases including infectious bursal disease [15], Newcastle disease [16, 17], infectious laryngotracheitis disease [18] and avian influenza [19]. HVT vectors with two inserted genes were also recently launched.

The HVT virus shares similar characteristics in terms of in vivo [20] and in vitro [21] replication with *Gallid herpesvirus 2*, the Marek's Disease Virus (MDV). Thus, they share similar production platform and technologies. In vitro, HVT and its vectored derivatives are generally cultivated in adherent primary avian cells like CEF, leading to the use of monolayer technology for upstream process. The Marek's disease viruses are cell-associated, and therefore, the vaccine AI is the living infected cells formulated and frozen with liquid nitrogen under controlled conditions. No further downstream process is required. At an industrial scale, plastic or glass roller bottles (Fig. 1), as well as multilayer culture flasks, are used as culture vessels. Vaccine storage and distribution require liquid nitrogen tanks.

The different steps of production of HVT vector vaccines are described below:

1. Primary CEF preparation and cell culture: CEF are prepared by enzymatic digestion of embryos from specific pathogen-free (SPF) eggs as for canarypox virus production (see above). Roller bottles are seeded with the cells in a suitable medium.
2. Viral inoculation, culture and harvest: The inoculums used are composed of living

infected cells coming from freshly harvested culture or infected cells stored in liquid nitrogen. After several passages from MSV or WSV, inoculation of cells is performed. AI production can be realized in one or several passages. The post-inoculation culture duration depends on the MOI applied. The infected cells are harvested by treating the monolayer with an enzymatic solution. The infected cells are concentrated by centrifugation and resuspended using freezing media for storage in liquid nitrogen.

3. Freezing and storage: The cell suspension is distributed in glass ampoules which are sealed and frozen using an adapted freezing cycle before to be stored in a liquid nitrogen tank. A glass vial contains enough infected cells to vaccinate 1000 or several thousands of chickens.

4. Quality control: HVT-vectored vaccine is mainly tested like an HVT Marek's disease vaccine. The difference is in the identity test that needs to identify specifically the vector vaccine and its foreign gene. It may be done by immunofluorescence detecting the foreign gene expression or by a specific polymerase chain reaction (PCR). The potency test is based on the viral titre. The HVT titration is evaluated by counting the plaque forming unit, and when required, the titration for the foreign gene is usually performed by immunofluorescence with a specific monoclonal antibody.

5. Transport and vaccine preparation: The vaccine is transported in liquid nitrogen Dewar tanks to the hatcheries where the vaccine is used. Glass ampoules are thawed in a water bath, and infected cells are resuspended in a bag of diluent. The vaccine administration may be done by the *in ovo* to 18–19-day-old embryos or by the subcutaneous route to 1-day-old chicks. Vaccine administration should be done within the hour following vaccine preparation.

This process of production is complex and is facing many difficulties relative to:

1. Cell substrate: Primary CEF cells need to be used since there is today no suitable cell line

available. Hence, the egg quality and supply will have a direct impact on vaccine supply and sustainability. In addition, it is challenging to handle embryonated eggs in a controlled environment.

2. Culture vessels: It has not been possible to adapt HVT production in bioreactor so far due to the replication mechanism requiring high-density cell layer and the necessity to harvest live cells. The use of roller bottles or other alternative culture vessels may lead to multiple open-phase operations during different stages of production and inevitable heavy labour, which cause high contamination risks and limitations on productivities.

3. Infected cell freezing and storage: The cell-associated nature of HVT requires conserving live infected cells in liquid nitrogen. HVT virus may also be freeze-dried, but this cell-free-based vaccine has been shown to be less efficacious and more sensitive to maternally derived antibodies than the cell-associated vaccine [22]. The liquid nitrogen requirement complicates the storage and distribution of such vaccine. Break in the cold chain will have quickly an impact on the viral titre [23] and, so, on efficacy.

## 6 Manufacturing of Adenovirus Vectors

Two types of adenovirus vectors have been developed for animals: a replicative human adenovirus type 5 for oral vaccination of wildlife against rabies [24, 25] and a non-replicative human adenovirus type 5 (hAd5) for cattle vaccination to protect against foot and mouth disease (FMD) [26, 27]. The latter is the most recent conditionally licensed vaccine for animal health utilizing capsid and 3C protease-coding regions of FMD virus to protect cattle for a potential outbreak of the disease in the United States. Although conceptually interesting, examples of replication-competent adenovirus vector produced from a target species for bivalent protection in the same species were not able to overcome interfering antibodies in subsequent vaccinations or protect via the preferred route of oral administration [28]. Primarily non-replicative adenovirus vectors (such as those based on hAd5) are used for today's evaluations which may require modified cell lines for vector production or sequence "switches" to up- or downregulate a gene encoded. Vaccine design gives a hint to the complexity of the upstream, downstream, formulation and testing of adenovirus vector vaccines. This is a subsequent challenge in the case of vaccines targeting disease outbreaks or multi-dose presentations in which expensive raw materials, industrialization scale or antigen load for protection can be cost-prohibitive. Adenovirus vaccines have been shown to be effective with or without adjuvants and/or immunomodulators depending on the target disease. Combinations of multiple serotypes in one formulation and use in prime-boost regimes continue to be a strong area of focus for veterinary research.

The production and formulation of adenovirus vectors have been reviewed in detail by Altaras *et al.* [29] and Vellinga *et al.* [29, 30]. A summary of steps of production of adenovirus vector vaccines are described below:

1. Cells and media: Cell lines used to produce adenovirus vector vaccines are usually HEK293, PerC6 or A549 derivatives. All can be grown in adherent cultures for testing purposes and preferably suspension-based in serum-free media without animal origin components for the production of active ingredients. Cell density effect described with HEK293 at high densities greater than one million cells/ml can have a detrimental impact on vector productivity without proper analysis of cell health and supplementation. Many commercial serum-free mediums are available in liquid and powder formulation or development of custom media formulation. Cell growth is optimized in a medium to reach reproducible densities and doubling time in the fewest number of manipulations. Working cell banks are frozen in liquid nitrogen at regular passage intervals, with adequate densities, in ampules or cryobags.

2. Cell passaging: HEK293 derivatives are expanded from frozen cell banks with attention to recovery from liquid nitrogen, cell counts, imaging and trending of doubling time. Additional data points can be acquired with online sampling and metabolite analysis. Cells can be prompted into specific physiological phases for infection utilizing monitoring or media changes prior to the infection process.

3. Viral infection and harvest: Once viral working seed lots are produced, an optimal MOI for the adenovirus vector is established over a series of passages. The medium is usually changed prior to infection. In the cases of non-replicative vector, induction may be required ahead of infection and media change using hollow fibres or perfusion after infection depending on the vector design. Set points for bioreactor production conditions such as temperature, pH, dissolved oxygen and time of harvest should be optimized.

4. Downstream processing: At the time of harvest, cells may be lysed by mechanical methods or detergent. Endonucleases such as benzonase may be used to treat lysate prior to clarification. Concentration and/or diafiltration in selected buffers prior to sterile filtration and storage may occur.

5. Formulation with or without adjuvant follows and serials are tested accordingly.

6. In-process testing and quality control: As for any vaccine, each serial or sub-serial must satisfactorily meet specification requirements described in the marketing authorization. Potency is developed in parallel with the new product. Vector and gene product characterization can be complex for in-process and release testing. It is critical to test the genetic stability of the construct, the ratio of viral particles to infectious unit (P/I), the presence of defective particles and replication-competent adenoviruses (generated by recombination between the vector and the homologous E1 sequences present in the manufacturing cell line) by different titration methods.

Commercialization is used to be the major concern for adenovirus vector veterinary vaccines, and platforms that have proven cost-effective approaches can be achieved although areas of challenge still exist. Care needs to be taken in development to consistently producing a large number of doses using conventional stainless or single-use stirred-tank bioreactors, feed or perfusion approaches to reach high vector productivity per culture volume in a minimal manufacturing footprint. With regard to testing, parallel tracks for test development and process development are important due to the complexity of the monoclonal antibodies, reagents and analysis needed for a well-characterized final product.

# 7 Future Directions and Potential for Other Applications

The state of the art in vaccine manufacturing has advanced tremendously over the years. The blooming biopharmaceutical industry created the opportunity for researchers and manufacturers to explore more robust and cost-effective production methods to express protein therapeutics and antibodies in fermentation and cell culture processes. These efforts have benefited the vaccine industry that adapted the significant bioprocess technology breakthroughs to fit the viral production processes. Future trends in vector vaccine production are described below for each step of the process.

## 7.1 Upstream Bioprocessing: Viral Antigen Production

With the development of molecular biology tools, improved microbial/mammalian expression efficiencies and advancement of single-use processing systems, the application of recombinant viral vector technology in vaccine development has progressed significantly over the years. The current technology wave in biologics manufacturing offers solutions to the industrialisation of viral vector vaccines at an affordable cost with high product quality. Below are some

examples of recent improvement and future directions:

1. **Raw Materials**: Cell adaptations to suspension and serum-free attempts have been widely explored in the vaccine production industry. The dependency of virus propagation on serum and animal origin raw materials has been a difficulty to vaccine manufacturers. The variability in quality, the supply and the regulations on the use of animal origin materials and its viral purity are currently the drivers to adapt production processes to serum-free, chemically defined media and the use of recombinant enzymes [31, 32]. Simultaneous attempts to adapt cells to suspension for high-cell-density application with perfusion systems have gained much attention to increasing the antigen yield for a more efficient process [33]. Removing the use of microcarriers and associated trypsin treatment process during cell expansion further simplifies the process and decreases the production cost, time and risks. The increasing awareness on the impacts of antibiotics resistance issue to the environment and public health also spurs manufacturers from using antibiotics in vaccine production, especially as a preservative in the final product. Much consideration is put into research and development of new products to produce viral constructs free of antibiotic-resistant gene, removal/alternatives to antibiotics in vaccine production and controlled treatment of production waste.

2. **Continuous Cell Line**: The requirement for the fermentation and cell culture process has increased with more demanding regulations enforced for biologics. Efforts on cell line development for vaccine production using continuous cell expression systems have changed the vaccine production industry significantly. Antigens are more and more produced using microbial, mammalian or insect cell cultures instead of primary cell lines and embryonic cultures [34, 35]. This reduces the risk of potential extraneous agents' contamination and overcomes the supply chain issues

with unstable quantity and quality supply of the embryos.

3. **Bioreactors Cultivation System**: The use of continuous cell line production allows the move from static vessel cultures such as roller bottles to bioreactor production vessels. Although the conventional roller bottle production is still practised for several atypical viruses such as HVT in poultry vaccine production, the advancement of bioreactor technology has penetrated the industry to replace the laborious and high-contamination-risk production method. Hollow fibres and fixed-bed bioreactors with large surface area for cell attachment and controlled aeration are commonly used to replace roller bottles to improve cell growth and viral propagation [36]. Adherent cells are also grown on microcarriers, suspended in the bioreactor system that provides better scalability as well as process controls as compared to static bioreactor cultures [32, 37–40]. Vaccines were primarily manufactured in stainless steel bioreactors. The stainless steel set-up offers flexibility in the bioreactor design that allows manipulation of the bioreactor parts to fit different processes. However, they are progressively replaced by single-use technology.

4. **Single-Use Technology (SUT)**: SUT offers solution to overcome the stainless steel drawbacks with increase in production efficiency and capability, fast turnover rate between batches, absence of complex cleaning validations, reduced manufacturing footprints and reduced capital expenditure investment and COGs. Introduction of SUT in vaccine development gives light to the possibility of single-use end-to-end production with a continuous closed production system reducing the risks of contamination. The single-use systems such as wave bags and bioreactors are designed differently with respective advantages and have demonstrated comparable efficiency if not better than the stainless steel systems [37, 41]. High-cell-density inoculum is viable with the application of perfusion technology coupled with bioreactor system, replacing the tedious and time-consuming cell expansion prior to bioreactor inoculation [42]. Research and development are working on media development, and feeding strategy using the advanced single-use high-throughput miniature bioreactors coupled with analytical tools to support bioprocess design of experiments and to accelerate the product development stages. Supplying sufficient essential media component and nutrient supplementation at cell growth and viral propagation stages are imperative to support the high-cell-density continuous system [43, 44]. These high-throughput lab-scale single-use platforms for development studies have demonstrated scalability to pilot and production scales. The SUT also comes with several challenges, including the integrity of materials and risk of bag leakage and contamination.

5. **Continuous Bioprocessing**: The current manufacturing trend is to implement integrated processing systems for continuous flow modular automation from upstream to downstream processing, equipped with online process monitoring and feedback controls. Such technology is however not easy to set up for viral production.

6. **Analytical Tools**: Manufacturing control with process analytical technologies, pump systems and sensors is being developed. Raman spectrometry, biomass probe and mass spectrometry are examples of analytical tools that are useful in providing real-time biomass and bioprofile measurements for immediate feedback controls and analysis of production data.

## 7.2 Downstream Bioprocessing

The innovation in downstream purification is driven by the increasing demand to meet the upstream high titres as well as the stringent regulations governing the safety and efficacy of vaccines. The biggest hurdle in downstream bioprocessing is to develop cost-effective processes to develop high-quality products with short processing time, which is especially challenging in the animal health vaccine industry that deals with very low selling price per dose of vaccine. A big advantage of vector vaccines is that their production usually does not require complex downstream processing. The purification processes (including chromatography) aiming to remove host cell proteins, nucleic acids, toxins and various adventitious agents may not be needed for vector vaccine manufacturing. Progresses in purification technology over several decades have shown vast improvement in overall process capacity, chromatography efficiency and capability, robust membrane technology, disposable technology and high-throughput process and analytics developments. Some improvements are described below:

1. **Harvest and Clarification**: Cell lysis process by chemical or physical treatment (freeze-thaw, sonication) is typically carried out prior to clarification when dealing with cell-associated viruses or inefficient extracellular secretion of viral particles to release the antigen from the host cells. Clarification of the bulk harvest is carried out to remove cell debris and large aggregates, usually by centrifugation and/or filtration to ensure the efficiency of the subsequent purification steps. The commercially available dead-end depth filters and microfilters are now well designed to support adequate clarification of bulk harvest with different degree of turbidity and viscosity. Very often, the clarification step is carried out as the purification step prior to formulation in animal livestock live vector vaccines due to cost restrictions.

2. **Concentration and Purification**: The clarified supernatant may be concentrated and purified by ultrafiltration and/or diafiltration. The current advancement of membrane technology enables improved uniformity, controlled pore size and customization capability to enhance yield, purity, efficiency and affordability.

## 7.3 Formulation and Delivery

Viral vectored vaccines are generally freeze-dried as the lyophilisation process is well known to preserve the infectivity and stability of the vaccines. However, freeze-drying capability is one of the biggest bottlenecks in vaccine manufacturing as it involves low turnover rate due to the long freeze-drying cycle and turnover maintenance time. Improvements in formulation (stabilizers) and pharmaceutical processing technology are areas of research for many vaccine developers. Future perspectives in the alternative formulation include microencapsulation and nanotechnology that may allow mucosal administration and induction of mucosal immune response which are necessary for some pathogens [45–47]. However, attempts to formulate live viruses including viral vectors with polymer matrices to form microspheres, nanoparticles and hydrogels have been relatively scarce [48–50]. In particular, nanoparticle-vector hybrid delivery strategy utilises the advantage of both, the viral and nanomaterials to complement each other's drawbacks [50, 51]. Increasing the thermal stability of live vector vaccines is also an important area of research; it may be obtained with additives as shown for hAd5 [52].

Automatization, digitalization, data management and artificial intelligence can be applied at each step of vector vaccine production to secure production, minimize downtime, and reduce COGs.

## 8 Summary

The development of production process is a key step before the licensing and commercialization of veterinary vector vaccines. The process development may be the cause of R&D project failure if it does not allow to produce the vector vaccine at acceptable quality, safety, potency and, last but not least for veterinary vaccines, cost. The quality of the process will impact the quality and supply of the vaccine. However, once a process for a particular vector is well established, it can be reused for the production of other products based on the same vector. Challenges in process development include the selection of continuous cell line allowing the production of vector viruses without the substance of animal origins and antibiotics and growing in suspension in bioreactors that may now be of single use. Freeze-drying is also a delicate and time-consuming step in the manufacturing of live vector vaccines, and any alternative formulation solutions allowing the production of a stable vaccine in another form are needed.

# References

1. Baron MD, Iqbal M, Nair V. Recent advances in viral vectors in veterinary vaccinology. Curr Opin Virol. 2018;29:1–7.
2. Brochier B, Kieny MP, Costy F, Coppens P, Bauduin B, Lecocq JP, et al. Large-scale eradication of rabies using recombinant vaccinia-rabies vaccine. Nature. 1991;354(6354):520–2.
3. Poulet H, Minke J, Pardo MC, Juillard V, Nordgren B, Audonnet JC. Development and registration of recombinant veterinary vaccines. The example of the canarypox vector platform. Vaccine. 2007;25 (30):5606–12.
4. Swayne DE, Beck JR, Mickle TR. Efficacy of recombinant fowl poxvirus vaccine in protecting chickens against a highly pathogenic Mexican-origin H5N2 avian influenza virus. Avian Dis. 1997;41(4):910–22.
5. Taylor J, Christensen L, Gettig R, Goebel SJ, Bouquet JF, Mickle TR, et al. Efficacy of a recombinant fowl pox-based Newcastle disease virus vaccine candidate against velogenic and respiratory challenge. Avian Dis. 1996;40(1):173–80.
6. Weyer J, Rupprecht CE, Nel LH. Poxvirus-vectored vaccines for rabies – a review. Vaccine. 2009;27 (51):7198–201.
7. Volz A, Sutter G. Protective efficacy of modified vaccinia virus Ankara in preclinical studies. Vaccine. 2013;31(39):4235–40.
8. Kyriakis CS, De Vleeschauwer A, Barbe F, Bublot M, Van Reeth K. Safety, immunogenicity and efficacy of poxvirus-based vector vaccines expressing the haemagglutinin gene of a highly pathogenic H5N1 avian influenza virus in pigs. Vaccine. 2009;27 (16):2258–64.
9. Tartaglia J, Perkus ME, Taylor J, Norton EK, Audonnet JC, Cox WI, et al. NYVAC: a highly attenuated strain of vaccinia virus. Virology. 1992;188(1):217–32.
10. Maki J, Guiot AL, Aubert M, Brochier B, Cliquet F, Hanlon CA, et al. Oral vaccination of wildlife using a vaccinia-rabies-glycoprotein recombinant virus vaccine (RABORAL V-RG((R))): a global review. Vet Res. 2017;48(1):57.
11. Giotis ES, Montillet G, Pain B, Skinner MA. Chicken embryonic-stem cells are permissive to poxvirus recombinant vaccine vectors. Genes (Basel). 2019;10(3)
12. Jordan I, Northoff S, Thiele M, Hartmann S, Horn D, Howing K, et al. A chemically defined production process for highly attenuated poxviruses. Biologicals. 2011;39(1):50–8.
13. Vazquez-Ramirez D, Genzel Y, Jordan I, Sandig V, Reichl U. High-cell-density cultivations to increase MVA virus production. Vaccine. 2018;36(22):3124–33.
14. Witter RL, Sharma JM, Offenbecker L. Turkey herpesvirus infection in chickens: induction of lymphoproliferative lesions and characterization of vaccinal immunity against Marek's disease. Avian Dis. 1976;20(4):676–92.
15. Bublot M, Pritchard N, Le Gros FX, Goutebroze S. Use of a vectored vaccine against infectious bursal disease of chickens in the face of high-titred maternally derived antibody. J Comp Pathol. 2007;137 (Suppl 1):S81–4.
16. Morgan RW, Gelb J Jr, Schreurs CS, Lutticken D, Rosenberger JK, Sondermeijer PJ. Protection of chickens from Newcastle and Marek's diseases with a recombinant herpesvirus of turkeys vaccine expressing the Newcastle disease virus fusion protein. Avian Dis. 1992;36(4):858–70.
17. Palya V, Kiss I, Tatar-Kis T, Mato T, Felfoldi B, Gardin Y. Advancement in vaccination against Newcastle disease: recombinant HVT NDV provides high clinical protection and reduces challenge virus shedding with the absence of vaccine reactions. Avian Dis. 2012;56(2):282–7.
18. Johnson DI, Vagnozzi A, Dorea F, Riblet SM, Mundt A, Zavala G, et al. Protection against infectious laryngotracheitis by in ovo vaccination with commercially available viral vector recombinant vaccines. Avian Dis. 2010;54(4):1251–9.
19. Kapczynski DR, Esaki M, Dorsey KM, Jiang H, Jackwood M, Moraes M, et al. Vaccine protection of chickens against antigenically diverse H5 highly pathogenic avian influenza isolates with a live HVT vector vaccine expressing the influenza hemagglutinin gene derived from a clade 2.2 avian influenza virus. Vaccine. 2015;33(9):1197–205.
20. Witter RL, Nazerian K, Solomon JJ. Studies on the in vivo replication of Turkey herpesvirus. J Natl Cancer Inst. 1972;49(4):1121–30.
21. Witter RL, Nazerian K, Purchase HG, Burgoyne GH. Isolation from turkeys of a cell-associated herpesvirus antigenically related to Marek's disease virus. Am J Vet Res. 1970;31(3):525–38.
22. Sharma JM, Graham CK. Influence of maternal antibody on efficacy of embryo vaccination with cell-associated and cell-free Marek's disease vaccine. Avian Dis. 1982;26(4):860–70.

23. Halvorson DA, Mitchell DO. Loss of cell-associated Marek's disease vaccine titer during thawing, reconstitution, and use. Avian Dis. 1979;23(4):848–53.

24. Knowles MK, Nadin-Davis SA, Sheen M, Rosatte R, Mueller R, Beresford A. Safety studies on an adenovirus recombinant vaccine for rabies (AdRG1.3-ONRAB) in target and non-target species. Vaccine. 2009;27(47):6619–26.

25. Knowles MK, Roberts D, Craig S, Sheen M, Nadin-Davis SA, Wandeler AI. In vitro and in vivo genetic stability studies of a human adenovirus type 5 recombinant rabies glycoprotein vaccine (ONRAB). Vaccine. 2009;27(20):2662–8.

26. Barrera J, Brake DA, Schutta C, Ettyreddy D, Kamicker BJ, Rasmussen MV, et al. Versatility of the adenovirus-vectored foot-and-mouth disease vaccine platform across multiple foot-and-mouth disease virus serotypes and topotypes using a vaccine dose representative of the AdtA24 conditionally licensed vaccine. Vaccine. 2018;36(48):7345–52.

27. Sreenivasa BP, Mohapatra JK, Pauszek SJ, Koster M, Dhanya VC, Tamil Selvan RP, et al. Recombinant human adenovirus-5 expressing capsid proteins of Indian vaccine strains of foot-and-mouth disease virus elicits effective antibody response in cattle. Vet Microbiol. 2017;203:196–201.

28. Fischer L, Tronel JP, Pardo-David C, Tanner P, Colombet G, Minke J, et al. Vaccination of puppies born to immune dams with a canine adenovirus-based vaccine protects against a canine distemper virus challenge. Vaccine. 2002;20(29–30):3485–97.

29. Altaras NE, Aunins JG, Evans RK, Kamen A, Konz JO, Wolf JJ. Production and formulation of adenovirus vectors. Adv Biochem Eng Biotechnol. 2005;99:193–260.

30. Vellinga J, Smith JP, Lipiec A, Majhen D, Lemckert A, van Ooij M, et al. Challenges in manufacturing adenoviral vectors for global vaccine product deployment. Hum Gene Ther. 2014;25(4):318–27.

31. Genzel Y, Fischer M, Reichl U. Serum-free influenza virus production avoiding washing steps and medium exchange in large-scale microcarrier culture. Vaccine. 2006;24(16):3261–72.

32. Rourou S, van der Ark A, van der Velden T, Kallel H. A microcarrier cell culture process for propagating rabies virus in Vero cells grown in a stirred bioreactor under fully animal component free conditions. Vaccine. 2007;25(19):3879–89.

33. Perrin P, Madhusudana S, Gontier-Jallet C, Petres S, Tordo N, Merten OW. An experimental rabies vaccine produced with a new BHK-21 suspension cell culture process: use of serum-free medium and perfusion-reactor system. Vaccine. 1995;13(13):1244–50.

34. Josefsberg JO, Buckland B. Vaccine process technology. Biotechnol Bioeng. 2012;109(6):1443–60.

35. Tree JA, Richardson C, Fooks AR, Clegg JC, Looby D. Comparison of large-scale mammalian cell culture systems with egg culture for the production of influenza virus a vaccine strains. Vaccine. 2001;19(25–26):3444–50.

36. Wu SC, Liau MY, Lin YC, Sun CJ, Wang CT. The feasibility of a novel bioreactor for vaccine production of classical swine fever virus. Vaccine. 2013;31(6):867–72.

37. Genzel Y, Olmer RM, Schafer B, Reichl U. Wave microcarrier cultivation of MDCK cells for influenza virus production in serum containing and serum-free media. Vaccine. 2006;24(35–36):6074–87.

38. Kong D, Chen M, Gentz R, Zhang J. Cell growth and protein formation on various microcarriers. Cytotechnology. 1999;29(2):151–8.

39. Silva AC, Delgado I, Sousa MF, Carrondo MJ, Alves PM. Scalable culture systems using different cell lines for the production of Peste des Petits ruminants vaccine. Vaccine. 2008;26(26):3305–11.

40. Wu SC, Liu CC, Lian WC. Optimization of microcarrier cell culture process for the inactivated enterovirus type 71 vaccine development. Vaccine. 2004;22(29–30):3858–64.

41. Chu L, Robinson DK. Industrial choices for protein production by large-scale cell culture. Curr Opin Biotechnol. 2001;12(2):180–7.

42. Genzel Y, Vogel T, Buck J, Behrendt I, Ramirez DV, Schiedner G, et al. High cell density cultivations by alternating tangential flow (ATF) perfusion for influenza a virus production using suspension cells. Vaccine. 2014;32(24):2770–81.

43. Drews M, Paalme T, Vilu R. The growth and nutrient utilization of the insect cell line *Spodoptera frugiperda* Sf9 in batch and continuous culture. J Biotechnol. 1995;40(3):187–98.

44. Ikonomou L, Bastin G, Schneider YJ, Agathos SN. Design of an efficient medium for insect cell growth and recombinant protein production. In Vitro Cell Dev Biol Anim. 2001;37(9):549–59.

45. Belyakov IM, Ahlers JD. What role does the route of immunization play in the generation of protective immunity against mucosal pathogens? J Immunol. 2009;183(11):6883–92.

46. Russell-Jones GJ. Oral vaccine delivery. J Control Release. 2000;65(1–2):49–54.

47. Peek LJ, Middaugh CR, Berkland C. Nanotechnology in vaccine delivery. Adv Drug Deliv Rev. 2008;60(8):915–28.

48. Rauw F, Gardin Y, Palya V, Anbari S, Gonze M, Lemaire S, et al. The positive adjuvant effect of chitosan on antigen-specific cell-mediated immunity after chickens vaccination with live Newcastle disease vaccine. Vet Immunol Immunopathol. 2010;134(3–4):249–58.

49. Zhao K, Zhang Y, Zhang X, Li W, Shi C, Guo C, et al. Preparation and efficacy of Newcastle disease virus DNA vaccine encapsulated in chitosan nanoparticles. Int J Nanomedicine. 2014;9:389–402.

50. Selot R, Marepally S, Kumar Vemula P, Jayandharan GR. Nanoparticle coated viral vectors for gene therapy. Curr Biotechnol. 2016;5(1):44–53.

51. Kasala D, Yoon AR, Hong J, Kim SW, Yun CO. Evolving lessons on nanomaterial-coated viral vectors for local and systemic gene therapy. Nanomedicine (Lond). 2016;11(13):1689–713.

52. Pelliccia M, Andreozzi P, Paulose J, D'Alicarnasso M, Cagno V, Donalisio M, et al. Additives for vaccine storage to improve thermal stability of adenoviruses from hours to months. Nat Commun. 2016;7:13520.

# Regulatory Strategies and Factors Affecting Veterinary Viral Vector Development

Michel Bublot, Virginie Woerly, Qinghua Wang, and Hallie King

## Abstract

The development of a vector vaccine starts after the establishment of the proof of concept in the research phase. It includes process and analytical tools development, safety, and efficacy testing required to build the dossier that is submitted to regulatory authorities to get market authorization (MA). Licensing requirements depend on the country in which MA is requested. The basics of the licensing process in the USA, European Union, and China are reviewed. Critical factors taken into consideration for the quality part include the ability to produce the vaccine at an acceptable level of quality and cost of goods and with minimal or absence of replication-competent virus for replication-defective vectors. Critical factors linked to safety taken into considerations are the genetic stability, shed and spread, reversion to virulence, recombination with vaccine or wild-type strains, and changes in tropism. A risk assessment of potential impact on human health and environment is built based on data obtained with the vaccine candidate and scientific literature on the vector and inserted sequences. Critical efficacy factors include the minimum protective dose and the potential interference of previous passive (maternally derived immunity) or active immunity and/or that of coadministration with other vaccines. The market authorization decision is based on a risk/benefit balance analysis.

**Keywords**

Quality · Safety · Efficacy · Registration · Risk assessment · Recombination

M. Bublot (✉) · V. Woerly
Boehringer Ingelheim Animal Health, R&D, Saint Priest, France
e-mail: michel.bublot@boehringer-ingelheim.com;
Virginie.WOERLY@boehringer-ingelheim.com

Q. Wang
Boehringer Ingelheim Animal Health, Regulatory Affairs, Beijing, China
e-mail: QingHua.Wang@boehringer-ingelheim.com

H. King
Boehringer Ingelheim Animal Health, R&D, Athens, GA, USA
e-mail: Hallie.King@boehringer-ingelheim.com

## Abbreviations

| | |
|---|---|
| AI | Avian influenza |
| APHIS | Animal and Plant Health Inspection Service (USA) |
| CEVD | Committee for Evaluation for Veterinary Drugs (China) |
| 9CFR | Title 9 of the Code of Federal Regulations |
| COGS | Cost of good sold |
| CVB | Center for Veterinary Biologics (USA) |

T. Vanniasinkam et al. (eds.), *Viral Vectors in Veterinary Vaccine Development*,
https://doi.org/10.1007/978-3-030-51927-8_13

CVDE Center for Veterinary Drug Evaluation (China)
CVMP Committee for Medicinal Products for Veterinary Use (European Union)
EC European Commission
EMA European Medicines Agency
EU European Union
FONSI Finding of non-significant impact
GMM Genetically modified microorganism
GMO Genetically modified organism
HVT Herpesvirus of turkey
IVDC Institute of Veterinary Drug Control (China)
MA Market authorization
MARA Ministry of Agriculture and Rural Affairs (China)
MDA Maternally derived antibodies
MDV Marek's disease virus
MPD Minimum protective dose
MSV Master seed virus
QC Quality control
RA Risk assessment
SIF Summary Information Format
SKU Stock-keeping unit
SPF Specific pathogen-free
SOP Standard Operating Procedures
USDA United States Department of Agriculture

**Learning Objectives**
After reading this chapter, you should be able to:

1. State that the regulatory requirements for vector vaccine licensing depends on the country in which the vaccine is being licensed
2. Explain that the registration dossier contains three major parts: the quality (including process of production, quality control, and specifications), safety, and efficacy
3. Explain how results of numerous safety studies will allow to make a risk assessment of potential impact of the vector vaccine on the environment (including other animals) and human health that will be a key component on the risk/benefit analysis of the vaccine
4. Recognize that the developed process should allow the consistent release of a vaccine of adequate quality and at a titer above the

minimum release dose, which is derived from the minimum protective dose
5. Recognize that anti-vector and/or anti-insert antigens passive and active immunity may potentially interfere with vector vaccine efficacy

# 1 Introduction

The generation and testing of many veterinary vector vaccines have been published, but only a very few were successfully licensed and reached the market. Once the proof of concept of a vector vaccine candidate is demonstrated at the research level, the development work starts. Results of development studies are compiled in the vaccine dossier that will be evaluated by regulatory authorities. The vaccine dossier contains basically three parts: [1] quality including the process of production, in-process control, quality control (QC), and specifications, [2] safety including a risk assessment, and [3] efficacy. The dossier is submitted to the regulatory authorities to get market authorization (MA). Requirements for licensing depend on the country in which the dossier is applied; differences between European Union, the USA, and China are outlined. Failures in development may occur at different stages. Key points to take into consideration in the development and the building of the dossier to license veterinary vector vaccines are described. The process and QC development have already been covered in the chapter entitled "Manufacturing and Control of Viral Vectored Vaccines: Challenges," and therefore, this chapter will mainly focus on regulatory, safety, and efficacy requirements.

# 2 Regulatory Requirements

The use and licensing of vector vaccines depend on two main regulatory requirements: the first one linked to the contained use and deliberate release into the environment of genetically modified organisms (GMO) and the second one to the licensing process of veterinary vaccines to get the MA. The regulatory requirements and procedures vary from country to country. However, there is a trend at harmonizing technical requirements for veterinary product registration by the International Cooperation on Harmonisation of Technical Requirements for Registration of Veterinary Medicinal Products (VICH), which is a trilateral (EU-Japan-USA) program officially launched in April 1996 (https://www.vichsec.org/en/). General and specific requirements for veterinary vaccines may also be found in the World Organization for Animal Health (Office International des Epizooties) Manual of Diagnostic Tests and Vaccines for Terrestrial animals (https://www.oie.int/en/standard-setting/terrestrial-manual/). The basic principles and process of registration in three major countries or regions are explained below.

## 2.1    United States of America

The Animal and Plant Health Inspection Service (APHIS), which is part of the United States Department of Agriculture (USDA), regulates veterinary biologics including vaccines. The evaluation is performed by APHIS Center for Veterinary Biologics (CVB). Standard requirements for licensing veterinary vaccines are outlined in the Title 9 of the Code of Federal Regulations (9CFR). In addition, Veterinary Services Memoranda and CVB Notices provide guidance and details regarding specific aspects of 9CFR requirements. The proposed manufacturing and testing facilities must also meet the requirements as outlined in the 9CFR. The licensing process of veterinary vector vaccines in the USA involves a continuous exchange of information between the applicant and an assigned CVB licensing reviewer. First, a license application package is submitted by the applicant; it includes a brief description of the vaccine, a preliminary outline of production, blueprints, and pivotal safety and efficacy protocols that will be reviewed prior to initiating licensing studies. CVB will assign a product code and official name for the vaccine, which they send to the applicant along with any comments on the study protocols. Studies are

conducted by the applicant or external partners, and each time a report is finalized, it can be submitted to the CVB reviewer for approval. A Summary Information Format (SIF) document that details the GMO design, construction, and biological properties is created and used to assess the risk associated with the manufacturing and release of the biological organism(s). Live vector vaccines are biotechnology-derived biologics belonging to Category III. The SIF contains multiple sections that can be submitted separately. Documents containing Parts I (general information and objective for the use of the proposed product) and II (description of the regulated biological agent) must be submitted prior to conducting animal testing needed for licensure. Parts I and II will be evaluated by CVB in order to get approval from their Institutional Biosafety Committee (IBC) to work with the GMO in the CVB laboratory and to recommend an appropriate containment for the safety and efficacy studies performed by the applicant. The master seed virus (MSV) is then submitted to CVB for confirmatory testing of identity and purity. Once all safety studies are completed, Part III of the SIF including the biological properties or virulence for the regulated biological agent used for MSV is added to the initial document. Part III includes a risk assessment (RA) conducted using data obtained with the construct as well as information coming from peer-reviewed scientific literature on similar agents. The proposal to conduct experimental vaccination under field study conditions is also included. This detailed SIF + RA will support the documentation required for publication of a Federal Register Notice of availability of an RA and an environmental assessment for public comment. This publication and assessment of public comments occur before issuing the Finding of No Significant Impact (FONSI) for release of the vector vaccine in the environment, which is required before starting field safety trials. Once CVB MSV confirmatory control is completed and considered satisfactory, authorization is given to move MSV to the production facility. Seed lots and three successive prelicensing serials (batches) are then produced and quality controlled. These three batches are submitted to CVB with appropriate SOPs and reagents for confirmatory testing. Once all laboratory safety, efficacy, and potency test validation reports are accepted, the three prelicensing serials are released and can be used in the field trial. After all the required documents including the field trial report have been approved by CVB, MA can be obtained usually within a few weeks. Recently, a possibility to be granted a "categorical exclusion" for vector vaccines has been outlined in APHIS Memorandum 800.215. This may be applicable for well-characterized vectors for which at least two products are already licensed by the applicant that differ only in the insertion of genetic sequence coding for another pathogen's immunogen and that have high safety records. This procedure may allow the issuance by CVB of a FONSI based on the SIF with RA without having the notice published in the Federal Register. The process to get field trial authorization will therefore be faster. Licensed veterinary biological product information is available online (https://www. aphis.usda.gov/aphis/ourfocus/animalhealth/veteri nary-biologics/ct_vb_licensed_products).

## 2.2    European Union

The contained use of genetically modified microorganisms (GMMs) is regulated in the EU by Directive 2009/41/EC. The new GMMs need to be declared to the National Authority, in the country where they are manipulated, who will then evaluate the user assessment of the contained uses as regards to the risks to human health and the environment that those contained use may pose. This assessment shall result in the final classification of the contained use in one of the four classes (one to four; see Table 1). The classification establishes the minimum biosafety level of containment for this GMM use in notified confined laboratory or animal facility premises. The deliberate release into the environment of GMOs is regulated by Directive 2001/18/EC. A notification is submitted by the applicant to the competent authority of the Member State where such a GMO is to be tested for the first time (field trial). This notification contains information

relating to the GMO, to the conditions of release and the receiving environment, and to the interactions between the GMO and the environment. It also includes an environmental RA evaluating the risk to human health and the environment, which the deliberate release of the GMO may pose. The RA is based on internal data generated with the GMO and literature review. Labelling, specific conditions of use and handling, information to the public and safeguard clause, monitoring plan, control, waste treatment, and the emergency response plans are part of the notification. An independent national body (for instance in France, the High Council for Biotechnology) will evaluate the notification in regard to the scientific information as well as other aspects such as ethical and possibly societal and economical aspects.

European veterinary vaccine registration requirements are described in different official texts such as the European Pharmacopoeia, EU Rules governing the veterinary medicinal products (for instance, Directive 2001/82/EC and its amendments, Directives 2004/28/EC and 2009/9/EC), EMA Guidelines (for instance, EMEA/CVMP/004/04 relative to live recombinant vector vaccines for veterinary use), as well as national regulations. New or updated documents are published on a regular basis, in particular Regulation (EU) 2019/6 on veterinary medicinal products that will repeal Directive 2001/82/EC in 2022. Today, veterinary vector vaccines, as products derived from biotechnology, are licensed in the EU using the so-called centralized procedure. The major advantage of this procedure is that there is one single regulatory registration assessment that allows valid MA for the 27 European Commission (EC) Member States, as well as Iceland, Norway, and Liechtenstein. The registration is managed administratively by the European Medicines Agency (EMA) located in Amsterdam. The scientific evaluation is performed by the Committee for Medicinal Products for Veterinary Use (CVMP) with the help of designated rapporteur and co-rapporteur from two different European countries. The general principle of registration in Europe, contrary to the USA, for example, is to

**Table 1** EU classification of the contained uses of GMMs in four classes, which will result in the assignment of containment levels (based on Directive 2009/41/EC)

| Class | Description |
|---|---|
| 1 | Activities of no or negligible risk, that is to say activities for which level 1 containment is appropriate to protect human health and the environment |
| 2 | Activities of low risk, that is to say activities for which level 2 containment is appropriate to protect human health and the environment |
| 3 | Activities of moderate risk, that is to say activities for which level 3 containment is appropriate to protect human health and the environment |
| 4 | Activities of high risk, that is to say activities for which level 4 containment is appropriate to protect human health and the environment |

submit to the agency the entire dossier once completed. Following the dossier review led by the rapporteur and co-rapporteur, the CVMP sends a list of questions to the applicant. The applicant then submits corresponding answers, and after CVMP review, a second list of questions may evolve needing further clarification. If major points need to be clarified, CVMP may request the applicant to provide further explanation at an oral hearing. If satisfied by the written answers and/or oral hearing, the CVMP gives a positive opinion leading to MA decision by the EC. The entire regulatory process from dossier submission to market authorization usually takes 18 months. Summary of product characteristics and European public assessment reports of licensed vector vaccine in the EU are available online (https://www.ema.europa.eu/en).

## 2.3 China

The Chinese process is the longest of the three regulatory processes. Like in the USA and the EU, the development of vector vaccines in China has two main parts: one linked to the use of genetically modified organism (GMO) and the other linked to the licensing of veterinary vaccines. The GMO Safety Certificate needs to be obtained first before starting the official veterinary vaccine registration process that will lead to the MA. Whereas the veterinary vaccine registration process depends on different bodies including the Institute of Veterinary Drug Control (IVDC), the GMO evaluation process depends on the Science Technology and Education Bureau. Both agencies belong to the Ministry of Agriculture and Rural Affairs (MARA) of People's Republic of China (China).

The GMO process applicable to veterinary vector vaccines is the same as for any live GMO such as genetically modified crop; it includes three stages (Tier 1, 2, and 3) of safety field testing after the initial safety testing under laboratory conditions. The GMO information and reports of safety lab testing are submitted together with a first field trial protocol to get the authorization to conduct the Tier 1 field trial. Once the authorization is obtained, this field trial is implemented, and reported results are then submitted with the protocol proposal for the Tier 2 trial and subsequently of similar manner for the Tier 3 trial. The number of animals involved increases in these successive trials. Once all trials are completed and reported, the GMO Safety Certificate may be applied. However, for veterinary vector vaccines, the Tier 3 and even Tier 2 trials may be skipped if the GMO has good records of safety. During the GMO safety evaluation process, the GMO and parental virus (negative control) are submitted together with the GMO identification analytical tool to the competent lab for eventual confirmatory testing. The process is quite time-consuming since it will take around 3–5 years and sometimes longer to get the GMO Safety Certificate, once the laboratory studies are completed.

If the product is developed in China by a local R&D team, the vaccine lab efficacy and safety studies for new veterinary vaccine registration purpose can be initiated at the same time as the GMO process. However, for registration of imported product, the GMO Safety Certificate needs to be obtained before going to the registration step. The normal process of vaccine registration is quite slow, but there are faster tracks such as the "green channel" if the developed vaccine protects against one of the 16 first priority diseases and if it allows the distinction between vaccine and wild-type infection or if it corresponds to an urgent need or the "emerging case" in the case of outbreak of a critical devastating disease such as African swine fever (http://www.moa.gov.cn/nybgb/2017/201711/201 802/t20180201_6136237.htm). The dossier is submitted to the Center of Veterinary Drug Evaluation (CVDE), which is part of MARA/IVDC. It is then reviewed by several regulatory bodies including the Committee for Evaluation for Veterinary Drugs (CEVD). In parallel, three batches of vaccine samples need to be tested by IVDC. The dossier is frequently rejected and needs to be resubmitted several times with additional requested information. Once the dossier is submitted, the registration time is very variable, but it is rarely shorter than 20 months and often longer than 3 years.

## 3 Information on the Vector Genetic Construct Generation, Control, and Master Seed Production

Whatever the country (USA, EU, or China), full information on the origin and production of the parental virus, the cloning and construction of plasmids or other cloning vehicles used to generate the construct, and the details on its generation, testing, and passage history up to the MSV level are required (see, for instance, EMEA/CVMP/ 004/04 guideline).

Each genetic component such as the foreign gene and expression signals (promoter, polyadenylation signal, or other) needs to be described in details. The sequence of the entire vector virus genome is usually not requested, but the one corresponding to the inserted sequence and flanking vector sequences is required. In addition, the techniques such as polymerase chain reaction, Southern blotting, sequencing, Western blotting, and/or immunostaining used to characterize and identify the construct need to be detailed. In the USA, the MSV is submitted to the official laboratory (CVB) for confirmatory testing. In China, seed lots, MSV, and WSV are submitted to IVDC for archiving before getting the registration approval. No virus/product submission is required for licensing in EU. The genetic and phenotypic (foreign gene expression) stability analysis is reported after cell culture passages and after passing in target animals. The genomic part analyzed for stability is usually the foreign inserted sequence as well as neighbor vector sequences, but authorities may request a deeper whole genome analysis. The in vitro genetic stability is a key factor investigated early in development (or even in the research phase).

## 4 Quality: Manufacturing Process, Quality Control, Specifications

The development of the production process is described in detail in the chapter entitled "Manufacturing and Control of Viral Vectored Vaccines: Challenges." The registration dossier contains the outline of production describing in detail the raw/starting materials, cells and virus seeds and methods to produce and control intermediate passage stocks (such as working seed virus), active ingredient, and final product.

Specifications are a key component of the production file since they contain a list of tests, references to analytical procedures, and appropriate acceptance criteria, which are numerical limits, ranges, or other criteria for the tests described. They establish the set of criteria, to which the vector vaccine testing results should comply, to be considered acceptable for its intended use on the market. These ranges of values are determined during development and determined by the safety, efficacy, analytical technique validation and stability data [1]. Specifications are critical quality standards that are proposed and justified by the manufacturer and approved by regulatory authorities as conditions of approval.

Three successive prelicensing batches of vector vaccine active ingredient and finished product need to be successfully produced and released based on specifications to verify the batch-to-batch consistency. In China, the active ingredient is formulated immediately after production. Confirmatory testing of the three batches is performed by authorities in the USA and in China, but not in the EU. The stability of both the active ingredient (not for China) and finished product is followed to confirm the specifications. Stability of final product needs to be proven for an additional 3 months after the claimed shelf life.

A key factor that may affect vector vaccine development is the cost of goods sold (COGS) for a vaccine dose that may be compatible with a reasonable selling price, especially for production animals for which return on investment of vaccination is taken into account by customers. The minimum release dose is usually calculated from the minimum protective dose, taking into account the potential loss during stability and the variability of the potency test analytical technique (viral titration). This minimum release dose and consequently the COGS of replication-defective vectors are generally higher than that of replication-competent vectors. Thus, the latter are typically licensed for low-cost production animals such as chickens. A maximum release dose is also set up in the EU and China. This dose will define the dose to be tested in safety studies. As an example, for replication-deficient vectors such as E1-deleted human adenovirus 5 grown in complementing 293 cells, the absence in a batch of replication-competent adenovirus above a certain threshold (for instance, 1 in $3 \times 10^{10}$ particles based on FDA guidelines for Ad5 vectors) is key to guarantee the safety and

avoid the risk of spread and shed [2]. The use of another cell line may alleviate this risk [3].

## 5 Safety Evaluation

The safety testing of vector vaccines involves both in vivo and in vitro testing. Some key studies and areas of investigation are detailed below.

### 5.1 Residual Virulence of the Seed and Vaccine Safety Testing in Target Species

Most safety studies are performed in specific pathogen-free (SPF) target species if available. It includes testing of the residual virulence of the construct, which is usually performed with ten times dose of the MSV or an early passage from MSV. Testing conditions vary depending on the vector backbone virus as described in documents specific for a type of virus, such as in Pharmacopoeia or 9CFR, or according to approved test protocols. For instance, in EU, for Marek's disease virus (MDV)-based vector vaccines, the evaluation is performed during 120 days after administration by the recommended route, at the end of which no Marek's disease gross lesions should be present, thus demonstrating absence of residual pathogenicity of the strain.

The safety of the final vaccine is evaluated using the maximum release dose administered by each recommended route to animals of the minimum age of administration. Animals are observed daily for local and systemic reactions for at least 14 days after administration. The safety of the administration of an overdose (at least ten times the maximum release dose) needs to be evaluated similarly. If the vaccination scheme includes booster vaccination, the safety of repeated dose(s) is tested as well. If the vaccine is recommended for use in pregnant animals or laying birds, its impact on reproductive performance and/or pregnancy is investigated. For constructs that may affect the immunological functions, the impact on immune response must be tested.

Safety studies of some vector vaccines may potentially be performed in animals with anti-vector maternally derived antibodies (MDA) as well since such MDA have been shown to be harmful in some cases such as adenovirus-vectored HIV vaccine candidate [4].

At the end of development, field safety trials are conducted in the target animals to confirm the safety in future use conditions. Requirements for field trials vary depending on the regulatory agency and the country. Field trials are conducted with batches (at least three in China and two in the USA) produced according to the manufacturing process described in the MA application. In the USA, only safety field trials are requested, but in Europe and China, both field efficacy and safety may be investigated in the same or different field trials.

## 5.2    Reversion to Virulence

Five successive passages of the strain (usually MSV) are performed in target animals (most sensitive class, age, sex, and serological status), and no evidence of increase of virulence must be seen. Virus isolation is performed at each passage to confirm the presence of virus. The safety of the virus recovered at the last passage is evaluated in comparison with the initial inoculum for at least 21 days. The genetic/phenotypic stability of the vector vaccine candidate after in vivo passages is usually performed on recovered viruses from this study. Non-replicative vectors could become undetectable after the first or second passage; if so, a repeat of the passage is needed to confirm the loss of the GMO. This is a critical study to perform early in development, especially for replicating vectors, since viruses passed many times in cell culture have a tendency of regaining in vivo fitness and possibly virulence after successive passages in animals and thus not appropriate for the market [5]. Vector vaccine attenuation by virulence gene deletion and/or codon de-optimization [6] or codon pair bias [7] may decrease the risk of reversion to virulence.

## 5.3    Tissue Tropism or Internal Dissemination Within Vaccinated Animal

The tropism or bio-distribution of vector vaccine may be modified by the expression of the foreign gene, especially if the gene product is expressed at the surface of the virion or if an immunomodulator is expressed. Such insert-induced changes in tropism and/or virulence have been documented at least for herpesvirus [8] and poxvirus [9–12] vectors. The deletion of a vector gene may also potentially increase the virulence as observed with equine herpesvirus 1 [13, 14] and vaccinia virus [15]. It is therefore important to compare the distribution and persistence of the vector vaccine in different tissues of the host animal in parallel to that of the parental vector in order to identify potential tropism changes brought by the genetic modification of the parental virus.

## 5.4    Shed and Spread

One key component of replicative vector vaccine safety that will have a major impact on the RA of the GMO is its ability to shed into the environment and to spread to other animals. Studies looking at virus shedding from the different mucosal tissues and skin are conducted in the target species. Unvaccinated sensitive animals are raised in close contacts to vaccinated animals in order to evaluate horizontal transmission. Vertical transmission may be investigated as well if recommended for use in breeders. To our knowledge, no vector vaccine able to easily spread from vaccinated to naïve contact animals have been licensed so far in Europe and in the USA. Newcastle disease virus (NDV) readily spreads to naïve birds, and thus, it can be assumed that NDV vector vaccines against influenza licensed for chickens in China and in Mexico may probably spread to unvaccinated chickens. Shed and spread studies may also need to be performed in non-target species that are known to be susceptible to the vector virus. The case of herpesvirus of turkey (HVT) vector for chickens is an interesting example since HVT does not easily spread from vaccinated to unvaccinated chickens, but it may potentially spread from vaccinated chickens to naïve turkeys and from infected turkeys to naïve turkeys. However, it does not spread from vaccinated chickens to HVT-infected turkeys, and since HVT infection is widely distributed in turkey flocks, the risk of transmission of the HVT vector from chickens to turkeys is considered extremely low. Additional studies in target- and non-target-susceptible (domestic or wild) species sharing the same ecosystem as vaccinated animals may sometimes be needed to support the RA of the vaccine for environment, especially with new vectors not already licensed. Finally, the risk of transmission from vaccinated target animals to humans needs to be assessed using scientific literature and eventually appropriate animal models. The shed and spread of replication-abortive vectors should be low and/or absent.

## 5.5    Recombination/Reassortment

The potential for intratypic or intertypic recombination and/or reassortment with other vaccine strains or with wild-type pathogenic strains must be investigated thoroughly. An assessment of this issue has been recently published by the Brighton Collaboration [15, 16]. Reassortment is an exchange of viral genomic segments that occurs in a cell coinfected by at least two viruses leading to new viruses that may have biological properties different from the parental viruses. It is an issue for RNA viruses with segmented genome such as those belonging to *Arenaviridae, Birnaviridae, Bunyavirales, Orthomyxoviridae, Picobirnaviridae,* and *Reoviridae.* These viruses are not favored candidates for vector development. Recombination is an exchange of genome sequences of variable length that may naturally occur when two different viruses are infecting the same cell in vitro or in vivo. Some viruses (DNA viruses, retroviruses, positive stranded RNA viruses) are more prone to recombination than others (negative strand RNA viruses). Once cells are coinfected with two viruses from the same

species, the percentage of intratypic recombinant viruses may be quite high as observed for herpesviruses [17] and, to a lesser extent, poxviruses [18]. Although much less frequent, intertypic recombination may occasionally occur [19–22]. The probability of occurrence of recombination between vector vaccine and circulating wild-type or conventional live vaccine viruses has to be assessed. Indeed, since recombination naturally occurs among and between some live vaccine and wild-type viruses, it may likely also happen with some vector vaccines. An important part of the RA will focus on possible outcomes of such recombination. The major issue is if the recombination generates a hybrid virus that has undesirable properties affecting transmission or virulence [16]. Recombination events between conventional live vaccine and wild-type virulent viruses were detected by sequence analysis [23], but to our knowledge, there is no report of creating a more virulent or transmissible hybrid virus than the parental wild-type virus. In contrast, recombination between two attenuated strains generating a virulent hybrid virus has been reported at least for herpes simplex [24, 25], pseudorabies virus [26–30], and infectious laryngotracheitis virus [31]. Indeed, when two attenuated strains have attenuated mutations/deletions/insertions located in different parts of their genome, recombination events can recreate a hybrid virus with genome similar to virulent virus. This critical issue needs to be carefully discussed in the RA especially when replicative vectors have intrinsic recombination properties and when live vaccines based on the same species of virus are widely used in the same target species population. For instance, a vaccine based on a herpesvirus vector genetically attenuated by the deletion of a gene playing a key role in virulence but nonessential for its replication (the Meq gene of MDV would be a good example [32–34]) may potentially revert to virulence by acquiring the missing gene by recombination with another vaccine strain. The RA will therefore take into consideration the presence of such vaccine strain in the environment, including possibly in the vaccinated animals, and the possible outcome of recombination with the vector vaccine.

Additional experimental in vitro and/or in vivo studies may sometimes be needed to support the risk assessment as it was done with the yellow fever vector (ChimeriVax) platform [35]. This issue of recombination is a critical factor for vector vaccine development and should be assessed early in development and even in research when choosing and generating the vector vaccine candidate. Ways to decrease such risk include the choice of a vector with lower intrinsic recombination properties of the parental virus, attenuation by codon deoptimization [6] or codon pair bias [7] to decrease sequence relatedness of vector and wild-type virus, the choice of a non-replicative vector such as canarypox virus or human adenovirus 5, and/or the choice of a vector with restricted host range and its use in a target species not naturally infected or vaccinated with similar viruses (e.g., canarypox for mammals).

## 5.6    Safety in Non-target Species

The safety of vector vaccines is assessed in non-target species that share the same ecosystem as vaccinated animals or that may be accidentally vaccinated. Scientific knowledge on the vector host range will orient the design of in vitro and/or in vivo studies to be performed to evaluate this risk. As described above for HVT, shed and spread of a vector able to infect different species should be studied in non-target species as well. The choice of a vector with a narrow host range will facilitate the RA in non-targeted species. For viruses with a known wide tropism such as vaccinia, it is preferable to delete genes responsible for this broad host range and/or to use replication-incompetent vaccinia [36].

## 5.7    Survival in the Environment

The persistence of infectivity of the vaccine in the environment is based on scientific publications on the vector, but it may also require in vitro testing to support the RA. For instance, in vitro testing is required in the USA to confirm survivability has not been altered by the genetic modification.

## 5.8    Ecotoxicity Evaluation, Risk Assessment

A risk assessment (RA) quantifying the risk for human health and environment is necessary. The RA is used to evaluate the overall risk/benefit ratio on which regulatory authorities base their decisions to authorize the deliberate release in field trials and for licensing such new vector vaccines. Data of safety studies performed with the candidate as well as scientific knowledge on the vector virus are key to support the RA. Information required to perform the RA includes details on the GMO construct, its generation and biological properties, characteristics affecting GMO survival, multiplication and dissemination, interactions with the environment including safety in target and non-target species, genetic transfer capability, genetic stability, residual virulence, possibility to revert to virulence, and competitive advantage in relation to parental virus. GMO monitoring techniques used to follow the released GMO, measures to limit GMO spread, and emergency response plans in case of unexpected spread or adverse reactions are described in the submitted document. RA will be based on this information and is defined as the likelihood of an adverse event occurring and the consequences if that adverse event occurs. An adverse event is defined as safety hazards to animals, public health, or the environment. A safety hazard is defined as a danger, risk, or peril; absence of predictability associated with an event; or an expected or unpredicted event. RA is made by identifying the different hazards and evaluating the likelihood and consequences of each hazard occurring. The final assessment is made using a matrix in which the likelihood and consequence of hazard are scored. Each regulatory authority may have its own matrix; the rating of degree of certainty is also taken into consideration in the USA APHIS matrix. Risk management recommendations are based on the RA and determine means of reducing or eliminating safety risks to animals, public health, or the environment. As an example, key considerations for a risk/benefit assessment of vector vaccine based on the vesicular stomatitis virus have been published [37].

# 6 Efficacy Evaluation

The first pivotal efficacy studies to perform are those aiming to establish minimum protective dose (MPD) against the target disease and from which the minimum release dose will be derived as explained before. They are performed in SPF animals vaccinated with different doses of a representative vaccine batch by each targeted administration route using the challenge model described in the most relevant pharmacopoeia or 9CFR documents, if existing. These studies may set up the onset of immunity for diseases targeted by the insert and possibly also by the vector, if the vector itself is also providing protection against a disease. Back-titration of the vaccine at the time of vaccination and challenge virus at the time of challenge is required by 9CFR. An indirect (by serology on serum sampled before challenge) or direct (vaccine virus isolation or nucleic acid detection) proof of vaccine take may be required (Europe and China). For recombinant vector vaccines for which no claim is made for the vector, the anti-vector immune response induced after vaccination may need to be documented (Europe). Additional studies with a later challenge may be performed to determine the duration of immunity/protection allowed by the vaccine.

The possibility to boost the induced immunity against the vector antigen and/or the foreign antigen(s) within a claimed/intended vaccination schedule should be investigated if booster vaccinations are deemed necessary. The effect of preexisting immunity to the vector and/or the foreign antigen(s) expressed by the vector should be studied for licensing in the EU and in China, but not necessarily in the USA. The minimum release dose can be calculated taking into account the MPD, precision of titration technique (potency test), and stability data for the claimed shelf life (+ 3 months). In the USA, for some vectors, including HVT, there is a standard overage applied that is three times the MPD at release.

Interference of the following factors on the immunogenicity of vector vaccines may need to be investigated: maternally derived (passive) immunity, active immunity, and coadministration with other vaccines. The impact of maternally derived immunity (mostly MDA) is requested to be evaluated for the EU and China licensing, especially if the vaccine is targeted for newborn or young animals. For vector vaccines, both anti-vector and anti-insert MDA interferences are typically investigated. For replicative poxvirus [38] and paramyxovirus [39] vectors in poultry, the anti-insert interference was shown to be more severe than the anti-vector one, but it may not be the case for other vectors such as those based on adenovirus [40]. The level of MDA interference may depend on the type of vector and the immunization route. Immunogenicity of persisting vectors such as herpesvirus of turkey (HVT) is less affected than non-persisting vectors such as Newcastle disease virus (NDV) vector [41]. Mucosal administration route with a replication-competent vector may better overcome MDA interference than parenteral one but not systematically [42]. The anti-insert MDA impact may be observed in the lower antibody response induced after the first vector vaccine administration; however, rapid and high humoral response increase after a boost with another vaccine or the same vector vaccine (in case of non-replicative vector) clearly shows that the immune system has been primed by the first vector vaccine administration despite MDA presence [38, 43]. Since the humoral immune response is usually impacted but not necessarily the protection, vaccination/challenge studies may be needed to study MDA interference with vector vaccines. A lower protection in animals with MDA does not necessarily mean that the vaccine is not licensable, especially if the MDA interference is not worse when compared to that of a classical live or inactivated benchmark vaccine [44, 45].

Active immunity against the vector may interfere on vector vaccine immunogenicity. This was demonstrated, for instance, for replicative fowlpox virus vector: previous vaccination of chicken with fowlpox decreases the avian influenza (AI) protection induced by a fowlpox-vectored AI vaccine [46]; this has been demonstrated with HVT vector as well [47]. For non-replicative vector that must be administered several times in the same animal to get optimal and long-term immunity, the anti-vector induced by the first administration may potentially interfere with subsequent administration, as shown with adenovirus vectors [40]. Interestingly, no such interference has been observed with the canarypox ALVAC vector; this may be due to the absence of detectable neutralizing antibodies after administration [44]. Active immunity against the insert before vector vaccine administration usually has no negative impact; in contrast, the vector vaccine will induce a boosted immune response in a primed immune system [48–50].

There may also be interference when a vector vaccine is administered with the empty vector (when the vector is itself a vaccine) or when two vector vaccines based on the same vector but targeting two different diseases are coadministered [51–53]. The vector vaccine may be part of a vaccination program in which other vaccines may be administered at the same time. Lack of interference studies will need to be conducted to verify compatibility with these other vaccines.

To be successful, the route of administration of a vector vaccine needs to be convenient. For poultry, it should be applicable at the hatchery (in vivo, subcutaneously or by spray) or using a mass administration (spray or drinking water) in the farms. Intramuscular route is the easiest route for ruminants, swine, and horses. Intranasal administration to dogs and cats is not easy to perform; parenteral or oral vaccination should be preferred.

In addition to regulatory requirements, the widely admitted concept to reduce, refine, or replace animals (3Rs) is increasingly applied to the development and quality control of veterinary vaccines, especially for release test of vaccine batches. Indeed, safety testing is usually performed by administering an overdose of each produced batch in the most sensitive target species. Some authorities like in the EU allow waiving the batch safety test when consistency of production is well established. Similarly, in vitro methods to test for extraneous viruses in vaccines may replace in the future the currently used in vivo methods [54]. Similar initiative is ongoing in the USA, for instance, with the 2017 USDA update on Veterinary Services Memorandum 800.116. The development of new vaccines should include the 3Rs principles.

In certain situations, vector vaccine may not be efficacious enough to be licensed as a stand-alone vaccine. However, the priming with a vector vaccine followed by a heterologous boost with a classical or another vector vaccine may sometimes be highly immunogenic [38, 55]. There is today a lack of guidelines from the authorities on how to license such a combination, but specific rules may come in the future to allow licensing of vector vaccine as priming and/or boosting vaccination.

When the same vector has already been licensed against different diseases, there should be a simplified way of licensing new vaccines based on the same vector. This trend is already in place in the USA with the categorical exclusion possibility (see above), and this approach could be applied in the future in other countries or regions.

# 7 Future Directions

As explained in this chapter, regulatory requirements depend on the licensing country or region. Although harmonization is ongoing via VICH on specific topics, progresses are slow, but further changes toward harmonized requirements are expected in the future.

# 8 Summary

The development of veterinary viral vector vaccines is a long process that requires expertise in many different fields (regulatory affairs, clinical testing, process and analytical development, manufacturing and quality control, quality assurance, molecular biology) and has a non-negligible cost. It is therefore important to initiate early in development the key and most challenging studies for the vector vaccine candidate to satisfy the regulatory requirements for quality, safety, and efficacy at an acceptable cost of production. Conducting studies in this manner allows a rapid increase to the probability of success of development.

# References

1. Minor P. Considerations for setting the specifications of vaccines. Expert Rev Vaccines. 2012;11(5):579–85.
2. Vellinga J, Smith JP, Lipiec A, Majhen D, Lemckert A, van Ooij M, et al. Challenges in manufacturing adenoviral vectors for global vaccine product deployment. Hum Gene Ther. 2014;25 (4):318–27.
3. Kovesdi I, Hedley SJ. Adenoviral producer cells. Viruses. 2010;2(8):1681–703.
4. Sekaly RP. The failed HIV Merck vaccine study: a step back or a launching point for future vaccine development? J Exp Med. 2008;205(1):7–12.
5. Hanley KA. The double-edged sword: how evolution can make or break a live-attenuated virus vaccine. Evolution (N Y). 2011;4(4):635–43.
6. Mueller S, Papamichail D, Coleman JR, Skiena S, Wimmer E. Reduction of the rate of poliovirus protein synthesis through large-scale codon deoptimization causes attenuation of viral virulence by lowering specific infectivity. J Virol. 2006;80(19):9687–96.
7. Coleman JR, Papamichail D, Skiena S, Futcher B, Wimmer E, Mueller S. Virus attenuation by genome-scale changes in codon pair bias. Science. 2008;320 (5884):1784–7.
8. Taylor G, Rijsewijk FA, Thomas LH, Wyld SG, Gaddum RM, Cook RS, et al. Resistance to bovine respiratory syncytial virus (BRSV) induced in calves by a recombinant bovine herpesvirus-1 expressing the attachment glycoprotein of BRSV. J Gen Virol. 1998;79(Pt 7):1759–67.
9. Sharma DP, Ramsay AJ, Maguire DJ, Rolph MS, Ramshaw IA. Interleukin-4 mediates down regulation of antiviral cytokine expression and cytotoxic T-lymphocyte responses and exacerbates vaccinia virus infection in vivo. J Virol. 1996;70(10):7103–7.
10. Taylor G, Stott EJ, Wertz G, Ball A. Comparison of the virulence of wild-type thymidine kinase (tk)-deficient and tk+ phenotypes of vaccinia virus recombinants after intranasal inoculation of mice. J Gen Virol. 1991;72(Pt 1):125–30.
11. Johnson TR, Fischer JE, Graham BS. Construction and characterization of recombinant vaccinia viruses co-expressing a respiratory syncytial virus protein and a cytokine. J Gen Virol. 2001;82(Pt 9):2107–16.
12. Jackson RJ, Ramsay AJ, Christensen CD, Beaton S, Hall DF, Ramshaw IA. Expression of mouse interleukin-4 by a recombinant ectromelia virus suppresses cytolytic lymphocyte responses and overcomes genetic resistance to mousepox. J Virol. 2001;75(3):1205–10.
13. Van de Walle GR, May ML, Sukhumavasi W, von Einem J, Osterrieder N. Herpesvirus chemokine-binding glycoprotein G (gG) efficiently inhibits neutrophil chemotaxis in vitro and in vivo. J Immunol. 2007;179(6):4161–9.
14. von Einem J, Smith PM, Van de Walle GR, O'Callaghan DJ, Osterrieder N. In vitro and in vivo characterization of equine herpesvirus type 1 (EHV-1) mutants devoid of the viral chemokine-binding glycoprotein G (gG). Virology. 2007;362(1):151–62.
15. Alcami A, Smith GL. A mechanism for the inhibition of fever by a virus. Proc Natl Acad Sci U S A. 1996;93 (20):11029–34.
16. Condit RC, Williamson AL, Sheets R, Seligman SJ, Monath TP, Excler JL, et al. Unique safety issues associated with virus-vectored vaccines: potential for and theoretical consequences of recombination with wild type virus strains. Vaccine. 2016;34(51):6610–6.
17. Thiry E, Meurens F, Muylkens B, McVoy M, Gogev S, Thiry J, et al. Recombination in alphaherpesviruses. Rev Med Virol. 2005;15(2):89–103.
18. Qin L, Evans DH. Genome scale patterns of recombination between coinfecting vaccinia viruses. J Virol. 2014;88(10):5277–86.
19. Hirai K, Yamada M, Arao Y, Kato S, Nii S. Replicating Marek's disease virus (MDV) serotype 2 DNA with inserted MDV serotype 1 DNA sequences in a Marek's disease lymphoblastoid cell line MSB1-41C. Arch Virol. 1990;114(3–4):153–65.
20. Meurens F, Keil GM, Muylkens B, Gogev S, Schynts F, Negro S, et al. Interspecific recombination between two ruminant alphaherpesviruses, bovine herpesviruses 1 and 5. J Virol. 2004;78(18):9828–36.
21. Pagamjav O, Sakata T, Matsumura T, Yamaguchi T, Fukushi H. Natural recombinant between equine herpesviruses 1 and 4 in the ICP4 gene. Microbiol Immunol. 2005;49(2):167–79.
22. Maidana SS, Craig PO, Craig MI, Ludwig L, Mauroy A, Thiry E, et al. Evidence of natural interspecific recombinant viruses between bovine alphaherpesviruses 1 and 5. Virus Res. 2017;242:122–30.
23. He L, Li J, Peng P, Nie J, Luo J, Cao Y, et al. Genomic analysis of a Chinese MDV strain derived from vaccine strain CVI988 through recombination. Infect Genet Evol. 2020;78:104045.
24. Brandt CR, Grau DR. Mixed infection with herpes simplex virus type 1 generates recombinants with increased ocular and neurovirulence. Invest Ophthalmol Vis Sci. 1990;31(11):2214–23.
25. Kintner RL, Allan RW, Brandt CR. Recombinants are isolated at high frequency following in vivo mixed ocular infection with two avirulent herpes simplex virus type 1 strains. Arch Virol. 1995;140(2):231–44.
26. Henderson LM, Katz JB, Erickson GA, Mayfield JE. In vivo and in vitro genetic recombination between

conventional and gene-deleted vaccine strains of pseudorabies virus. Am J Vet Res. 1990;51 (10):1656–62.

27. Henderson LM, Levings RL, Davis AJ, Sturtz DR. Recombination of pseudorabies virus vaccine strains in swine. Am J Vet Res. 1991;52(6):820–5.

28. Katz JB, Henderson LM, Erickson GA. Recombination in vivo of pseudorabies vaccine strains to produce new virus strains. Vaccine. 1990;8 (3):286–8.

29. Katz JB, Henderson LM, Erickson GA, Osorio FA. Exposure of pigs to a pseudorabies virus formed by in vivo recombination of two vaccine strains in sheep. J Vet Diagn Investig. 1990;2(2):135–6.

30. Glazenburg KL, Moormann RJ, Kimman TG, Gielkens AL, Peeters BP. In vivo recombination of pseudorabies virus strains in mice. Virus Res. 1994;34(2):115–26.

31. Lee SW, Markham PF, Coppo MJ, Legione AR, Markham JF, Noormohammadi AH, et al. Attenuated vaccines can recombine to form virulent field viruses. Science. 2012;337(6091):188.

32. Chang S, Ding Z, Dunn JR, Lee LF, Heidari M, Song J, et al. A comparative evaluation of the protective efficacy of rMd5deltaMeq and CVI988/ Rispens against a vv+ strain of Marek's disease virus infection in a series of recombinant congenic strains of White Leghorn chickens. Avian Dis. 2011;55(3):384–90.

33. Lee LF, Heidari M, Zhang H, Lupiani B, Reddy SM, Fadly A. Cell culture attenuation eliminates rMd5DeltaMeq-induced bursal and thymic atrophy and renders the mutant virus as an effective and safe vaccine against Marek's disease. Vaccine. 2012;30 (34):5151–8.

34. Su S, Cui N, Zhou Y, Chen Z, Li Y, Ding J, et al. A recombinant field strain of Marek's disease (MD) virus with reticuloendotheliosis virus long terminal repeat insert lacking the meq gene as a vaccine against MD. Vaccine. 2015;33(5):596–603.

35. McGee CE, Tsetsarkin KA, Guy B, Lang J, Plante K, Vanlandingham DL, et al. Stability of yellow fever virus under recombinatory pressure as compared with chikungunya virus. PLoS One. 2011;6(8):e23247.

36. Vanderplasschen A, Pastoret PP. The uses of poxviruses as vectors. Curr Gene Ther. 2003;3 (6):583–95.

37. Clarke DK, Hendry RM, Singh V, Rose JK, Seligman SJ, Klug B, et al. Live virus vaccines based on a vesicular stomatitis virus (VSV) backbone: standardized template with key considerations for a risk/benefit assessment. Vaccine. 2016;34 (51):6597–609.

38. Richard-Mazet A, Goutebroze S, Le Gros FX, Swayne DE, Bublot M. Immunogenicity and efficacy of fowlpox-vectored and inactivated avian influenza vaccines alone or in a prime-boost schedule in chickens with maternal antibodies. Vet Res. 2014;45:107.

39. Lambrecht B, Lardinois A, Vandersleyen O, Steensels M, Desloges N, Mast J, et al. Stronger interference of Avian Influenza than Newcastle Disease Virus specific maternal derived antibodies with a recombinant NDV-H5 vaccine. Avian Dis. 2015;In press.

40. Ndi OL, Barton MD, Vanniasinkam T. Adenoviral vectors in veterinary vaccine development: potential for further development. World J Vaccines. 2013;3:111–21.

41. Bertran K, Lee DH, Criado MF, Balzli CL, Killmaster LF, Kapczynski DR, et al. Maternal antibody inhibition of recombinant Newcastle disease virus vectored vaccine in a primary or booster avian influenza vaccination program of broiler chickens. Vaccine. 2018;36 (43):6361–72.

42. Fischer L, Tronel JP, Pardo-David C, Tanner P, Colombet G, Minke J, et al. Vaccination of puppies born to immune dams with a canine adenovirus-based vaccine protects against a canine distemper virus challenge. Vaccine. 2002;20(29–30):3485–97.

43. Minke JM, Toulemonde CE, Dinic S, Cozette V, Cullinane A, Audonnet JC. Effective priming of foals born to immune dams against influenza by a canarypox-vectored recombinant influenza H3N8 vaccine. J Comp Pathol. 2007;137(Suppl 1):S76–80.

44. Poulet H, Minke J, Pardo MC, Juillard V, Nordgren B, Audonnet JC. Development and registration of recombinant veterinary vaccines. The example of the canarypox vector platform. Vaccine. 2007;25 (30):5606–12.

45. Pardo MC, Tanner P, Bauman J, Silver K, Fischer L. Immunization of puppies in the presence of maternally derived antibodies against canine distemper virus. J Comp Pathol. 2007;137(Suppl 1):S72–5.

46. Swayne DE, Beck JR, Kinney N. Failure of a recombinant fowl poxvirus vaccine containing an avian influenza hemagglutinin gene to provide consistent protection against influenza in chickens preimmunized with a fowl pox vaccine. Avian Dis. 2000;44 (1):132–7.

47. Bublot M, Pritchard N, Le Gros F-X, Goutebroze S. Use of a vectored vaccine against infectious bursal disease of chickens in the face of high-titred maternally derived antibody. J Comp Path. 2007;137:S81–4.

48. Grosenbaugh DA, Backus CS, Karaca K, Minke JM, Nordgren RM. The anamnestic serologic response to vaccination with a canarypox virus-vectored recombinant West Nile virus (WNV) vaccine in horses previously vaccinated with an inactivated WNV vaccine. Vet Ther. 2004;5(4):251–7.

49. Grosenbaugh DA, Leard T, Pardo MC. Protection from challenge following administration of a canarypox virus-vectored recombinant feline leukemia virus vaccine in cats previously vaccinated with a killed virus vaccine. J Am Vet Med Assoc. 2006;228(5):726–7.

50. El Garch H, Minke JM, Rehder J, Richard S, Edlund Toulemonde C, Dinic S, et al. A West Nile virus (WNV) recombinant canarypox virus vaccine elicits

WNV-specific neutralizing antibodies and cell-mediated immune responses in the horse. Vet Immunol Immunopathol. 2008;123(3–4):230–9.

51. Slacum G, Hein R, Lynch P, editors. The compatibility of HVT recombinants with other Marek's disease vaccines. Sacramento: 58th Western Poultry Disease Conference; 2009.

52. Hein RG, editor. Issues of the poultry recombinant viral vector vaccines which may cause a negative effect on the economic benefits of those vaccines. Cancun: XVII International Congress of the World Veterinary Poultry Association; 2011.

53. Kumar N, Sharma S, Barua S, Tripathi BN, Rouse BT. Virological and immunological outcomes of coinfections. Clin Microbiol Rev. 2018;31(4).

54. Woodland R. European regulatory requirements for veterinary vaccine safety and potency testing and recent progress towards reducing animal use. Proc Vaccinol. 2011;5:151–5.

55. Niqueux E, Guionie O, Amelot M, Jestin V. Prime-boost vaccination with recombinant H5-fowlpox and Newcastle disease virus vectors affords lasting protection in SPF Muscovy ducks against highly pathogenic H5N1 influenza virus. Vaccine. 2013;31(38):4121–8.

# Emerging Viral-Vectored Technology: Future Potential of Capripoxvirus and African Swine Fever Virus as Viral Vectors

Shawn Babiuk

**Abstract**

The ability to generate recombinant viruses allows for the development of more effective live-attenuated vaccines. Furthermore, these vaccines can also be used a viral vector to induce immune responses against expressed protective antigens to generate multivalent vaccines. Capripoxviruses and African swine fever virus are becoming increasingly important due to their spread into new regions. This chapter will describe these viruses, the clinical disease they cause, and the impact of the disease as well as how recombinant viruses are constructed. The current state of the art for capripoxvirus and African swine fever viruses as vaccines and viral vectors will be presented. Finally, future areas of research for improving capripoxvirus and African swine fever virus vectors will be discussed.

**Keywords**

Capripoxvirus · Sheep pox · Goat pox · Lumpy skin disease · African swine fever · Vector · Homologous recombination · Vaccine · DIVA

**Learning Objectives**

S. Babiuk (✉)
Canadian Food Inspection Agency, National Centre for Foreign Animal Disease, Winnipeg, MB, Canada

Department of Immunology, University of Manitoba, Winnipeg, MB, Canada
e-mail: shawn.babiuk@canada.gc.ca

After reading this chapter, you should be able to:

- Describe the importance of capripoxvirus and African swine fever virus
- Explain how recombinant DNA viruses can be generated
- Describe features that are important for a viral vector
- Discuss the rationale for the development of multivalent vaccines
- Explain the principle of Differentiating Infected from Vaccinated Animals (DIVA)

# 1 Chapter Introduction

## 1.1 Background of Capripoxviruses

Sheep pox, goat pox, and lumpy skin disease virus are the three members of the genus *Capripoxvirus* in the family *Poxviridae*. Capripoxviruses have similar morphology to orthopoxviruses, which have closed hairpin loops at their termini and share many similar genes. Capripoxviruses are double-stranded DNA viruses with genome sizes of approximately 150 kbp which encode 147 putative genes for sheep pox and goat pox with lumpy skin disease having an additional nine genes which are disrupted and nonfunctional in sheep pox and goat pox [1, 2]. There are no serotypes of capripoxviruses due to their genetic similarity.

© Springer Nature Switzerland AG 2021
T. Vanniasinkam et al. (eds.), *Viral Vectors in Veterinary Vaccine Development*,
https://doi.org/10.1007/978-3-030-51927-8_14

Capripoxviruses are thermostable similar to other poxviruses which allows the virus to contaminate the environment for prolonged periods of time [3].

Sheep pox and goat pox viruses have been observed since ancient times and are endemic throughout most of Central and Northern Africa with the absence of goat pox in Morocco. Sheep pox and goat pox are endemic throughout the Middle East and Turkey with outbreaks reported in Greece and Bulgaria between 2013 and 2015. The disease is endemic in many parts of Asia, including Russia, Mongolia, Kazakhstan, Kyrgyzstan, Afghanistan, Pakistan, India, Bangladesh, Nepal, China, Vietnam, and Chinese Taipei [4]. The determination of sheep pox and goat pox is based on the host that the virus has been isolated from. Many isolates of sheep pox and goat pox demonstrate a host preference; however, there are some isolates which can infect both species [5].

Lumpy skin disease virus was historically defined as a disease affecting cattle in Africa. Lumpy skin disease virus has a limited host range of cattle and water buffalo. There has been one occurrence where lumpy skin disease virus infected sheep and was misidentified as sheep pox; since at the time when it was isolated, there was no molecular methods available to properly identify the virus as lumpy skin disease [6]. It was first identified in 1929 in sub-Saharan Africa, where it has spread throughout most of Africa, including Madagascar, with exception of the Northern African countries of Libya, Tunisia, Algeria, and Morocco. Before 2012, lumpy skin disease virus only caused sporadic outbreaks in the Middle East, which did not lead to lumpy skin disease virus to become endemic. A lumpy skin disease outbreak was reported in the Golan Heights bordering Syria in 2012. From 2013 to 2015, the disease spread in Israel, Lebanon, Jordan and into new bordering regions including Turkey, Saudi Arabia, Kuwait, Iraq, Iran, Azerbaijan, and Cyprus. In 2015, lumpy skin disease was reported in Greece as well as the Caucasus region of Russia, Dagestan, and Chechnya [4]. In 2016, lumpy skin disease virus spread into Bulgaria, Serbia, Montenegro, Republic of Macedonia, Kosovo, Albania, Georgia, Kazakhstan, and further into regions of Russia [7].

## 1.2 Clinical Importance of Capripoxviruses

Sheep pox and goat pox affect sheep and goats, respectively, and the virus generally demonstrates a host preference for either sheep or goats although some isolates can infect both species. The disease is spread by contact with infected animals or premises. The clinical signs of sheep pox and goat pox are similar, causing a severe disease characterized by fever, nasal discharge, and skin lesions as well as lesions on internal organs such as the lung and gastrointestinal tract [8]. The disease causes significant economic losses to farmers due to reduced productivity from decreased weight gain, milk production, and damage to wool and hides. The morbidity and mortality rates can approach 100% in naïve animals causing a major loss to producers. Because of the high morbidity and mortality, sheep pox and goat pox devastate the livelihoods of poor small-scale farmers in endemic regions.

Lumpy skin disease virus causes variable morbidity around 50% with mortality usually not exceeding 1–3%. The main mode of transmission of lumpy skin disease is through mechanical transmission by insects and ticks. The disease, following experimental infection, is also variable, with the clinical signs observed in cattle similar to what is observed with the natural infection in the field [9]. Production losses include lower reproductive rates due to abortions and infertility as well as reduced weight gain due to long-term morbidity. Losses from damage to hides caused by the scars from the skin lesions as well as reduction in draft power contribute to overall economic losses from lumpy skin disease. Lumpy skin disease virus infection also decreases milk production and can lead to complications of mastitis [7].

An outbreak of any capripoxvirus disease in a non-endemic country causes major issues with trade. With sheep pox and goat pox viruses, the disease can be brought under control through a slaughter policy to stamp out the disease. Unfortunately, for lumpy skin disease virus, stamping out is not effective, and vaccination must be done to control the disease [10]. This is due to the spread of lumpy skin disease being transmitted by vectors through mechanical transmission. Following vaccination, it is difficult to demonstrate freedom of disease. This is illustrated in European countries which have not yet demonstrated freedom of disease for trade purposes years following the outbreak since live-attenuated vaccines are used.

## 1.3    Construction of Capripoxvirus Vectors

There are several different capripoxviruses which can be used as viral vectors. The most commonly used capripoxvirus vectors are existing capripoxvirus vaccines which are already attenuated. However, attenuation can be further enhanced but more importantly better controlled by deleting specific genes. There are many potential different genes and/or combinations of genes that could be used to attenuate capripoxviruses. This is illustrated by the multitude of different capripoxvirus vaccines developed in different countries and laboratories through serial passage of the virus. These genes can potentially be identified by full genome sequencing of virulent capripoxviruses and attenuated capripoxvirus vaccines, followed by genetic comparison between vaccines and virulent isolates. Due to the genetic similarities between capripoxviruses, it has been proposed that a single vaccine could be developed to protect sheep, goats, and cattle from sheep pox, goat pox, and lumpy skin disease virus [11]. While this is theoretically true, for lumpy skin disease virus, vaccines based on lumpy skin disease virus seem to be most effective for controlling lumpy skin disease, but vaccines based on goat pox vaccines have demonstrated to provide protection against lumpy skin disease [12]. In contrast, sheep pox-based vaccines are not as effective for lumpy skin disease virus in cattle [13]. For

sheep pox and goat pox, there are a variety of sheep pox and goat pox vaccines that can protect sheep and goats. In addition, vaccines derived from lumpy skin disease virus can also protect sheep and goats as illustrated by the KS-1 sheep and goat pox vaccine, which was developed from a lumpy skin disease virus isolated from a sheep [6]. Additionally, an IL-10 gene-deleted lumpy skin disease virus has been demonstrated to protect sheep and goats from virulent sheep pox and goat pox [14].

The choice of capripoxvirus to be used is determined by the geographic region where the vaccine will be used. The reason for this is that lumpy skin disease virus-vectored vaccines or sheep pox- or goat pox-vectored vaccines would not be used in regions where these viruses are not endemic. The generation of capripoxvirus vectors is done using homologous recombination to either insert a gene of interest or delete a specific viral gene to attenuate the virus. This is done using a plasmid which encodes the flanking sequences of the regions where the homologous recombination will occur and will result in the insertion of an enhanced green fluorescent protein (EGFP) reporter gene for visualization of recombinant virus, as well as the E. coli guanine phosphoribosyl transferase (gpt) gene as a positive selection marker. If a specific antigen encoding gene is desired to be expressed, this gene is also placed in the plasmid between the flanking regions for insertion with a vaccinia virus promoter to allow its expression. In order to remove the selection markers, they will be flanked by two vaccinia 7.5 K early/late promoters, which upon removal of selective pressure will enable their removal through recombinant deletion [15]. Different insertion sites such as the thymidine kinase [16], IL-10 gene homologue [14], and interferon-gamma receptor-like gene [17] have been demonstrated to be useful insertion sites for foreign genes. Following infection of susceptible cells such as OA3.ts with the capripoxvirus of interest and transfection with the transfer plasmid with the reporter and selection genes as well as a foreign antigen gene or genes, viruses undergo homologous recombination with display GFP and be able to grow under selection media. These viruses are then plaque purified and grown for three rounds of selection and evaluated for the presence of wild-type virus. This is done using PCR with primers designed to differentiate between wild-type and recombinant viruses on plaque-purified viruses. Once it is determined that the plaque-purified virus is the recombinant virus of interest and is free from wild-type virus, the removal of the selection markers can be achieved by propagating the virus in non-selection media. This will allow homologous recombination to occur and remove the marker genes, allowing plaque purification of the recombinant virus which does not have the marker genes. The recombinant virus is then sequenced to confirm the genes are correctly inserted, and the virus is evaluated for the expression of the inserted antigen genes using immunostaining or Western blotting.

An additional system used to generate recombinant poxviruses is the Cre (Cyclization Recombination Enzyme)/LoxP system that allows the selection markers to be easily removed through the Cre/loxP site-specific recombination system [18]. In order to increase the efficacy of generating recombinant capripoxviruses, Clustered Regularly Interspaced Short Palindromic Repeats associated nuclease 9 system (CRISPR-Cas9) could be used; although CRISPR-Cas9 has not been demonstrated for capripoxviruses at this time, it has been used in orthopoxvirus genome editing [19].

It is likely possible that capripoxvirus vectors can be generated using synthetic biology similar to how horsepox virus was generated through gene synthesis and rescue using Shope fibroma virus (SFV) [20]. The advantage of using a synthetic approach is that it is possible to generate a viral vector construct with multiple genomic changes simultaneously or to generate several different combinations of genomic changes in the viral vector including multiple gene inserts to code for multiple antigens. This approach could provide a vector to protect against a wide array of diseases.

## 1.4 Application of Capripoxvirus Vectors

One of the greatest benefits of using capripoxvirus vectors is that these vectors can elicit immunity using only a single vaccination. The principle that capripoxviruses could be used to deliver protective antigens from other pathogens was first demonstrated in 1994 using the hemagglutinin (H) protein from rinderpest inserted into the thymidine kinase gene of the KS-1 (LSDV) vaccine which was able to protect cattle against a rinderpest challenge [21]. The KS-1 vaccine expressing the H rinderpest protein and an additional construct expressing the fusion (F) gene of rinderpest were both able to protect goats against the related peste des petits ruminants virus [22]. The duration of immunity elicited by a KS-1 (LSDV) vaccine expression, both the F and H proteins, from rinderpest was evaluated in cattle, and protection against rinderpest was approximately 50% after 2 years, demonstrating the duration of immunity against expressed antigens is suitable for vaccines [23]. The F [24] and H [25] proteins from peste des petits ruminants in the KS-1 vaccine were able to protect goats against a virulent peste des petits ruminants challenge. The KS-1 vaccine expressing bluetongue virus (BTV) antigens, including VP7 [26] VP2, NS1, and NS3 provided partial protection following challenge in sheep and goats [27]. The Rift valley fever glycoproteins Gn and Gc have been expressed in the KS-1 vector and demonstrated to elicit both humoral and cellular immunity and partial protection against Rift valley fever virus in cattle [28]. A similar KS-1 construct with Gn and Gc demonstrated neutralizing antibodies and protection in sheep against Rift valley fever challenge [29]. The role of preexisting immunity was evaluated using a KS-1-vectored peste des petits ruminants F and H constructs. The results demonstrated that preexisting immunity against capripoxvirus decreased the efficacy of the vectored vaccines resulting in partial protection against peste des petits ruminants [30].

The Chinese AV41 capripoxvirus (goat pox) vaccine expressing either the F or H proteins from peste des petits ruminants was evaluated in sheep and goats to induce neutralizing antibodies against peste des petits ruminants. The results of these studies demonstrated that AV41 vaccine expressing the H protein was able to generate neutralizing antibodies and that the antibody responses could be enhanced using a second vaccination. In addition, the role of preexisting capripoxvirus immunity was evaluated, and it was observed that preexisting immunity decreased the efficacy of the AV41-vectored vaccine; however, the impact of preexisting immunity can be overcome by using a secondary immunization [31]. Further evaluation of the AV41 vector comparing constructs with the peste des petits ruminants proteins F, H, and combination of F and H demonstrated that the combination of both F and H conferred better protection compared to F or H antigens alone [32]. The brucella outer membrane protein 25 has been expressed in the goat-pox AV41 vaccine, and immunity was demonstrated in mice [33]. The Echinococcus granulosus EG95 antigen has also been expressed in the goat pox AV41 vaccine, although it has not been evaluated for protection in animal studies [34].

There have been a few studies which have demonstrated attenuation of capripoxviruses through gene deletion of different genes. The sheeppox-019 Kelch like protein gene was deleted from a virulent sheep pox virus isolated in Kazakhstan, and this deletion was demonstrated to attenuate the sheep pox virus in sheep [35]. Further attenuation of the AV41 vaccine has been demonstrated through deletion of the TK as well as open reading frames 8–18, demonstrating that large regions of capripoxvirus can be deleted while still allowing the virus to replicate in vitro while still being an effective vaccine in goats against a virulent AV40 challenge [36].

Using a virulent lumpy skin virus, the open reading frame 005 IL-10 gene as well as the open reading frame 008 interferon-gamma receptor genes were deleted and evaluated in cattle, and although improved immunity was generated by these constructs, there were severe postvaccinal reactions observed, and two cattle inoculated with the interferon-gamma receptor-deleted lumpy skin disease virus developed skin lesions [17]. These results indicated that the IL-10 gene deletion was able to partially attenuate lumpy skin disease virus in cattle; however, the virus is not attenuated enough to be a useful vector in cattle. The same IL-10 gene-deleted lumpy skin disease virus was demonstrated to be safe and protect against virulent sheep pox and goat-pox in sheep and goats [14]. The F protein from peste des petits ruminants and Gn/Gc glycoproteins from Rift Valley fever virus were inserted in this lumpy skin disease virus [36] and evaluated for protection against peste des petits ruminants in sheep as well as Rift valley fever virus in sheep and goats. This construct was able to protect against peste des petits ruminants as well as Rift Valley fever virus, with neutralizing antibodies against Rift Valley fever virus observed following vaccination (manuscript in preparation).

## 1.5 Summary of the Application of Capripoxvirus Vectors

Currently, the only licensed vaccines used against sheep pox, goat-pox, and lumpy skin disease virus are different live-attenuated capripoxvirus vaccines. There have been no viral vectors encoding capripoxvirus antigens that have been demonstrated to be effective against any capripoxvirus. This is due to the complexity of the capripoxvirus proteome and the lack of understanding of the protective antigens for capripoxvirus. Due to the large geographic distribution of capripoxviruses and the use of live-attenuated vaccines to control these diseases, it makes economic sense to use capripoxvirus-vectored vaccines to control several different diseases in sheep and goats as well as in cattle. Since capripoxviruses are genetically stable DNA large viruses, they have the ability to be used as viral vectors since large and/or multiple genes can be inserted into their genome. Multivalent capripoxvirus vaccines have benefits of being thermostable, allowing their distribution in regions without a cold chain. In addition, they have bene demonstrated to be effective using a single dose with an annual revaccination [37].

## 1.6    Future Directions and Potential for Other Applications for Capripoxviruses as Vectors

Since capripoxviruses are foreign animal diseases in many regions, vaccines based on capripoxvirus would not be used in countries free from these viruses. Since different geographic regions have different capripoxviruses, it is required to develop capripoxvirus vectors based on which virus (es) are present. For this reason, a viral vector platform using lumpy skin disease virus as well as sheep pox viral vectors is required. Although currently used, live-attenuated lumpy skin disease virus and sheep pox and goat pox vaccines are considered generally safe and effective, their safety profile could be improved. This is especially the case with lumpy skin disease vaccines where they can cause injection site reactions. It is known that intradermal inoculation of virulent sheep pox, goat pox, and lumpy skin disease viruses into their natural hosts will result in very large severe skin lesions. It has been demonstrated that deletion of the catalytically inactive homologs of cellular Cu-Zn superoxide dismutase (SOD1) gene in Shope fibroma virus can decrease the injection site reactions in rabbits [38]. The role of SOD1 in capripoxvirus has not been evaluated in sheep, goats, and cattle; however, it is possible that SOD1 deletion could decrease injection site reactions.

Even though capripoxvirus-vectored vaccines have been demonstrated to be effective following a single immunization to both capripoxviruses and the vaccine-encoded antigen, the antibody responses measured are low. This is likely due to the strong cellular immunity generated by poxviruses as well as the numerous immunomodulatory proteins encoded by the viruses which influence immune responses generated by the vector and encoded antigens. It may be possible to enhance antibody responses in capripoxvirus

vectors by deleting possible immunomodulatory proteins.

For capripoxvirus-vectored vaccines to be used in non-endemic countries, a vaccine that can **D**ifferentiate **I**nfected from **V**accinated **A**nimals (DIVA) would need to be developed. Having a DIVA vaccine would also be a useful tool for endemic countries so they could run serological epidemiology studies to understand and reduce the disease burden as well as to develop an effective eradication campaign, leading to eventually acquire disease-free status. Currently, there are no capripoxvirus vaccines that have DIVA capability. A DIVA vaccine can be generated by simply deleting a gene from the virus and then developing a companion diagnostic serology test which can identify animals which have been infected as these animals will develop antibodies against the viral protein found in the wild-type virus but not from the vaccine. Although the principle of DIVA is simple, implementation of the DIVA can be difficult. In order to improve the chance of developing a DIVA vaccine, it is best to start with the diagnostic serological test. The reason for this is that there is no point in generating gene-deleted viruses where the deleted gene will not be useful for a serological test. Even if an antigen is identified to be a useful antigen for a diagnostic ELISA, it is possible that the antigen is an essential protein for the viral vector. If this is the case, it will be impossible to generate the gene deletion in the vector. Currently, there is a double-core antigen ELISA which can detect antibodies following capripoxvirus infections [39]. Unfortunately, these antigens are essential proteins for the virus and not suitable candidates to generate a DIVA vaccine. Capripoxviruses generate neutralizing antibodies, and an ELISA has been developed using inactivated whole virus as antigen [40], demonstrating that there are other antigens present which could be used in an ELISA. Further work is required to identify a suitable antigen to generate a DIVA vaccine for capripoxviruses.

Several of the previously described capripoxvirus-vectored vaccines would be a DIVA vaccine for the disease against the inserted protective antigens. For example, with peste des petits ruminants, the capripoxvirus-vectored vaccine would generate antibodies against the H antigen which could be used to evaluate vaccine efficacy. Animals following vaccination would have antibodies against H antigen and, following infection with peste des petits ruminants, would generate antibodies against nucleoprotein (N) antigen. By using two diagnostic ELISA tests, one specific for H and one specific for N, it could determine if animals have been vaccinated and not infected or vaccinated and infected. The same principle can be used for Rift valley fever virus where animals vaccinated using a capripoxvirus-vectored Gn/Gc vaccine would elicit antibodies against Gn/Gc but not to nucleoprotein. Following infection with Rift valley fever, antibodies would be elicited against nucleoprotein.

Capripoxviruses have a great potential to be used as multivalent vaccines. Currently, for sheep and goats, it has been demonstrated that a combination of sheep pox, goat pox, Rift valley fever, and peste des petits ruminants vaccine can be effective in sheep and goats (manuscript in preparation). Further demonstration of additional protective antigens for other diseases would further increase the utility of capripoxvirus-vectored vaccines. It is currently unknown how many different antigens can be expressed in a capripoxvirus vector while still being able to elicit protective immunity. It is likely that more than two different antigens can be expressed to generate an effective multivalent vaccine. Further work to evaluate the duration of immunity of capripoxvirus-vectored vaccines is required.

Since sheep pox and goat pox are transmitted through contact with infected animals and the environment and shed virus at mucosal sites, it is possible that capripoxvirus vectors could be administered by intranasal administration. Unfortunately, there have been no studies with capripoxvirus-vectored vaccines that have evaluated the intranasal route for administration.

Even through capripoxviruses have been demonstrated to be a useful vector experimentally, currently, there are no commercially available capripoxvirus-vectored vaccines despite the use of several different live-attenuated vaccines

for lumpy skin disease, sheep pox, and goat pox. Since these diseases primarily affect farmers in developing regions, there are little economic incentives to develop these vaccines. The major issue in getting these vaccines to be used in the field is the lack of vaccine companies willing to spend the required resources for licensing of these vaccines.

## 2 Chapter Introduction

### 2.1 Background of African Swine Fever Virus

African swine fever virus is the sole member of family *Asfarviridae* genus *Asfivirus* which has a double-stranded DNA virus with a genome ranging from 170 to 194 kbps with hairpin loops and encoding over 160 proteins [41]. There have been 23 genotypes characterized based on the p72 gene [42, 43]. African swine fever virus is a non-zoonotic disease with host specificity for suids. The natural hosts for African swine fever virus are wild pigs in Africa including warthogs, bush pigs, and giant forest hogs which can have unapparent infections. African swine fever virus can be transmitted through soft ticks of the *Ornithodoros* genus which maintain the reservoir of African swine fever virus in their natural hosts. Domestic pigs and wild boars are highly susceptible to African swine fever virus [44]. The virus is spread through the oral/nasal route by direct and indirect contact with infected pigs or pig products. This is due to the stability of African swine fever virus which can persist on contaminated boots, clothing, and equipment as well as in the meat of infected animals and pork products, such as sausages and cured meats [45]. In Europe, African swine fever virus is primarily spread through the movement of infected wild boar [46].

African swine fever virus was first described in Kenya in 1921 and identified warthogs as carriers of the virus which do not display clinical disease [47]. African swine fever virus has subsequently been reported in most sub-Saharan African countries where the disease was recognized to be present in the natural hosts in East and Southern Africa. African swine fever virus has spread into central, West Africa, and Madagascar.

African swine fever virus was a disease of Africa until it was introduced into Portugal in 1957 and rapidly eradicated. In 1960, ASF was reintroduced into Portugal and spread into Spain where it was widely present during the 1970s and 1980s and resulted in outbreaks in France (1964, 1967, 1977), Belgium (1985), the Netherlands (1986), and Italy (1967, 1980). African swine fever virus caused outbreaks in the Western Hemisphere from feeding ASF-contaminated food waste to pigs in Cuba (1971, 1978–1980), the Dominican Republic (1978–1981), Haiti (1979–1984), and Brazil (1978–1981) [48]. The disease was eradicated following all of these outbreaks, with the exception of Sardinia where ASF remains endemic today [49].

In 2007, ASF spread into Georgia where delays in the control of the disease resulted in the spread of ASF to neighboring regions such as Armenia and Azerbaijan and into Russia. African swine fever virus continued to spread into Ukraine (2012); Belarus (2013); Latvia, Lithuania, Estonia, and Poland (2014); Romania and Czech Republic (2017); Hungary, Belgium, and China (2018); and Vietnam (2019) [50].

## 2.2 Clinical Importance of African Swine Fever Virus

The clinical signs of African swine fever are high fever, weakness, loss of appetite, hemorrhages in the skin, internal bleeding, vomiting and diarrhea, and death. High viral loads can be found in the blood as well as oral and nasal secretions [51]. There are different levels of virulence with African swine fever viruses having peracute, acute, subacute, and chronic forms [52]. Mortality from highly virulent African swine fever viruses can approach 100% leading to severe losses for producers. The disease spreads rapidly in backyard farms with low biosecurity. The ASF epidemic in China is causing catastrophic consequences on national pork production since

China is the largest pork producer in the world [53].

Although it has been previously demonstrated that African swine fever virus can be eradicated, it remains difficult to eradicate the disease once established in a region. The establishment of African swine fever in wild boars throughout Europe, Russia, and China is making the control of the disease more difficult. Similar to capripoxvirus, a disease outbreak of African swine fever virus in a non-endemic country causes major issues with trade.

## 2.3 Construction of African Swine Fever Virus Vectors

The generation of recombinant African swine fever viruses is very similar to the generation of recombinant capripoxviruses. These viruses are generated using homologous recombination to insert and remove genes. Differences with African swine fever virus and recombinant viruses are generated in susceptible cells such as primary swine macrophages or COS-1 cells. In addition, the transfer plasmid will also be different using the flanking regions of the African swine fever virus as well as the P72 promoter from African swine fever virus used for selection markers and foreign antigen gene insertions. Selection of recombinant African swine fever virus is done using either plaque purification or successive rounds of limiting dilution purification, similar to the generation of recombinant capripoxviruses. Unfortunately, homologous recombination is not highly efficient with homologous recombination occurring at less than one recombinant virus for every $10^6$ wild-type viruses. The use of CRISPR-Cas9 to generate recombinant African swine fever viruses has been demonstrated to be much more efficient compared to homologous recombination at approximately one recombinant virus for every $10^2$ wild-type viruses [54].

It will be more difficult to generated African swine fever virus vectors compared to capripoxviruses using synthetic biology since it would be difficult or impossible to identify a helper virus to rescue the synthetic genome since they are the sole member of their genus.

## 2.4 Application of African Swine Fever Virus Vectors

Currently, there is no licensed vaccine for African swine fever virus. African swine fever virus live-attenuated vaccines generated by serial passage in primary bone marrow or blood macrophage cell cultures were first used in Portugal and Spain in the 1960s. These vaccines caused unacceptable postvaccination reactions as well as chronic forms of African swine fever and are no longer used [55]. However, their use first demonstrated the proof of principle that it is possible to develop a vaccine against African swine fever virus despite the fact that these vaccines were not completely safe. Further evaluation of naturally attenuated African swine fever virus NHV/P68 was demonstrated to be able to protect pigs from virulent L60 African swine fever virus [56]. The naturally attenuated African swine fever virus OURT88/3 followed by administration of virulent OURT88/1 virus conferred protection against Benin 97/1 isolate and Uganda 1965 isolate [57]. These studies provided the concept that protective immunity could be generated using live attenuated African swine fever viruses. Attenuation of African swine fever virus occurs when the virus is propagated in Vero cells leading to major deletions since the genome is not stable in these cells [58].

Currently, there are several different attenuated African swine fever viruses created through gene deletion described. Deletion of the thymidine kinase gene of the Georgia 2007 resulted in attenuation of the virus. However, this gene-deleted virus was not able to protect against ASF challenge [59]. The DP148R gene deleted in the Benin 97/1 virus caused mild clinical signs in pigs following administration and was able to protect against homologous challenge [60]. Deletion of MGF360 and MGF530/505 gene families in Benin 97/1 attenuated the virus, and protection was demonstrated against homologous challenge following two immunizations

[61]. The MGF360 and MGF505 genes deleted in Georgia 2007 resulted in attenuation. This virus was able to protect against homologous challenge following a single intramuscular administration of $10^2$ or $10^4$ 50% hemadsorbing doses (HAD50) [62]. African swine fever virus gene-deleted 9GL Georgia 2007 is attenuated and can protect pigs; however, the difference in the dose between a lethal infection and attenuation is only two logs [63]. This vaccine is therefore not considered safe and cannot be used. To further attenuate the 9GL gene-deleted virus, additional genes were deleted and evaluated for attenuation and protection. An African swine fever virus with deleted 9GL and MGF360/505 genes in Georgia 2007 was demonstrated to be highly attenuated in swine but did not confer protection against ASF [64]. A double gene-deleted African swine fever virus with the 9GL and UK genes in Georgia 2007 was attenuated and able to provide protection against homologous ASF challenge [65]. A live-attenuated CD2v gene deleted from the BA71 African swine fever virus (BA71ΔCD2) was generated in COS-1 cells. This BA71ΔCD2 virus was demonstrated to be attenuated and able to protect against homologous BA71 and heterologous Georgia 2007/1 challenge [66]. Several gene-deleted African swine fever viruses were constructed using the low-virulent NH/P68 virus deleting the A238L, A224L, A276R, and EP153R genes. The NH/P68ΔA238L, NH/P68ΔA224L, and NH/P68ΔEP153R protected against homologous Lisbon 60 challenge but not Arm07 challenge [67].

## 2.5 Summary of the Application of African Swine Fever Virus Vectors

African swine fever virus is in the early stages of development as a vaccine vector. Although it has not been yet demonstrated to be useful as a vaccine vector, African swine fever virus has several characteristics which would make it a useful vector. Similar to capripoxviruses, it is likely that a live-attenuated African swine fever virus would be a useful vaccine. Immunity following a single administration with low doses of an attenuated African swine fever virus had been demonstrated. The large DNA genome in African swine fever virus allows the ability to insert large and/or multiple genes similar to capripoxviruses. The thermostability of African swine fever virus is an important feature of these viruses, and the natural routes of infection make oral delivery for vaccination a possibility to elicit mucosal immunity to the vaccine. These features make African swine fever virus a potentially useful vector to control important diseases of swine.

## 2.6 Future Directions and Potential for Other Applications for African Swine Fever as Vectors

The development of a safe and effective live-attenuated African swine fever virus vaccine is still in early stages although proof of principle has been demonstrated. Improvements in the safety and efficacy of live-attenuated African swine fever virus are being attempted through gene deletion of different virulence genes either by a single deletion or a combination of deletions. The complexity of the virus makes it difficult to determine which genes are critical for attenuating the virus while still being able to elicit a protective immune response. Several different research groups are currently evaluating different gene-deleted African swine fever viruses as potential vaccines.

Once a live-attenuated African swine fever virus vaccine has been developed, this could potentially be used to elicit immune responses to a protective antigen from a different virus generating a multivalent vaccine. It has not currently been demonstrated that protective antigens can be expressed in an African swine fever virus vector and that these antigens can elicit protective immunity. For instance, the protective antigen E2 protein from classical swine fever virus [68] could be used in an African swine fever viral vector to protect swine from African swine fever virus and classical swine fever virus. The physical properties of African swine fever virus have many features of a useful viral vector. These features include the ability for large/multiple genes to be expressed. African swine fever virus has a very narrow host range of members of the family Suidae, making these vectors safe since they will not infect humans or other animal species. The thermostability of the African swine fever virus is an additional positive feature to allow vaccine distribution without a cold chain. Since African swine fever virus can be transmitted through the oral/nasal route, this potentially would allow African swine fever virus vectored vaccines to be administered by either the oral or intranasal route allowing for the induction of mucosal immunity. The ability to deliver a vectored vaccine orally is critical for being able to administer vaccines to wild boars as illustrated by the classical swine fever vaccination program [69]. Live-attenuated African swine fever viruses are able to generate immunity following a single immunization at reasonable doses of virus 10 [2]. The duration of immunity elicited by live-attenuated African swine fever virus-vectored vaccines to African swine fever virus as well as the expressed antigen requires further study.

For African swine fever virus-vectored vaccines to be used in non-endemic countries, a DIVA vaccine would need to be developed. Currently, there has been no proof of principle of a DIVA vaccine for African swine fever virus. As discussed in the previous section with capripoxviruses, to generate a DIVA vaccine, it is best to start with a validated diagnostic assay. For African swine fever virus, there are several validated/candidate indirect and competitive ELISA assays based on different viral antigens P72, P54, and P30 as well as others in development [70]. It is likely that P72, P54, and P30 antigens are essential antigens for African swine fever virus; however, this has not been demonstrated. If these antigens are essential proteins, it may be possible to modify the protein epitope which binds to the monoclonal antibody in the diagnostic ELISA. This virus would then have DIVA capability since it would not generate antibodies that would be detected by the diagnostic test. Since African swine fever virus is a complicated virus with many antigens, there are likely several other antigens that could be nonessential antigens for African swine fever virus that could be developed into a diagnostic assay.

The greatest impediment for African swine fever virus-vectored vaccines will likely be how to commercially produce these vaccines. Unfortunately, currently, there is no cell line that can be used to propagate African swine fever virus without adaptation. Hence, for diagnostic isolation protocols, use either primary porcine alveolar macrophages or monocyte and macrophage cultures. The development of a cell line to produce the African swine fever virus vectors is a major hurdle for African swine fever virus vaccines and the use of African swine fever virus as a vector.

# References

1. Tulman ER, Afonso CL, Lu Z, Zsak L, Sur JH, Sandybaev NT, Kerembekova UZ, Zaitsev VL, Kutish GF, Rock DL. The genomes of sheeppox and goatpox viruses. J Virol. 2002;76:6054–61.

2. Tulman ER, Afonso CL, Lu Z, Zsak L, Kutish GF, Rock DL. Genome of lumpy skin disease virus. J Virol. 2001;75:7122–30.

3. Babiuk S, Bowden TR, Boyle DB, Wallace DB, Kitching RP. Capripoxviruses: an emerging worldwide threat to sheep, goats and cattle. Transbound Emerg Dis. 2008;55:263–72. https://doi.org/10.1111/j.1865-1682.2008.01043.x.

4. Tuppurainen ESM, Venter EH, Shisler JL, Gari G, Mekonnen GA, Juleff N, Lyons NA, De Clercq K, Upton C, Bowden TR, Babiuk S, Babiuk LA. Review: capripoxvirus diseases: current status and opportunities for control. Transbound Emerg Dis. 2017;64:729–45. https://doi.org/10.1111/tbed.12444.

5. Babiuk S, Bowden TR, Parkyn G, Dalman B, Hoa DM, Long NT, Vu PP, Bieudo X, Copps J, Boyle DB. Yemen and Vietnam capripoxviruses demonstrate a distinct host preference for goats compared with sheep. J Gen Virol. 2009;90:105–14. https://doi.org/10.1099/vir.0.004507-0.

6. Tuppurainen ES, Pearson CR, Bachanek-Bankowska-K, Knowles NJ, Amareen S, Frost L, Henstock MR, Lamien CE, Diallo A, Mertens PP. Characterization of sheep pox virus vaccine for cattle against lumpy skin disease virus. Antivir Res. 2014;109:1–6. https://doi.org/10.1016/j.antiviral.2014.06.009.

7. Tuppurainen ESM, Babiuk S, Klement E. Lumpy skin disease: Springer; 2018.

8. Bowden TR, Babiuk SL, Parkyn GR, Copps JS, Boyle DB. Capripoxvirus tissue tropism and shedding: a quantitative study in experimentally infected sheep and goats. Virology. 2008;371:380–93.

9. Babiuk S, Bowden TR, Parkyn G, Dalman B, Manning L, Neufeld J, Embury-Hyatt C, Copps J, Boyle DB. Quantification of lumpy skin disease virus following experimental infection in cattle. Transbound Emerg Dis. 2008;55:299–307. https://doi.org/10.1111/j.1865-1682.2008.01024.x.

10. EFSA Panel on Animal Health and Welfare. Urgent advice on lumpy skin disease. EFSA J. 2016;14:e04573. https://doi.org/10.2903/j.efsa.2016.4573.

11. Kitching RP. Vaccines for lumpy skin disease, sheep pox and goat pox. Dev Biol (Basel). 2003;114:161–7.

12. Gari G, Abie G, Gizaw D, Wubete A, Kidane M, Asgedom H, Bayissa B, Ayelet G, Oura CA, Roger F, Tuppurainen ES. Evaluation of the safety, immunogenicity and efficacy of three capripoxvirus vaccine strains against lumpy skin disease virus. Vaccine. 2015;33:3256–61. https://doi.org/10.1016/j.vaccine.2015.01.035.

13. Ben-Gera J, Klement E, Khinich E, Stram Y, Shpigel NY. Comparison of the efficacy of Neethling lumpy skin disease virus and x10RM65 sheep-pox live attenuated vaccines for the prevention of lumpy skin disease – the results of a randomized controlled field study. Vaccine. 2015;33:4837–42. https://doi.org/10.1016/j.vaccine.2015.07.071.

14. Boshra H, Truong T, Nfon C, Bowden TR, Gerdts V, Tikoo S, Babiuk LA, Kara P, Mather A, Wallace DB, Babiuk S. A lumpy skin disease virus deficient of an IL-10 gene homologue provides protective immunity against virulent capripoxvirus challenge in sheep and goats. Antivir Res. 2015;123:39–49. https://doi.org/10.1016/j.antiviral.2015.08.016.

15. Boshra H, Cao J, Babiuk S. Generation of recombinant capripoxvirus vectors for vaccines and gene knockout function studies. Methods Mol Biol. 2016;1349:151–61. https://doi.org/10.1007/978-1-4939-3008-1_10.

16. Wallace DB, Viljoen GJ. Immune responses to recombinants of the South African vaccine strain of lumpy skin disease virus generated by using thymidine kinase gene insertion. Vaccine. 2005;23:3061–7.

17. Kara PD, Mather AS, Pretorius A, Chetty T, Babiuk S, Wallace DB. Characterisation of putative immunomodulatory gene knockouts of lumpy skin disease virus in cattle towards an improved vaccine. Vaccine. 2018;36:4708–15. https://doi.org/10.1016/j.vaccine.2018.06.017.

18. Rintoul JL, Wang J, Gammon DB, van Buuren NJ, Garson K, Jardine K, Barry M, Evans DH, Bell JC. A selectable and excisable marker system for the rapid creation of recombinant poxviruses. PLoS One. 2011;6:e24643. https://doi.org/10.1371/journal.pone.0024643.

19. Okoli A, Okeke MI, Tryland M, Moens U. CRISPR/Cas9—advancing orthopoxvirus genome editing for

vaccine and vector development. Viruses. 2018;10:50. https://doi.org/10.3390/v10010050.

20. Noyce RS, Lederman S, Evans DH. Construction of an infectious horsepox virus vaccine from chemically synthesized DNA fragments. PLoS One. 2018;13: e0188453. https://doi.org/10.1371/journal.pone. 0188453.

21. Romero CH, Barrett T, Chamberlain RW, Kitching RP, Fleming M, Black DN. Recombinant capripoxvirus expressing the hemagglutinin protein gene of rinderpest virus: protection of cattle against rinderpest and lumpy skin disease viruses. Virology. 1994;204:425–9.

22. Romero CH, Barrett T, Kitching RP, Bostock C, Black DN. Protection of goats against peste des petits ruminants with recombinant capripoxviruses expressing the fusion and haemagglutinin protein genes of rinderpest virus. Vaccine. 1995;13:36–40.

23. Ngichabe CK, Wamwayi HM, Ndungu EK, Mirangi PK, Bostock CJ, Black DN, Barrett T. Long term immunity in African cattle vaccinated with a recombinant capripox-rinderpest virus vaccine. Epidemiol Infect. 2002;128:343–9.

24. Berhe G, Minet C, Le Goff C, Barrett T, Ngangnou A, Grillet C, Libeau G, Fleming M, Black DN, Diallo A. Development of a dual recombinant vaccine to protect small ruminants against peste-des-petits-ruminants virus and capripoxvirus infections. J Virol. 2003;77:1571–7.

25. Diallo A, Minet C, Berhe G, Le Goff C, Black DN, Fleming M, Barrett T, Grillet C, Libeau G. Goat immune response to capripox vaccine expressing the hemagglutinin protein of peste des petits ruminants. Ann N Y Acad Sci. 2002;969:88–91.

26. Wade-Evans AM, Romero CH, Mellor P, Takamatsu H, Anderson J, Thevasagayam J, Fleming MJ, Mertens PP, Black DN. Expression of the major core structural protein (VP7) of bluetongue virus, by a recombinant capripox virus, provides partial protection of sheep against a virulent heterotypic bluetongue virus challenge. Virology. 1996;220:227–31.

27. Perrin A, Albina E, Bréard E, Sailleau C, Promé S, Grillet C, Kwiatek O, Russo P, Thiéry R, Zientara S, Cêtre-Sossah C. Recombinant capripoxviruses expressing proteins of bluetongue virus: evaluation of immune responses and protection in small ruminants. Vaccine. 2007;25:6774–83.

28. Wallace DB, Ellis CE, Espach A, Smith SJ, Greyling RR, Viljoen GJ. Protective immune responses induced by different recombinant vaccine regimes to Rift Valley fever. Vaccine. 2006;24:7181–9.

29. Soi RK, Rurangirwa FR, McGuire TC, Rwambo PM, DeMartini JC, Crawford TB. Protection of sheep against Rift Valley fever virus and sheep poxvirus with a recombinant capripoxvirus vaccine. Clin Vaccine Immunol. 2010;17:1842–9. https://doi.org/10. 1128/CVI.00220-10.

30. Caufour P, Rufael T, Lamien CE, Lancelot R, Kidane M, Awel D, Sertse T, Kwiatek O, Libeau G,

Sahle M, Diallo A, Albina E. Protective efficacy of a single immunization with capripoxvirus-vectored recombinant peste des petits ruminants vaccines in presence of pre-existing immunity. Vaccine. 2014;32:3772–9. https://doi.org/10.1016/j.vaccine. 2014.05.025.

31. Chen W, Hu S, Qu L, Hu Q, Zhang Q, Zhi H, Huang K, Bu Z. A goat poxvirus-vectored peste-des-petits-ruminants vaccine induces long-lasting neutralization antibody to high levels in goats and sheep. Vaccine. 2010;28:4742–50. https://doi.org/10.1016/j. vaccine.2010.04.102.

32. Fakri F, Bamouh Z, Ghzal F, Baha W, Tadlaoui K, Fihri OF, Chen W, Bu Z, Elharrak M. Comparative evaluation of three capripoxvirus-vectored peste des petits ruminants vaccines. Virology. 2018 Jan;514:211–5. https://doi.org/10.1016/j.virol.2017. 11.015.

33. Sun Z, Liu L, Zhang H, Li Y, Wei F, Li Z, Wang P, Fu Q, Ren Y, Zhang Y, Guo Z, Chen C. Expression and functional analysis of Brucella outer membrane protein 25 in recombinant goat pox virus. Mol Med Rep. 2019;19:2323–9. https://doi.org/10.3892/mmr. 2019.9868.

34. Liu F, Fan X, Li L, Ren W, Han X, Wu X, Wang Z. Development of recombinant goatpox virus expressing Echinococcus granulosus EG95 vaccine antigen. J Virol Methods. 2018;261:28–33. https:// doi.org/10.1016/j.jviromet.2018.08.002.

35. Balinsky CA, Delhon G, Afonso CL, Risatti GR, Borca MV, French RA, Tulman ER, Geary SJ, Rock DL. Sheeppox virus kelch-like gene SPPV-019 affects virus virulence. J Virol. 2007;81:11392–401.

36. Zhu Y, Li Y, Bai B, Fang J, Zhang K, Yin X, Li S, Li W, Ma Y, Cui Y, Wang J, Liu X, Li X, Sun L, Jin N. Construction of an attenuated goatpox virus AV41 strain by deleting the TK gene and ORF8-18. Antivir Res. 2018;157:111–9. https://doi.org/10.1016/j. antiviral.2018.07.008.

37. Boshra H, Truong T, Nfon C, Gerdts V, Tikoo S, Babiuk LA, Kara P, Mather A, Wallace D, Babiuk S. Capripoxvirus-vectored vaccines against livestock diseases in Africa. Antivir Res. 2013;98:217–27. https://doi.org/10.1016/j.antiviral.2013.02.016.

38. Teoh ML, Turner PV, Evans DH. Tumorigenic poxviruses up-regulate intracellular superoxide to inhibit apoptosis and promote cell proliferation. J Virol. 2005;79:5799–811.

39. Bowden TR, Coupar BE, Babiuk SL, White JR, Boyd V, Duch CJ, Shiell BJ, Ueda N, Parkyn GR, Copps JS, Boyle DB. Detection of antibodies specific for sheeppox and goatpox viruses using recombinant capripoxvirus antigens in an indirect enzyme-linked immunosorbent assay. J Virol Methods. 2009;161:19–29. https://doi.org/10.1016/j.jviromet. 2009.04.031.

40. Babiuk S, Wallace DB, Smith SJ, Bowden TR, Dalman B, Parkyn G, Copps J, Boyle DB. Detection of antibodies against capripoxviruses using an

inactivated sheeppox virus ELISA. Transbound Emerg Dis. 2009;56:132–41. https://doi.org/10.1111/j.1865-1682.2009.01067.x.

41. Alonso C, Borca M, Dixon L, Revilla Y, Rodriguez F, Escribano JM. ICTV report consortium. ICTV virus taxonomy profile: Asfarviridae. J Gen Virol. 2018;99:613–4. https://doi.org/10.1099/jgv.0.001049.

42. Lubisi BA, Bastos AD, Dwarka RM, Vosloo W. Molecular epidemiology of African swine fever in East Africa. Arch Virol. 2005;150:2439–52.

43. Gallardo C, Fernández-Pinero J, Pelayo V, Gazaev I, Markowska-Daniel I, Pridotkas G, Nieto R, Fernández-Pacheco P, Bokhan S, Nevolko O, Drozhzhe Z, Pérez C, Soler A, Kolvasov D, Arias M. Genetic variation among African swine fever genotype II viruses, eastern and central Europe. Emerg Infect Dis. 2014;20:1544–7. https://doi.org/10.3201/eid2009.140554.

44. Blome S, Gabriel C, Beer M. Pathogenesis of African swine fever in domestic pigs and European wild boar. Virus Res. 2013;173:122–30. https://doi.org/10.1016/j.virusres.2012.10.026.

45. Petrini S, Feliziani F, Casciari C, Giammarioli M, Torresi C, De Mia GM. Survival of African swine fever virus (ASFV) in various traditional Italian dry-cured meat products. Prev Vet Med. 2019;162:126–30. https://doi.org/10.1016/j.prevetmed.2018.11.013.

46. Bosch J, Rodríguez A, Iglesias I, Muñoz MJ, Jurado C, Sánchez-Vizcaíno JM, de la Torre A. Update on the risk of introduction of African Swine fever by wild boar into disease-free European Union countries. Transbound Emerg Dis. 2017;64:1424–32. https://doi.org/10.1111/tbed.12527.

47. Montgomery RE. On a form of swine fever occurring in British East Africa (Kenya Colony). J Comp Pathol Ther. 1921;34:159–91.

48. Costard S, Wieland B, de Glanville W, Jori F, Rowlands R, Vosloo W, Roger F, Pfeiffer DU, Dixon LK. African swine fever: how can global spread be prevented? Philos Trans R Soc Lond Ser B Biol Sci. 2009;364:2683–96. https://doi.org/10.1098/rstb.2009.0098.

49. Mur L, Atzeni M, Martínez-López B, Feliziani F, Rolesu S, Sanchez-Vizcaino JM. Thirty-five-year presence of African swine fever in Sardinia: history, evolution and risk factors for disease maintenance. Transbound Emerg Dis. 2016;63:e165–77. https://doi.org/10.1111/tbed.12264.

50. World Organisation for Animal Health, Wahid Database (OIE WAHID) Interface. Disease information.

51. Guinat C, Reis AL, Netherton CL, Goatley L, Pfeiffer DU, Dixon L. Dynamics of African swine fever virus shedding and excretion in domestic pigs infected by intramuscular inoculation and contact transmission. Vet Res. 2014;45:93. https://doi.org/10.1186/s13567-014-0093-8.

52. Schulz K, Staubach C, Blome S. African and classical swine fever: similarities, differences and epidemiological consequences. Vet Res. 2017;48:84. https://doi.org/10.1186/s13567-017-0490-x.

53. Wang T, Sun Y, Qiu HJ. African swine fever: an unprecedented disaster and challenge to China. Infect Dis Poverty. 2018;7:111. https://doi.org/10.1186/s40249-018-0495-3.

54. Borca MV, Holinka LG, Berggren KA, Gladue DP. CRISPR-Cas9, a tool to efficiently increase the development of recombinant African swine fever viruses. Sci Rep. 2018;8:3154. https://doi.org/10.1038/s41598-018-21575-8.

55. Manso-Ribeiro J, Nunes-Petisca JL, Lopez-Frazao F, Sobral M. Vaccination against ASF. Bull Off Int Epizoot. 1963;60:921–37.

56. Leitão A, Cartaxeiro C, Coelho R, Cruz B, Parkhouse RM, Portugal F, Vigário JD, Martins CL. The non-haemadsorbing African swine fever virus isolate ASFV/NH/P68 provides a model for defining the protective anti-virus immune response. J Gen Virol. 2001;82:513–23.

57. King K, Chapman D, Argilaguet JM, Fishbourne E, Hutet E, Cariolet R, Hutchings G, Oura CA, Netherton CL, Moffat K, Taylor G, Le Potier MF, Dixon LK, Takamatsu HH. Protection of European domestic pigs from virulent African isolates of African swine fever virus by experimental immunisation. Vaccine. 2011;29:4593–600. https://doi.org/10.1016/j.vaccine.2011.04.052.

58. Krug PW, Holinka LG, O'Donnell V, Reese B, Sanford B, Fernandez-Sainz I, Gladue DP, Arzt J, Rodriguez L, Risatti GR, Borca MV. The progressive adaptation of a georgian isolate of African swine fever virus to vero cells leads to a gradual attenuation of virulence in swine corresponding to major modifications of the viral genome. J Virol. 2015;89:2324–32. https://doi.org/10.1128/JVI.03250-14.

59. Sanford B, Holinka LG, O'Donnell V, Krug PW, Carlson J, Alfano M, Carrillo C, Wu P, Lowe A, Risatti GR, Gladue DP, Borca MV. Deletion of the thymidine kinase gene induces complete attenuation of the Georgia isolate of African swine fever virus. Virus Res. 2016;213:165–71. https://doi.org/10.1016/j.virusres.2015.12.002.

60. Reis AL, Goatley LC, Jabbar T, Sanchez-Cordon PJ, Netherton CL, Chapman DAG, Dixon LK. Deletion of the African swine fever virus gene DP148R does not reduce virus replication in culture but reduces virus virulence in pigs and induces high levels of protection against challenge. J Virol. 2017;91:e01428–17. https://doi.org/10.1128/JVI.01428-17.

61. Reis AL, Abrams CC, Goatley LC, Netherton C, Chapman DG, Sanchez-Cordon P, Dixon LK. Deletion of African swine fever virus interferon inhibitors from the genome of a virulent isolate reduces virulence in domestic pigs and induces a protective response. Vaccine. 2016;34:4698–705. https://doi.org/10.1016/j.vaccine.2016.08.011.

62. O'Donnell V, Holinka LG, Gladue DP, Sanford B, Krug PW, Lu X, Arzt J, Reese B, Carrillo C, Risatti GR, Borca MV. African swine fever virus Georgia isolate harboring deletions of MGF360 and MGF505 genes is attenuated in swine and confers protection against challenge with virulent parental virus. J Virol. 2015;89:6048–56. https://doi.org/10.1128/JVI.00554-15.

63. O'Donnell V, Holinka LG, Krug PW, Gladue DP, Carlson J, Sanford B, Alfano M, Kramer E, Lu Z, Arzt J, Reese B, Carrillo C, Risatti GR, Borca MV. African swine fever virus Georgia 2007 with a deletion of virulence-associated gene 9GL (B119L), when administered at low doses, leads to virus attenuation in swine and induces an effective protection against homologous challenge. J Virol. 2015;89:8556–66. https://doi.org/10.1128/JVI.00969-15.

64. O'Donnell V, Holinka LG, Sanford B, Krug PW, Carlson J, Pacheco JM, Reese B, Risatti GR, Gladue DP, Borca MV. African swine fever virus Georgia isolate harboring deletions of 9GL and MGF360/505 genes is highly attenuated in swine but does not confer protection against parental virus challenge. Virus Res. 2016;221:8–14. https://doi.org/10.1016/j.virusres.2016.05.014.

65. O'Donnell V, Risatti GR, Holinka LG, Krug PW, Carlson J, Velazquez-Salinas L, Azzinaro PA, Gladue DP, Borca MV. Simultaneous deletion of the 9GL and UK genes from the African swine fever virus Georgia 2007 isolate offers increased safety and protection against homologous challenge. J Virol. 2016;91: e01760–16. https://doi.org/10.1128/JVI.01760-16.

66. Monteagudo PL, Lacasta A, López E, Bosch L, Collado J, Pina-Pedrero S, Correa-Fiz F, Accensi F, Navas MJ, Vidal E, Bustos MJ, Rodríguez JM, Gallei A, Nikolin V, Salas ML, Rodríguez F. BA71ΔCD2: a new recombinant live attenuated African swine fever virus with cross-protective capabilities. J Virol. 2017;91:e01058–17. https://doi.org/10.1128/JVI.01058-17.

67. Gallardo C, Sánchez EG, Pérez-Núñez D, Nogal M, de León P, Carrascosa ÁL, Nieto R, Soler A, Arias ML, Revilla Y. African swine fever virus (ASFV) protection mediated by NH/P68 and NH/P68 recombinant live-attenuated viruses. Vaccine. 2018;36:2694–704. https://doi.org/10.1016/j.vaccine.2018.03.040.

68. König M, Lengsfeld T, Pauly T, Stark R, Thiel HJ. Classical swine fever virus: independent induction of protective immunity by two structural glycoproteins. J Virol. 1995;69:6479–86.

69. Rossi S, Staubach C, Blome S, Guberti V, Thulke HH, Vos A, Koenen F, Le Potier MF. Controlling of CSFV in European wild boar using oral vaccination: a review. Front Microbiol. 2015;6:1141. https://doi.org/10.3389/fmicb.2015.01141.

70. Gallardo C, Blanco E, Rodríguez JM, Carrascosa AL, Sanchez-Vizcaino JM. Antigenic properties and diagnostic potential of African swine fever virus protein pp62 expressed in insect cells. J Clin Microbiol. 2006;44:950–6.